Advances in Intelligent Systems and Computing

Volume 1030

The series "Advances in Intelligent Systems and Computing" contains publications on theory, applications, and design methods of Intelligent Systems and Intelligent Computing. Virtually all disciplines such as engineering, natural sciences, computer and information science, ICT, economics, business, e-commerce, environment, healthcare, life science are covered. The list of topics spans all the areas of modern intelligent systems and computing such as: computational intelligence, soft computing including neural networks, fuzzy systems, evolutionary computing and the fusion of these paradigms, social intelligence, ambient intelligence, computational neuroscience, artificial life, virtual worlds and society, cognitive science and systems, Perception and Vision, DNA and immune based systems, self-organizing and adaptive systems, e-Learning and teaching, human-centered and human-centric computing, recommender systems, intelligent control, robotics and mechatronics including human-machine teaming, knowledge-based paradigms, learning paradigms, machine ethics, intelligent data analysis, knowledge management, intelligent agents, intelligent decision making and support, intelligent network security, trust management, interactive entertainment, Web intelligence and multimedia.

The publications within "Advances in Intelligent Systems and Computing" are primarily proceedings of important conferences, symposia and congresses. They cover significant recent developments in the field, both of a foundational and applicable character. An important characteristic feature of the series is the short publication time and world-wide distribution. This permits a rapid and broad dissemination of research results.

** **Indexing: The books of this series are submitted to ISI Proceedings, EI-Compendex, DBLP, SCOPUS, Google Scholar and Springerlink** **

More information about this series at http://www.springer.com/series/11156

Srikanta Patnaik · Andrew W. H. Ip ·
Madjid Tavana · Vipul Jain
Editors

New Paradigm in Decision Science and Management

Proceedings of ICDSM 2018

 Springer

Editors
Srikanta Patnaik
Siksha 'O' Anusandhan University
Bhubaneswar, India

Madjid Tavana
Department of Business Systems
and Analytics
La Salle University
Philadelphia, PA, USA

Andrew W. H. Ip
Department of Industrial
and Systems Engineering
The Hong Kong Polytechnic University
Kowloon, Hong Kong

Vipul Jain
Victoria Business School,
School of Management
Victoria University of Wellington
Wellington, New Zealand

ISSN 2194-5357   ISSN 2194-5365   (electronic)
Advances in Intelligent Systems and Computing
ISBN 978-981-13-9329-7   ISBN 978-981-13-9330-3   (eBook)
https://doi.org/10.1007/978-981-13-9330-3

This Springer imprint is published by the registered company Springer Nature Singapore Pte Ltd.
The registered company address is: 152 Beach Road, #21-01/04 Gateway East, Singapore 189721,
Singapore

Preface

The International Conference on Decision Science and Management (ICDSM-2018) was organized by the Interscience Institute of Management and Technology (IIMT), Bhubaneswar, India, on December 22–23, 2018. ICDSM-2018 has provided an ideal platform for researchers, academicians, software developers, and industry experts to share their research works, findings, and new research directions with each other. The conference focuses on research and developmental activities in the field of decision science and its significance in various fields of management. A good response has been received for this conference, and the selected papers out of the total number of submissions have been divided into three major parts, i.e., data science and analytics, decision science and management, and finally ICT and innovative applications.

The first part, data science and analytics, is an interdisciplinary field as it provides a strong combination of algorithms with mathematics and statistics. The underlying techniques are capable of gathering a large amount of data from different sources and working with huge unstructured datasets and apply efficient techniques to extract meaningful, effective, and accurate insights using data mining, neural networks, and machine learning techniques. These insights further help in identifying hidden patterns and future trends and making actionable decisions for solving significant problems and create impact. Data science and analytics when combined with business management concepts can be used to solve some of the most crucial problems. Many of the submissions fall into this section of the proceeding.

The second part is decision science and management. Decision science is a multidisciplinary area that involves existing as well as emerging fields such as computer science, information technology, mathematics, statistics, and business management for solving several crucial problems of management processes. Various management processes and systems are being used to process data collected from heterogeneous sources for extracting information and provide significant insights. These insights support managers for enhancing their understanding of important problems in hand and make crucial decisions in uncertain and critical situations. Decision science thus adds value to various processes of business management sectors such as evaluation process while hiring new candidates to assist senior managements in dealing with vulnerable circumstances by keeping track of influential changes occurring in other sectors. Further, it provides new direction to the advancement in key areas such as adoption of Internet of things, distributed computing, human resource developments, resource management, marketing strategies, supply chain management, logistics, business operations, financial markets, thus leading the integration of data science and decision science techniques in business sector, a potential research area. We have received a good number of submissions in this section.

The third part is ICT and innovative applications which explore various application areas of industry as well as other sectors that utilize both data science and decision science for solving problems. Information and communication technology (ICT) is an extension of information technology and involves all the infrastructure and components that enable modern computing including computer systems, mobile devices, networking components, and applications. It allows organizations (including business firms to government bodies) and individuals to interact with each other in the digital world. This section of the conference has also received lots of attention. The conference not only encouraged the submission of unpublished original articles in the fields of data science, decision science, and management issues but also considered the cutting-edge applications in organizations and firms while scrutinizing the relevant papers.

Kowloon, Hong Kong   Andrew W. H. Ip
Philadelphia, USA   Madjid Tavana
Bhubaneswar, India   Srikanta Patnaik
Wellington, New Zealand   Vipul Jain

Acknowledgements

The contributions covered in this proceeding are the outcome of the contributions from more than one hundred researchers. We are thankful to the authors, paper contributors of this volume, and the departments which supported the event a lot.

We are thankful to the Editor-in-Chief of the Springer Book series on *Advances in Intelligent Systems and Computing* Prof. Janusz Kacprzyk for his support to bring out the maiden volume of the conference, i.e., ICDSM-2018.

We are also very much thankful to Dr. Aninda Bose, Senior Editor, Hard Sciences, Springer Nature publishing, and his team for his constant encouragement support and time-to-time monitoring.

We express our deep sense of gratitude to Prof. Taosheng Wang, Dean, School of Business, Hunan International Economics University, China, for addressing as the guest of honor coming from China, and Prof. Rabi Narayan Subudhi, School of Management, KIIT, for addressing the conference as Chief Speaker.

We would like to record our thanks to our academic partner Hunan International Economics University, China, who will be hosting the next edition of the conference. Last but not least, we are thankful to Interscience Institute of Management and Technology, Bhubaneswar, India, for hosting the event successfully.

We are sure that the readers shall get immense benefit and knowledge from this edited volume. We look forward to your valuable contribution and support to the next editions of ICDSM, which will be held in Hunan International Economics University, Changsha, China.

Andrew W. H. Ip
Madjid Tavana
Srikanta Patnaik
Vipul Jain

Contents

About the Editors

Dr. Srikanta Patnaik is a Professor at the Department of Computer Science and Engineering, Faculty of Engineering and Technology, SOA University, Bhubaneswar, India. Dr. Patnaik has published more than 60 research papers and book chapters, 2 textbooks, and has edited 12 books. He is the Editor-in-Chief of the International Journal of Information and Communication Technology and International Journal of Computational Vision and Robotics, both published by Inderscience, England, and Editor-in-Chief of the book series "Modeling and Optimization in Science and Technology" published by Springer, Germany.

Dr. Andrew W. H. Ip received his PhD from Loughborough University (UK), MBA from Brunel University (UK), MSc in Industrial Engineering from Cranfield University (UK) and LLB (Hons) from the University of Wolverhampton (UK). Dr. Ip is Principal Research Fellow of the Hong Kong Polytechnic University, Professor Emeritus and Adjunct Professor of Mechanical Engineering, University of Saskatchewan, Canada. He is a Visiting Professor at the Civil Aviation University of China, South China Normal University, and University of Electronic Science and Technology of China, and Honorary Industrial Fellow of the University of Warwick, Warwick Manufacturing Group. He is editor in chief of Enterprise Information Systems, Taylor and Francis (SCI indexed); founder and editor in chief of International Journal of Engineering Business Management, SAGE (ESCI and Scopus indexed).

Dr. Madjid Tavana is a Professor and Distinguished Chair of Business Analytics at La Salle University in Philadelphia, where he serves as Chairman of the Business Systems and Analytics Department. He also holds an Honorary Professorship in Business Information Systems at the University of Paderborn in Germany. Dr. Tavana is a Distinguished Research Fellow at the Kennedy Space Center, the Johnson Space Center, the Naval Research Laboratory at Stennis Space Center, and the Air Force Research Laboratory. He has published 15 books and over 250 research papers in scholarly academic journals, and is the Editor-in-Chief of several international journals.

Dr. Vipul Jain is currently working in the area of Operations and Supply Chain Management at Victoria Business School, Victoria University of Wellington, New Zealand. He has also worked as a French Government researcher for the French National Institute for Research in Computer Science and Control at Nancy, France. He is the Editor-in-Chief of the International Journal of Intelligent Enterprise and an Editorial Board Member for seven international journals.

Data Science and Analytics

Quality Index Evaluation Model Based on Index Screening Model

Fenglin Zhang, Mengpei Guo and Xingming Gao

Abstract The principal component analysis theory is quite mature in the simplification and comprehensive evaluation of real data, and the traditional principal component analysis only strives for the form that the index value is real or interval number. It lacks certain persuasion in the face of large uncertainty index values, but it lacks the simplification of the three-parameter interval number. Based on the principal component analysis theory that the index data is an interval number, the paper proposed the M-PCA (three-parameter interval principal component analysis) method to screen out the unrelated indicators. Principal component analysis does not consider the influence of the negative factor when it is used to reduce the dimension of the comprehensive index. There are some misunderstandings and defects in the application. In dealing with the weights of unrelated indicators, this paper considered the subjective theories to solve the weights of the new indicators, combined with the multi-attribute gray target decision model to evaluate quality indicators. Finally, it combined with examples and verified that the method is more scientific, reasonable, and effective.

Keywords Principal component analysis · Three-parameter interval · The weights of new indicators · Gray target decision · Quality indicators

F. Zhang (✉) · M. Guo · X. Gao (✉)
College of Economics and Management, Nanjing University of Aeronautics and Astronautics, 29 Jiangjun Road, Nanjing 211106, Jiangsu, China
e-mail: zflnuaa@163.com

X. Gao
e-mail: 1136624944@qq.com

M. Guo
e-mail: 772028639@qq.com

© Springer Nature Singapore Pte Ltd. 2020
S. Patnaik et al. (eds.), *New Paradigm in Decision Science and Management*,
Advances in Intelligent Systems and Computing 1030,
https://doi.org/10.1007/978-981-13-9330-3_1

1   Introduction

Quality is the lifeline of an enterprise. The quality of products is decisive for the development of enterprises. Nowadays, a reasonable evaluation of quality indicators can help companies more scientifically understand their current level of quality. How to evaluate the form of the quality index value as the three-end interval number will be a difficult problem for researchers to study in different environments and conditions. To evaluate the problem, the comprehensive evaluation of the indicator is an important link, and the scientific and reasonable method has a greater impact on the final result. Multi-attribute evaluation is a systematic analysis method for studying the problem of uncertainty evaluation. It is the problem of selecting, ranking, or evaluating multiple events and multiple countermeasures. Its purpose is to improve the decision-making process and systematically evaluate and analyze quality indicators. Multi-attribute evaluation methods such as analytic hierarchy process, TOPSIS method, and fuzzy decision-making method have attracted the attention of many scholars since their introduction. However, the above studies directly deal with the original indicator values, ignoring the interrelationship between the various quality indicators. The degree of preprocessing of the evaluation index is related to the quality and weakness of the quality evaluation and the complexity of the calculation, especially for the case where the quality index value is not a real number, such as the number of intervals or the number of three-parameter intervals.

Many scholars have conducted in-depth research on the interrelationship between evaluation indicators. Ait-Izem et al. [1] compared the four most common interval principal component analysis (PCA) methods and applied them to fault detection and isolation. Based on the reconstruction principle in the classical PCA method, the reconstruction of the interval is optimized, and a new criterion is derived to determine the structure of the interval PCA model. Liu et al. [2] proposed a new model. This new model is an interval-intuitive fuzzy principal component analysis. Based on the traditional principal component analysis, it treats decision makers and attributes as interval-intuitive fuzzy variables, and transforms these two kinds of variables through the proposed principal component analysis model. As an independent variable, Yu et al. [3] used principal component analysis to reduce the index of the stock price and predicted the stock price based on the generalized regression neural network model. Hou [4] proposed two improved methods, one is the principal component analysis method based on empirical statistics, and the other is the principal component analysis method based on interval matrix operation, which achieves the expected effect. Qi and Xiong [5] put forward two kinds of energy-saving evaluation methods for the central heating boiler room: qualitative and quantitative index.

In this paper, when processing the simplification of index data for the three-parameter interval data, it is based on the three-interval principal component analysis of the most probable value method (M-PCA—three-parameter interval principal component analysis) to filter out the indicators with low correlation and for the new indicator weight analysis. Finally, the multi-attribute gray target decision model is used to evaluate the quality indicators.

2 Three-Parameter Interval Principal Component Analysis

2.1 Principal Component Analysis

PCA is able to turn multiple indicators into fewer comprehensive indicators. The principle of comprehensive variables is the linear combination of multiple indicators.

2.2 Calculation Steps

(1) Constructing a sample matrix

$$
X = \begin{bmatrix} x_1^T \\ x_2^T \\ \vdots \\ x_m^T \end{bmatrix} = \begin{bmatrix} x_{11} & x_{12} & \cdots & x_{1n} \\ x_{21} & x_{22} & \cdots & x_{2n} \\ \vdots & \vdots & \vdots & \vdots \\ x_{m1} & x_{m2} & \cdots & x_{mn} \end{bmatrix} \tag{1}
$$

(2) Transform sample array X to $Q = \lfloor q_{ij} \rfloor_{m \times n}$

$$
q_{ij} = \begin{cases} x_{ij}, & \text{Positive intensity indicator} \\ 1/x_{ij}, & \text{Negative intensity indicator} \end{cases} \tag{2}
$$

(3) Normalize matrix Q to obtain a normalized matrix Z

$$
Z = \begin{bmatrix} z_1^T \\ z_2^T \\ \vdots \\ z_m^T \end{bmatrix} = \begin{bmatrix} z_{11} & z_{12} & \cdots & z_{1n} \\ z_{21} & z_{22} & \cdots & z_{2n} \\ \vdots & \vdots & \vdots & \vdots \\ z_{m1} & z_{m2} & \cdots & z_{mn} \end{bmatrix} \tag{3}
$$

$$
z_{ij} = \frac{q_{ij} - \bar{q}_j}{s_j} \tag{4}
$$

The standardization makes the variance of all indexes equal to 1, smoothing the variation of each indicator. The main component of the standardized data extraction contains information on the mutual influence of each index and does not accurately reflect all the information contained in the original data. To eliminate the influence of the dimension and magnitude of variables, this paper uses the "average" method.

$$
z_{ij} = \frac{q_{ij}}{\sum\limits_{j=1}^{m} q_{ij}} \tag{5}
$$

(4) Calculate the covariance matrix of the averaging matrix z

$$Y = \left[y_{ij} \right]_{\text{In } n} = \frac{Z^T Z}{m - 1} \tag{6}$$

(5) Solve the eigen value

$$|Y - \lambda I_n| = 0 \tag{7}$$

Solve to get n feature values $\lambda_1 \geq \lambda_2 \geq \ldots \geq \lambda_n \geq 0$.

(6) Determine the p-value. In general, the utilization rate of information is more than 85%.

$$\frac{\sum\limits_{j=1}^{p} \lambda_j}{\sum\limits_{j=1}^{m} \lambda_j} \geq 0.85 \tag{8}$$

Solving equations: $Yb = \lambda_j b$.

$$b_j^0 = \frac{b_j}{|b_j|} \tag{9}$$

(7) Find the p principal components of $z_i = (z_{i1}, z_{i2}, \ldots, z_{in})$

$$u_{ij} = z_i^T b_j^0, \, j = 1, 2, \ldots, p \tag{10}$$

Decision matrix:

$$U = \begin{bmatrix} u_1^T \\ u_2^T \\ \vdots \\ u_m^T \end{bmatrix} = \begin{bmatrix} u_{11} & u_{12} & \ldots & u_{1p} \\ u_{21} & u_{22} & \ldots & u_{2p} \\ \vdots & \vdots & \vdots & \vdots \\ u_{m1} & u_{m2} & \ldots & u_{mp} \end{bmatrix} \tag{11}$$

2.3 M-PCA Three-Parameter Interval Principal Component Analysis

In the actual decision problem, the interval range is made too large, which will increase the uncertainty of the decision result. It introduced three-parameter interval numbers and added the most likely number. This paper is based on the M-PCA to solve the three-parameter interval index principal component values.

2.4 Calculation Steps

(1) Construction matrix X:

$$X = \begin{bmatrix} x_1^T \\ x_2^T \\ \vdots \\ x_m^T \end{bmatrix} = \begin{bmatrix} \left[x_{11}^S, x_{11}^M, x_{11}^L\right] & \left[x_{12}^S, x_{12}^M, x_{12}^L\right] & \cdots & \left[x_{1n}^S, x_{1n}^M, x_{1n}^L\right] \\ \left[x_{21}^S, x_{21}^M, x_{21}^L\right] & \left[x_{22}^S, x_{22}^M, x_{22}^L\right] & \cdots & \left[x_{2n}^S, x_{2n}^M, x_{2n}^L\right] \\ \vdots & \vdots & \vdots & \vdots \\ \left[x_{m1}^S, x_{m1}^M, x_{m1}^L\right] & \left[x_{m2}^S, x_{m2}^M, x_{m2}^L\right] & \cdots & \left[x_{mn}^S, x_{mn}^M, x_{mn}^L\right] \end{bmatrix} \tag{12}$$

where x_{ij}^s represents the minimum value of the j index of the i product, x_{ij}^M represents the most likely value of the j index of the i product, and x_{ij}^L represents the maximum value of the j index of the i product.

(2) Constructing the most likely index value matrix X_M Perform a traditional covariance-based principal component analysis for Eq. (11) to obtain a decision matrix $u_{ij}^M = z_i^M b_j^0$, $j = 1, 2, \ldots, p$.

(3) Solving sample endpoint number decision matrix:

$$X^M = \begin{bmatrix} \left(x_1^M\right)^T \\ \left(x_2^M\right)^T \\ \vdots \\ \left(x_m^M\right)^T \end{bmatrix} = \begin{bmatrix} x_{11}^M & x_{12}^M & \cdots & x_{1n}^M \\ x_{21}^M & x_{22}^M & \cdots & x_{2n}^M \\ \vdots & \vdots & \vdots & \vdots \\ x_{m1}^M & x_{m2}^M & \cdots & x_{mn}^M \end{bmatrix} \tag{13}$$

$$u_{ij}^S = \min_{z_{ij}^S \le z_{ij} \le z_{ij}^L} b_j^0 \tag{14}$$

$$u_{ij}^L = \max_{z_j^S \le z_{ij} \le z_{ij}^L} b_j^0 \tag{15}$$

The linear weighted comprehensive evaluation method is established based on the weighted contribution ratio of the principal components, without considering the influence of negative factors, and there are some misunderstandings and defects in the application.

Because of measurement errors or difficult to measure, this value is not an exact real number and is therefore represented by a three-parameter interval number.

Therefore, it noted that x_{ij} is the index value of the product a under the attribute is b is $x_{ij} \in \left[x_{ij}^S, x_{ij}^M, x_{ij}^L\right] \left(0 \le x_{ij}^S \le x_{ij}^M \le x_{ij}^L, i = 1, 2, \ldots, m; j = 1, 2, \ldots, n\right)$. In order to eliminate the dimension and increase comparability, the data is averaged to obtain a normalized decision matrix $U = \left(u_{ij}\right)_{m \times n}$, where $u_{ij} \in \left(u_{ij}^S, u_{ij}^M, u_{ij}^L\right)$ is the three-parameter interval gray number, which represents the index value of the product a under the attribute b.

Definition 1

$u_j^+(\otimes) = \max\left\{\left(u_{ij}^S + u_{ij}^M + u_{ij}^L\right)/3\right\} i = 1, 2, \ldots, m, j = 1, 2, \ldots, n)$ the corresponding effect value is denoted $\left[\left(u_{ij}^S\right)^+, \left(u_{ij}^M\right)^+, \left(u_{ij}^L\right)^+\right]$, and

$$
\begin{aligned}
u^+(\otimes) &= \left\{u_1^+(\otimes), u_2^+(\otimes), \ldots, u_n^+(\otimes)\right\} \\
&= \left\{ \begin{array}{l}
\left[\left(u_{i1}^S\right)^+, \left(u_{i1}^M\right)^+, \left(u_{i1}^L\right)^+,\right], \left[\left(u_{i2}^S\right)^+, \left(u_{i2}^M\right)^+, \left(u_{i2}^L\right)^+,\right], \\
\ldots, \left[\left(u_{in}^S\right)^+, \left(u_{in}^M\right)^+, \left(u_{in}^L\right)^+,\right]
\end{array} \right\}
\end{aligned}
\tag{16}
$$

is called the optimal ideal effect vector.

2.5 The Determination of Subjective Index Weight Based on Analytic Hierarchy Process

Analytic hierarchy process (AHP) divides decision-making problems into different hierarchical structures and compares them at the same level, finds the weights of the same level, and finally analyzes the problems comprehensively.

2.6 Index Evaluation

Let $u_i = \{u_{i1}(\otimes), u_{i2}(\otimes), \ldots, u_{in}(\otimes)\}$ is the effect evaluation vector of scheme a_i, and $u^+(\otimes) = \{u_1^+(\otimes), u_2^+(\otimes), \ldots, u_n^+(\otimes)\} = \left\{\left[\left(u_{i1}^S\right)^+, \left(u_{i1}^M\right)^+, \left(u_{i1}^L\right)^+,\right], \left[\left(u_{i2}^S\right)^+, \left(u_{i2}^M\right)^+, \left(u_{i2}^L\right)^+,\right], \ldots, \left[\left(u_{in}^S\right)^+, \left(u_{in}^M\right)^+, \left(u_{in}^L\right)^+,\right]\right\}$ is the optimal ideal effect vector. We call

$$
\varepsilon_i^+ = 3^{-1/2}\left\{\sum_{j=1}^n \omega_j\left[\left(u_{ij}^S - \left(u_{ij}^S\right)^+\right)^2 + \left(u_{ij}^M - \left(u_{ij}^M\right)^+\right)^2 + \left(u_{ij}^L - \left(u_{ij}^L\right)^+\right)^2\right]\right\}^{1/2}
\tag{17}
$$

is the target center distance of the effect vector of scheme a_i. The smaller ε_i^+ is, the better the a_i scheme is. The bigger ε_i^+ is, the worse the a_i scheme is.

3 Analysis of Examples

When a unit purchases artillery weapons, there are four series of new types of guns a_i available for selection, and consider the following eight attributes: b_1 is the fire assault capacity coefficient; b_2 is the response capability index; b_3 is the mobility index; b_4 is the viability Index; b_5 is cost; b_6 is stability capability index; b_7 is index of risk resistance; and b_8 is an operational index. Recorded as the index value of the product under the attribute, the four types of artillery are new types of artillery. Due to measurement errors or difficulties in measurement, the value is not an exact real number and is therefore represented by a three-parameter interval number.

$$x_{ij} \in \left[x_{ij}^S, x_{ij}^M, x_{ij}^L\right]\left(x_{ij}^S \leq x_{ij}^M \leq x_{ij}^L, i = 1, 2, \ldots, 4; j = 1, 2, \ldots, 8\right).$$

Each scenario with respect to evaluation attributes is shown in the matrix X.

$$
\begin{bmatrix}
[26000, 26460, 27000] & [2, 3, 6, 4] & [18000, 18480, 19000] & [0.7, 0.75, 0.8] \\
[60000, 65400, 70000] & [3, 3.4, 4] & [16000, 16780, 19000] & [0.3, 0.36, 0.4] \\
[50000, 57600, 60000] & [2, 2, 7, 3] & [15000, 15360, 16000] & [0.7, 0.74, 0.8] \\
[40000, 44800, 50000] & [1, 1.8, 2] & [28000, 28680, 29000] & [0.4, 0.46, 0.5]
\end{bmatrix}
$$

$$
\begin{bmatrix}
[15000, 15560, 16000] & [0.4, 0.48, 0.5] & [4, 4.6, 6] & [6.5, 7, 7.5] \\
[27000, 27880, 28000] & [0.6, 0.68, 0.8] & [5, 5.8, 6] & [8, 8.5, 9] \\
[24000, 25380, 26000] & [0.6, 0.64, 0.7] & [6, 6.7, 7] & [7, 2, 8, 8] \\
[15000, 15620, 17000] & [0.5, 0.56, 0.6] & [3.6, 4, 4.8] & [6, 6.5, 7]
\end{bmatrix}
$$

$\varepsilon_1^+ = 0.086$, $\varepsilon_2^+ = 0.057$, $\varepsilon_3^+ = 0.103$, $\varepsilon_4^+ = 0.005$. Due to $\varepsilon_4^+ < \varepsilon_2^+ < \varepsilon_1^+ < \varepsilon_3^+$ the force should give priority to the fourth artillery. After a period of use and analysis of the use of other units, the force was found to be superior to other types of weapons. The results show that the proposed theory has a strong application value.

4 Conclusion

A novel M-PCA three-parameter interval principal component analysis method is proposed, which can extract mutually unrelated index attributes. Then, the multi-attribute target decision model is used to evaluate the quality indicators. Combined with examples, this method proved to be more scientific and effective. For the comprehensive consideration of the subjective and objective weights of the new principal component analysis will be the next research direction.

References

1. Ait-Izem, T., Harkat, M.F., Djeghaba, M., et al.: On the application of interval PCA to process monitoring: a robust strategy for sensor FDI with new efficient control statistics. J. Process Control **63**(1), 29–46 (2018)
2. Liu, B., Shen, Y., Zhang, W., et al.: An interval-valued intuitionistic fuzzy principal component analysis model-based method for complex multi-attribute large-group decision-making. Eur. J. Oper. Res. **245**(1), 209–225 (2015)
3. Yu, Z., Qin, W., Zhao, Z., Xiang.: The stock price prediction based on principal component analysis and generalized regression neural network. Stat. Decis. Mak. **34**(18), 168–171 (2018)
4. Hou, Z.: Research and application of principal component analysis based on interval number theory (2015)
5. Qi, J.F., Xiong, X.P.: Research to the central heating boiler room energy saving index evaluation system. Appl. Mech. Mater. **556–562**, 921–924 (2014)

Selection of Software Development Methodologies (SDMs) Using Bayesian Analysis

Himadri Bhusan Mahapatra, Vishal Chandra and Birendra Goswami

Abstract A software development methodology (SDM)—also called systems development methodology—is a formalized approach for the development of software. Although, there are many different SDMs. In this paper, we proposed a method for the selection of an appropriate SDM for a particular project using Bayesian analysis over various factors affecting the selection of SDM. Bayesian is widely used for decision making. There is a statistical model for various categories established by calculating the conditional probability density functions (PDF). We used some factors, i.e., cost, risk, size, and time duration. We classify categories with the maximum posterior probability within the Bayesian framework. Result of performance of the proposed model is shown by experimentation on the dataset using MATLAB and WINBUGS simulation.

Keywords Bayesian network · SDMs · Probability density function · Frequentist inference · Bayesian inferences

1 Introduction

According to IEEE, a software development methodology is defined as "A framework containing the processes, activities and tasks involved in the development, operation and maintenance of a software product, spanning the life of the system from the definition of its requirements to the termination of its use". It provides a

H. B. Mahapatra (✉)
Department of Computer Science, JRU, Ranchi, India
e-mail: himumahapatra@rediffmail.com

V. Chandra
Department of Computer Science and Engineering, RIT, Koderma, India
e-mail: vcvishalchandra@gmail.com

B. Goswami
Department of Computer Science, ICFAI University, Ranchi, India
e-mail: bgoswami@rediffmail.com

© Springer Nature Singapore Pte Ltd. 2020
S. Patnaik et al. (eds.), *New Paradigm in Decision Science and Management*,
Advances in Intelligent Systems and Computing 1030,
https://doi.org/10.1007/978-981-13-9330-3_2

flexible framework for enhancing the process. It enables effective communication, reuse, and process management [1, 2]. In last 50 years, there are various software development methodologies were developed by various researchers; few of them are: waterfall, iterative waterfall, prototyping model, spiral, evolutionary, incremental, time-boxing, RAD, V-model, etc. Any SDM framework is specific to the type of project we going to develop according to the time duration, complexity, mission-critical problem, resource required, client understandability, and also budget for the project. Therefore, it is the most crucial and most essential task to select the appropriate SDM for the success of project. The success of a software project mainly depends upon the selection of particular software development model for the particular project according to the requirements. Consideration of time and cost is most important when we are going to choose a particular SDM because they play a key role in project development. There are few major parameters which are come to the picture and required for the selection of specific SDM such as Cost, Requirement Specification, Time, Project Size, Project Type, and tolerance to change. The selection of SDM is either based upon the experience or heuristic way, and these approaches sometimes fail due to the involvement of uncertainty.

We have reviewed several literatures which are available on this topic. In [3], authors proposed a rule-based object-oriented framework for SDM selection. Another work proposed in [4], author present a model which was based on knowledge based known as ESPMS for SDM selection. In [5], authors proposed a model known as MODSET which was based on rule-based expert system combined with Likert scale measurement for SDM selection. In this paper, we proposed a method for the selection of an appropriate SDM for a particular project using Bayesian analysis over various factors affecting the selection of SDM. Bayesian analysis is widely used for decision making [6, 7]. There is a statistical model for various categories established by calculating the conditional probability density functions (PDFs). We used some factors' cost, risk, size, and time duration. We classify categories with the maximum posterior probability within the Bayesian framework. Result of performance of the proposed model is shown by experimentation on the dataset.

2 Proposed Work

Our proposed work is based on Bayesian probability distribution. There are various Bayesian inferences [8]; we used a model called Bayesian logistic regression model for the analysis of our dataset. Formula of logistic regression is given below, where **p** is the probability of posterior (likelihood) and **b** is prior distribution, and X_p is parameter.

$$p = \frac{e^{b_0 + b_1 X_1 + b_2 X_2 \ldots + b_p X_p}}{1 + e^{b_0 + b_1 X_1 \ldots + b_p X_p}}$$

MATLAB representation of logistic regression

```
logit = @ (b, x) exp (b (1) +b (2). *x)./(1+exp (b (1) +
b (2). *x))
prior
  prior1 = @(b1) normpdf(b1,0,20;
  likelihood @(b) prod (binopdf (iwf, total projects, logitp
  (b, cost))) ... * prior1(b (1)) * prior2(b (2));
```

Propose of this paper is to find the best software development model for a project considering four factors:-

1. Expected cost of project
2. Expected time duration of project
3. Expected risk of failure of project
4. Expected size of project (KLOC).

Expected cost of project:- When a new project starts, first the software development team has to make a rough estimation of the cost of project. This expected cost is determined by the previous experience of similar type of projects. This is estimated in terms of currency.

Expected time duration of project:- Second thing is to estimate how much time is required for completing the whole project. Development team has to determine the time duration to complete the project. This entire thing can be done with the help of the previously completed similar type of project. This factor is estimated in terms of months or weeks.

Expected risk of failure of project:- Every project is associated with a risk of failure of project. Software development team has to estimate the risk of failure of project. This is done by analyzing previously done similar type of projects. This factor is estimated in terms of percentage.

Expected size of project:- After doing all three processes, team decides technology and kilo line of code (KLOC). For example, suppose a new project handover to a team and team decides to do it in a particular language, then the team has to estimate how much lines of code required to do this based on the language. This is also done by using the past experience of similar types of projects. This is estimated in terms of kilo line of code (KLOC).

In this paper, we have compared three software development methodologies (SDMs):-

1. Iterative waterfall model
2. Prototype model
3. V-model.

Data are collected from various companies on software development using three models and four factors.

We draw the conclusion of the selection of SDM with two parameters:

1. Total number of projects
2. One of the four factors.

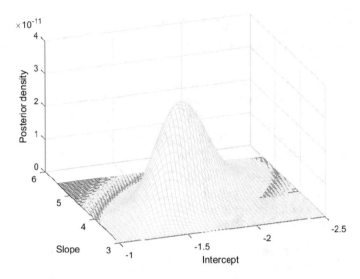

Fig. 1 Surface curve of iterative waterfall, prototype, v-model with cost and total numbers of projects

3 Proposed Model

In this work, we have given an algorithm for the proposed model which is described as follows:

```
model
{
for (i in 1: n) {
# Linear regression on logit
logit(p[i]) <- alpha + b.(factor*factor[i]) +
b. ((no. of projects) *(no. of projects[i]))
# Likelihood function for each data point
frac[i] ~ dbern(p[i])
}
alpha ~ dnorm(0.0,1.0E-4) # Prior for intercept
b. (no. of projects) ~ dnorm (0.0,1.0E-4) # Prior for slope of
number of projects
b. factor ~ dnorm(0.0,1.0E-4) # Prior for slope of factors
}
n=200
```

See Figs. 1, 2, 3 and 4 and Tables 1, 2, 3 and 4.

3.1 Experimental Result

We have collected 200 data from different companies. We used 200 iterations. Results are shown in below table (Tables 5 and 6).

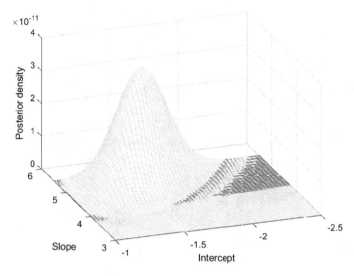

Fig. 2 Shows the surface curve of three SDMs, time, and total number of projects

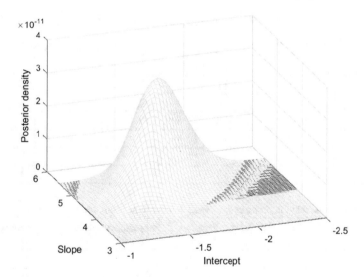

Fig. 3 Shows the surface curve of three SDMs, risk, and total number of projects

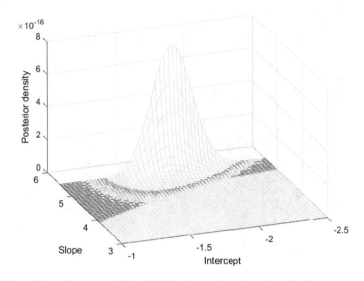

Fig. 4 Shows the surface curve of three SDMs, size, and total number of projects

Table 1 Shows the distribution of cost of projects and total number of projects and their relationship with three SDMs

Cost	Total no. of projects	Prototype model	Iterative waterfall model	V-model
2000	41	1	25	15
2500	42	2	22	18
3000	30	0	19	11
3100	35	6	19	7
5000	31	8	16	7
8000	21	8	9	4

Table 2 Shows the relation between time, total number of projects, and three SDMs

Time (Weeks)	Total no. of projects	Prototype model	Iterative waterfall model	V-model
1	41	0	36	5
3	42	2	20	20
7	30	1	20	9
12	35	6	19	7
15	31	18	5	8
24	21	12	4	5

Table 3 Shows the relation between risk, total number of projects, and three SDMs

Risk in %	Total no. of projects	Prototype model	Iterative waterfall model	V-model
0	41	0	37	4
5	42	2	30	10
8	30	21	4	5
10	35	20	0	15
15	31	24	0	7
20	21	18	0	3

Table 4 Shows the relation between size, total number of projects, and three SDMs

Size in KLOC	Total no. of projects	Prototype model	Iterative waterfall model	V-model
1000	41	0	39	2
1500	42	2	34	6
2300	30	6	16	8
2700	35	10	10	15
4000	31	22	1	8
5000	21	19	0	2

Table 5 Result (proposed model)

No. of projects	Cost	Time	Size	Risk	IWF model (p)	Prototype (p)	V-model (p)
20	2000	2	1000	0	0.91	0.25	0.43
27	1400	3	1400	2	0.78	0.31	0.52
15	4000	4	2000	5	0.64	0.55	0.42
30	3400	3	1600	4	0.79	0.54	0.38
40	4500	5	2300	2	0.49	0.71	0.54
55	6000	10	4000	10	0.26	0.89	0.67
10	1000	2	1200	0	0.79	0.54	0.39

4 Conclusion

On the basis of this research, we can conclude that there is some significant difference between frequentist inference method and Bayesian inference method for the selection of SDM for a particular project. We can also say that Bayesian approach is better than frequentist approach for the selection of SDM. Although proposed model has some degree of uncertainty as in other Bayesian model, it shows better result than other traditional approaches (frequentist). Bayesian method is more flexible and can handle more complex problem which contains uncertainty. The small dataset which

Table 6 Result (frequentist inferences)

No. of projects	Cost	Time	Size	Risk	IWF model (**p**)	Prototype (**p**)	V-model (**p**)
20	2000	2	1000	0	0.82	0.19	0.42
27	1400	3	1400	2	0.68	0.21	0.48
15	4000	4	2000	5	0.59	0.53	0.47
30	3400	3	1600	4	0.76	0.59	0.31
40	4500	5	2300	2	0.44	0.77	0.51
55	6000	10	4000	10	0.29	0.88	0.61
10	1000	2	1200	0	0.68	0.59	0.31

Where **p** is the probability of using that model in context of number of projects, cost, time, size and risk

we used shows significant result of enhancement in decision process for selection of SDM.

5 Future Scope

In our research, we used only four factors and three types of software development methodology. In future, we can increase the number of factors as well as the number of software development methodologies. For better result, we can also use large datasets. Again, we can also compare our model with other Bayesian inference models for better result.

References

1. Khurana, R.: Software Engineering Principles and Practices. Vikash Publication
2. Pressman, S.R.: Software Engineering: a Practitioner's Approach. 8th edn
3. Ayman, M., Ahmar, A.I.: Rule based expert system for selecting software development methodology. J Theor. Appl. Sci. (2010)
4. Khan, A.R., Rehman, Z.U., Amin, H.U.: Knowledge based system modelling for software process model selection, IJACSA **2**(2) (2011)
5. Verma, J., Bansal S., Pandey, H.: Develop framework for selecting best software development methodology, Int. J. Sci. Eng. Res. **5**(4) (2014)
6. Frank, R.: Bayesian networks with a logistic regression model for the conditional probabilities Int. J Approximate Reasoning **48** (2008)
7. Chulani, S., Boehm, B., Steece, B." Bayesian analysis of empirical software engineering cost models. IEEE Trans. Softw. Eng. **25** (1999)
8. Wandji, T.K., Sarkani, S., Eveleigh, T.H., Holzer T.H., Keiller, P.A.: Comparative analysis of Bayesian and classical approaches for software reliability measurement. In: International Symposium on Software Reliability Engineering Workshops (ISSREW), Pasadena, CA (2013)

Role of Social Media in Employer Branding—A Study on Selected Engineering Colleges (Private) in Bhubaneswar

Amita Panda

Abstract As social media is playing a transforming role among people around the world by facilitating a platform to share their feelings, memories, and happiness. Thus, we find the popularity of websites like Facebook and Twitter growing manifold. Even the corporate has been impacted by social media so much so that it acts as an opportunity for employer branding. Apart from that social media (like LinkedIn, Facebook, and Twitter) has also emerged as a new method of recruitment which is followed by most reputed IT enabled services (ITES) companies, institution, life insurance, hospital, etc. This paper explores the possibility of using social media by educational institutions for sourcing its employees, especially the professionals. The main purpose of this survey is to investigate thoroughly whether institutions are using branding in their recruitment process and how employer branding impacts the recruitment and retaining process in the engineering institutions and colleges of Bhubaneswar .

Keywords Social media · Facebook · LinkedIn · Recruitment · Employer branding

1 Introduction

In this aggressive market circumstances, the most important part of the organization is to face the challenges to attract and retain skilled and talented professionals. In order to face the challenges, the organizations should foresee competitors' strategy of attracting talent that leads to the organization to sustain in the market. The more talents are being retained and attracted by the organization that will create branding about themselves. Nowadays, the Internet plays a vital role which helps the employees to broadcast messages to outsiders. Apart from the Internet, social networking sites (SNS) like Facebook, LinkedIn, and Twitter play a vital role among the people

A. Panda (✉)
Astha School of Management, Bhubaneswar, India
e-mail: amitapandapati2011@gmail.com

© Springer Nature Singapore Pte Ltd. 2020
S. Patnaik et al. (eds.), *New Paradigm in Decision Science and Management*,
Advances in Intelligent Systems and Computing 1030,
https://doi.org/10.1007/978-981-13-9330-3_3

around the world. These SNS, LinkedIn, Twitter, and Facebook help the organization to recruit talented, high-potential, and experienced candidates. With the help of WhatsApp and other SNS, the organization can post the job openings and can take advantages of employee's Facebook profile, liking their post by their other friends which are in similar trades or professions. HR professionals also consider that SNS can be a great extent helpful to construct their image and create employer branding. For example, employee and alumni also permit to write over Facebook, Twitter, and their own site, which they can utilize for future advancements. Although social media is not much popular in non-government engineering colleges, they follow Facebook, Twitter, and LinkedIn for branding purpose also.

2 Objective

- To study the impact of social media on employer branding among the non-government engineering colleges;
- To find out the factors associated with employer branding in the above-said engineering colleges.

3 Methodology

The focus of this study is to explore the role of SNS, particularly Facebook and LinkedIn, in establishing a link between employers and potential/existing employees. The perspective of the employer is the central focus. A survey was conducted with sets of questions. The survey primarily consisted of statements. A total of 13 statements made upon a five-point Likert scale under the heads of the role of SNS toward the benefits of employer branding. The scale ranked from 1 (strongly disagree) to 5 (strongly agree). The study covered the non-government (or private) colleges operating in Bhubaneswar. The researcher interviewed five senior personnel of Silicon, ITER, Centurion University, Gandhi Institute of Technology with the structured questionnaire method and collected the data for further analysis.

4 Literature Review

4.1 Employer Branding

Burns et al. [1] "'states that one of the most appealing aspects of the semantic differential scale is the ability of the researcher to compute averages" and then plot a "profile of the brand or company image"'. Backhaus and Tikoo [2] "have found that effective employer branding leads to a competitive advantage, be of use for employees to internalize company values and supports in employee retention." Biswas and Suar [3] "mentioned in their study that although the concept of branding is well developed within the marketing literature, the concept of employer branding is still evolving, as practitioners' focus on employees' attractiveness to an employer, and literature on EB remain conceptual and result-oriented".

5 Cases of Private Institute

5.1 Case 1: Institute of Technical Education and Research, SOA University, Bhubaneswar

Institute of Technical Education and Research, Bhubaneswar is a technical college under Sikha 'O' Anusandhana University based in Bhubaneswar, which was founded in 1997. At this moment, ITER team in Bhubaneswar has more than 1000 members, including the top management of the company. The overall university ranking of SOA is 21 among all Indian universities. The case company has taken by the researcher to manifest the social media and its role in employer branding. As per the conversation with top management, the inference drawn is that ITER is recruiting non-teaching staff and teaching staff from advertisement, media, employees' references. The case company has not yet explored the social networking sites like (Facebook and LinkedIn) for recruitment. However, ITER has used social sites for campus recruitment as it helps in shaping formal communication with the companies'. Candidates can directly send their biodata or profile to the registrar at SOA University. Nevertheless, ITER has also benefited from Alumni through social networking sites.

Three key features of Institute of Technical Education and Research:

ITER has three key features which facilitate the usage of social media (online communication) within the organization. First of all, by inculcating strict academic process, ITER creates employer branding among the students as well as outsiders. Moreover, they are able to supervise the syllabus coverage, online question set, quiz test, and e-governance with the help of ERP software. Second of all, ITER encourages employees to participate in intra-social media at work. This means employees get rewarded with points for any activities they make on social media related to work. Lastly, ITER maintains transparency to intend learning culture among each employees and students.

ITER makes difference among its Competitors:

According to the head of HR, ITER is different from competitors by intensifying HR policy that gives assistance to attract talent from the labor market. Friendly environment, pay scale, increment structure and accommodation, and subsidized

Fig. 1 Outcomes of the analysis (ITER)

interest rate are differentiable factors that facilitate ITER to establish itself and create image among other technical institutions. ITER organizes so many research activities; more than 20 national seminars, one international conference, also do journals which construct the brand building affair.

Outcomes of the Case analysis through model representation:

See Fig. 1.

Analysis: (Case 1):

From the interview of personnel and questionnaire response, it was found that mostly the institute focuses on advertisement, employee reference, and open forum for recruiting teaching as well as non-teaching staff. They believe to build brand of an organization as an institution and an employer; it means to do events, national seminars, international conferences, teaching qualities, student campus selection, admission, etc. They are not active on social media for recruitment process; they strongly believe in Alumni, which creates a strong base on social media. Mostly, they prefer their own site rather than other SNS.

5.2 Case 2: Centurion University of Technology and Management (CUTM), Bhubaneswar

School of engineering and technology, Bhubaneswar, is a technical college under Centurion University of Technology and Management (CUTM), based in Bhubaneswar, which was established in the year 2005, by a group of intelligentsia with aspirations to provide high-quality education both nationally and internationally. There was a humongous challenge to establish Jagannath Institute for Technology and Management (JITM) in an anti-capitalist area. Despite all the challenges, JITM was successfully converted into Centurion University of Technology and Management in August 2010, which is one of its kind in that area, through an act of Odisha

Legislative Assembly. It became the first multi-sector State Private University in Odisha. At this moment, CUTM team in Bhubaneswar, has less than 1000 members, including the top management of the company. The case company has taken by the researcher to manifest the social media and its role in employer branding. As per the conversation with top management, the inference drawn that CUTM is recruiting non-teaching staff and teaching staff from social media like Facebook, LinkedIn, advertisement, media, employees' references. The case company gets much more helpful the social networking sites like (Facebook, LinkedIn) for recruitment. However, CUTM has used social sites for campus recruitment as it helps in shaping formal communication with the companies. Candidates can directly post their updated CV to Registrar at CUTM. Nevertheless, CUTM has also benefited from Alumni through social networking sites.

Three key features of Institute of Centurion University of Technology and Management (CUTM):

CUTM has three key features which facilitate the usage of social media (online communication) within the organization. First of all, by inculcating strict academic process, CUTM creates employer branding among the students as well as outsiders. To make a brand of their university, they believe in project work, employees achievements, daily event activity posted, seminars, conferences and so on are updates on their official site. Moreover, they are able to supervise the syllabus coverage, online question set, quiz test, and e-governance with the help of ERP software. Second of all, CUTM encourages their employees to voluntary participate in intra-social media at work. In result, employees get rewarded with points for any activities they make on social media related to work. And lastly, CUTM maintains transparency to intend learning culture among each employees and students.

CUTM makes difference among its Competitors:

According to the head of branding officer, CUTM is different from competitors by intensifying HR policy that gives assistance to attract talent from the labor market. Friendly environment, pay scale, increment structure and accommodation, and subsidized interest rate are differentiable factors that facilitate CUTM to establish itself and create image among other technical institutions and the university. CUTM organizes so many research activities; more than 15 national seminars, one international conference, and daily events also do journals which construct the brand building affair (Fig. 2).

Outcomes of the Case analysis through model representation:
Analysis: (Case 2):

From the interview, it was observed that CUTM strongly believes in social networking sites like Facebook, LinkedIn for recruitment and employer brand creation. Twitter is not that much effective as compared to others. The Alumni and daily-posted events on social sites make a good impact on their university branding. They believe in work and dedication. They maintain a friendly environment and an open communication system among juniors and superiors.

Fig. 2 Outcomes of the analysis (CUTM)

5.3 Case 3: Silicon Institute of Technology

Silicon Institute of Technology (SIT) is a technical college under Biju Patnaik University of Technology based in Bhubaneswar, which was founded in 2002. At this moment, Silicon team in Bhubaneswar has less than 300 members, including the top management of the company. Silicon Institute of Technology has got third rank among the engineering colleges in Bhubaneswar. The case company has taken by the researcher to manifest the social media and its role in employer branding. As per the conversation with top management, the inference drawn that SIT is recruiting non-teaching staff from job-posting websites like Naukri and Monster and teaching staff from advertisement and employees' references. The case company has not yet explored the social networking sites like (Facebook and LinkedIn) for recruitment. However, SIT has used social sites for campus recruitment as it helps in shaping formal communication with the companies. Nevertheless, Silicon has also benefited from Alumni through social networking sites.

Three key features of Silicon Institute of Technology:

SIT has three key features which facilitate the usage of social media (online communication) within the organization. First of all, by inculcating strict academic process, SIT creates employer branding among the students as well as outsiders. Moreover, they are able to supervise the syllabus coverage, online question set, quiz test, e-governance with help of ERP software. Second of all, SIT encourages employees to participate in intra-social media at work. This means employees get rewarded with points for any activities they make on social media related to work. Lastly, SIT maintains transparency to intend learning culture among each employees and students.

SIT makes difference among its Competitors:

According to the head of HR, SIT is different from competitors by intensifying HR policy that gives assistance to attract talent from the labor market. Friendly environment, pay scale, increment structure and accommodation, and subsidized interest rate are differentiable factors that facilitate SIT to establish itself and create image among other technical institutions. SIT organizes Tech-fest (three days of

Fig. 3 Outcomes of the analysis (SIT)

program) on the month of November, inter-college sports event like (Cricket, Volley ball and Basket ball) which construct the brand building affair.

Outcomes of the Case analysis through model representation:

See Fig. 3.

Analysis: (Case 3):

After the interview at SIT, it was found that employee is a big asset of the institution; the facilities they provide to the employee are really attractive; it creates its brand through **employee's word-of-mouth**. They are not that much actively following social networking site for employer branding rather they use SNS for post-campus (placement) follow-ups.

5.4 Case 4: Gandhi Engineering College (GEC) Bhubaneswar

Gandhi engineering college (GEC) is a technical college under Biju Patnaik University of Technology based in Bhubaneswar, which is a name to reckon within the field of technical education, was established in the year 2006, with due approval from AICTE, New Delhi and affiliation with BPUT, Odisha. At this moment, **GEC** team in Bhubaneswar has more than 300 members, including the top management of the company. The case company has taken by the researcher to manifest the social media and its role in employer branding. As per the conversation with top management, the inference drawn that GEC is recruiting non-teaching staff and teaching staff from Whatsapp, Instagram, Naukri, Monster, advertisement, media, and employees' references. The case company has explored the social networking sites like (Whatsapp and Instagram) for recruitment. However, GEC has used social sites for campus recruitment as it helps in shaping formal communication with the companies. Nevertheless, GEC has also benefited from Alumni through social networking sites.

Three key features of Gandhi engineering college:

GEC has main three key features which facilitate the usage of social media (online communication) within the organization. First of all, by inculcating strict academic

Fig. 4 Outcomes of the analysis (GEC)

process, GEC creates employer branding among the students as well as outsiders, by the use of SNS. Moreover, they are able to supervise the syllabus coverage, online question set, quiz test, and e-governance with help of ERP software. Second of all, GEC encourages employees to participate in intra-social media at work. This means employees get rewarded with points for any activities they make on social media related to work. Lastly, GEC maintains transparency to intend learning culture among each employees and students.

GEC makes difference among its Competitors:

According to the head of Academics, GEC is different from competitors due to its intense HR policy that gives assistance to attract talent from the employee market. Friendly environment, good pay scale, and attractive increment structure are differentiable factors that facilitates GEC to establish itself and create image among other technical institutions. GEC organizes so many research activities: national seminars, R&D activity, one international conference, and also do journals which construct the brand building affair.

Outcomes of the Case analysis through model representation:

See Fig. 4.

Analysis: (Case 4):

From the interview, it was observed that GEC strongly believes in social networking sites like Whatsapp, Instagram, Naukri, and Monster for recruitment and employer brand creation. Facebook, LinkedIn, and Twitter are not that much effective as compared to other SNS mentioned above. The Alumni and daily-posted videos of students and employees on social sites make a good impact on their institute branding. Their R&D is very strong. They believe in work and dedication. They maintain a good friendly environment and an effective communication network among superior and subordinates.

5.5   Case 5: Hi-Tech Institute of Technology (HIT), Bhubaneswar

Hi-Tech Institute of Technology (HIT), yet another feather added to the illustrious cap of Hi-Tech group of institutions for supporting the same adage. Having set milestones in Medical and Engineering education, the establishment of Hi-Tech Institute of Technology (HIT) in the year 2008, under Biju Patnaik University of Technology based in Bhubaneswar. At this moment, HIT team in Bhubaneswar has more than 500 members, including the top management of the company. The case company has taken by the researcher to manifest the social media and its role in employer branding. As per the conversation with top management, the inference drawn that HIT is recruiting non-teaching staff and teaching staff from Facebook, Whatsapp advertisement, media, and employees' references. The case company has not yet explored the social networking sites like (Twitter and LinkedIn) for recruitment. However, HIT has used social sites for campus recruitment as it helps in shaping formal communication with the companies. They do not focus on Alumni concept.

Key features of HIT:

HIT key features which facilitate the usage of social media (online communication) within the organization. First of all, by inculcating strict academic process, HIT creates employer branding among the students as well as outsiders. Moreover, they are able to supervise the syllabus coverage, online question set, quiz test, and e-governance with help of ERP software. Second of all, HIT encourages employees to participate in hygiene and safety awareness program which would helpful for students. Lastly, HIT maintains transparency to intend learning culture among each employees and students.

HIT makes difference among its Competitors:

According to the head of HR, HIT is different from competitors by intensifying HR policy that gives assistance to attract talent from the labor market. Friendly environment, pay scale, and increment structure are differentiable factors that facilitate HIT to establish itself and create image among other technical institutions.

Outcomes of the Case analysis through model representation:

See Fig. 5.

Analysis: (Case 5):

From my last interview, it was observed that this institution strongly believes in social networking sites like Whatsapp, Instagram, Naukri, and Monster for recruitment and employer brand creation. Facebook, LinkedIn, and Twitter are not that much effective as compared to other SNS mentioned above. The Alumni and daily-posted videos of students and employees on social sites make a good impact on their institute branding. Their R&D is very strong. They believe in work and dedication. They maintain a good friendly environment and a well-established communication system among all levels.

Fig. 5 Outcomes of the analysis (HIT)

6 Conclusion of the Study

In conclusion, the study suggests that the institutions are aware of social networking sites and they are using it for different purposes to connect with alumni, stakeholders and to connect with even campus recruiters. However, as for using it for employer branding, it has not yet shaped up very well. Other than two institutions CUTM and GEC, who are trying very well to take these social networking site services for employer branding activity, rest others have not yet thought over using it as a strong source of employer branding. However, through this particular study and these cases, we can conclude that there lies the high potential for using these social networking sites for creating employer branding and it should be used by educational institutions.

Acknowledgements It is hereby I have declared that all the information related to my survey is valid and ethical, which is collected by the authorized person on the institutions. Here, I have mentioned the details of all authorized persons in a simple manner:
ITER (SOA UNIVERSITY), Data collected in personal interview is ethical and authenticated by authorized person "Prof. Dr. Sujit Kumar Das (placement head), SOA University".
SILICON (BPUT), Data collected in personal interview is ethical and authenticated by authorized person "Prof. Saroj Kanta Misra (Director of Silicon), BPUT University".
Centurion University of Technology and Management (CUTM) Bhubaneswar, Data collected in personal interview is ethical and authenticated by authorized person "Prof. Sudhansu Nayak (placement head), CUTM University".
Gandhi engineering college (GEC) Bhubaneswar, Data collected in personal interview is ethical and authenticated by authorized person "Dr. Rati Ranjan Sabat (Principal), BPUT University".
Hi-Tech Institute of Technology (HIT) Bhubaneswar, Data collected in personal interview is ethical and authenticated by authorized person "Prof. Dr. Jayakrushna Moharana (Principal), BPUT University".

References

1. Burns, C.A., Veeck, A., Bush, F.R.: Marketing Research. Pearson Education Limited, Harlow (2017)
2. Backhaus, K., Tikoo, S.: Conceptualizing and researching employer branding. Career Dev. Int. **9**(5), 501–517 (2004)

3. Biswas, D., Suar, D.: Antecedents and consequences of employer branding. J Bus. Ethics 1–16 (2014)
4. The Study of Backhaus and Tikoo.: Employer Branding and its Effect on Organizational Attractiveness via the World WideWeb (2004), https://mafiadoc.com/
5. The Corporate Site.: www.facebook.com/jobvite, www.jobvite.com, visited and retrieved on 30 October 2018
6. The Corporate Site.: www.twitter.com/jobvite, www.jobvite.com, visited and retrieved on 30 October 2018
7. The Corporate Site.: www.linkedin.com/company/jobvite, www.jobvite.com, visited and retrieved on 30 October 2018

A Novel Approach for Artifact Removal from Brain Signal

Sandhyalati Behera and Mihir Narayan Mohanty

Abstract Brain signal is an essential physical characteristic of human being. The analysis of brain signal is most useful in many different fields including medical analysis, brain mapping, and crime analysis. Sometimes artifacts occur at the time of data acquisition. The removal of artifact from the signal is highly necessary for analysis and diagnosis. The task is very cumbersome. In this paper, authors have tried to remove it in a novel manner. Twenty-four channel EEG data is collected from which channel FP1 data found with artifact. The threshold is set first using statistical methods like kurtosis and MSE, and artifactual signals are identified. Further by visualization, the window is fixed for artifact signal. DWT is applied to remove the artifact and reconstructed the signal by taking that windowed signal. The inverse DWT provides a better result as compared to earlier method and shown in the result section.

Keywords Electroencephalogram (EEG) · Multiscale entropy (MSE) · Artifact removal · Kurtosis

1 Introduction

The sources of the generation of artifact define its classification. Artifact generated from the external sources is called external artifact. There are many causes of external artifact like electrode and lead movement. The artifacts generated from the human body during the time of recording of EEG signal are called physiological artifacts

S. Behera
Department of Electrical and Electronics Engineering, ITER, Sikhsha 'O' Anusandhan (Deemed to be University), Bhubaneswar, India
e-mail: sandhyalatibehera@soa.ac.in

M. N. Mohanty (✉)
Department of Electronics and Communication Engineering, ITER, Sikhsha 'O' Anusandhan (Deemed to be University), Bhubaneswar, India
e-mail: mihir.n.mohanty@gmail.com

© Springer Nature Singapore Pte Ltd. 2020
S. Patnaik et al. (eds.), *New Paradigm in Decision Science and Management*,
Advances in Intelligent Systems and Computing 1030,
https://doi.org/10.1007/978-981-13-9330-3_4

[1]. The ocular and muscle artifacts are generally two physiological artifacts. The eye blink, eye movement, and the movement of the eyeballs are the main causes of ocular artifact. The movement of the tongue is the main reason of muscle artifact [2]. The proposed method concerns with the ocular artifact.

The removal of artifact is the main challenge among the researchers. A number of time domain and frequency domain methods are used for artifact removal which include regression and filtering methods [3]. The regression method is based on subtracting artifact from the signal which creates undesired effect [4]. The filtering method lacks in selecting frequency band because artifact can occur at any frequency in EEG signal [5–9]. The time–frequency domain maps the time domain signal to frequency domain. The short-time Fourier transform (STFT) and wavelet transform are the main time–frequency domain methods which are suitable for non-stationary signal analysis. The wavelet transform uses a fixed window with fixed shape [10]. Due to this, wavelet transform is suitable for analyzing time-varying non-stationary signals. As EEG signal falls into this category, the wavelet transform applied for denoising the EEG signal [11].

An alternate method which uses higher-order statistics can detect the artifact in EEG due to small eye blinks [12]. The kurtosis which is a fourth-order moment is used to detect the transients in EEG, and a standard threshold is applied to detect artifact of specific spectral patterns [13]. The entropy which is considered as a degree of disorder of the system is also used as a statistical descriptor of the sharp changes within the EEG signal. The transients are the artifact in EEG signal has low entropy [14–16].

The proposed method uses kurtosis and multiscale entropy for identification of the artifact-contaminated channels in EEG signal. The artifact window is chosen visually, and then the wavelet transform is applied for denoising the window. This paper is organized as: Sect. 1 gives the brief introduction about the EEG artifact and some denoising methods, Sect. 2 describes about the proposed method, Sect. 3 result and discussion, and at the end conclusion.

2 Overview of Proposed Method

The proposed method framework is given in Fig. 1. This includes normalization of EEG data, detection of artifact channels, and removal of artifact. Kurtosis and multiscale entropy are used for detection of artifact and for removing artifact Wavelet transform is used.

2.1 Kurtosis

Kurtosis measures the degree of sharpness of the distribution. Highly positive kurtosis indicates the sharp distribution; negative kurtosis indicates the flat distribution. As

Fig. 1 Workflow of the
proposed method

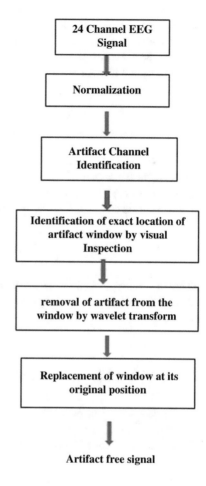

artifacts are the sharp changes in amplitude in EEG signal, kurtosis is used as a
measure to detect the sharp changes [17]. The kurtosis is calculated using the formula

$$k = m_4 - 3m_2^2 \tag{1}$$

where second-order moment $m_2 = E\{(x - m_1)^2\}$ and fourth-order moment $m_4 = E\{(x - m_1)^4\}$. E is the expectation function.

2.2 Multiscale Entropy (MSE)

The artifacts in EEG signal are of high probability, and its entropy is expected to be
low. The multiscale entropy is found by scaling the data in each channel, and then

the average is calculated for each scaled window [18, 19]. In the proposed method, each channel in EEG recordings has 2560 data samples. For calculating multiscale entropy, the scale factor is taken as 20 and the size of each window is 128.

The average of each scaled window is calculated as:

$$y_j^\tau = \frac{1}{\tau} \sum_{i=(j-1)\tau+1}^{j\tau} u_i; \, 1 \le j \le \frac{N}{\tau} \tag{2}$$

where τ is the scale factor, N is the number of data point in each channel in EEG signal, u_i is the signal of ith channel in EEG recordings, $\frac{N}{\tau}$ is the length of each scaled window, y_j^τ is the average of each scaled window.

The MSE is calculated for each window as:

$$\text{MSE}(m, r) = \log\left(\frac{B_r^m}{A_r^m}\right) \tag{3}$$

where m is taken as 2 and r is the tolerance which is 0.2 times the standard deviation of the data vectors. m is the length of matching templates; B and A track the template matches.

2.3 Artifact Channel Identification

For the identification of artifact channels, a statistical threshold is calculated by considering two-sided student's t-distribution with 96% CI of mean.

The threshold for MSE is calculated as:

$$t_{\text{lower}} = \bar{x} - \frac{s}{\sqrt{N}} \times t_{N-1} \tag{4}$$

where $\bar{x}$ is the mean of EEG signal, s is the standard deviation of EEG signal, t_{N-1} is taken as 2.201 and sample size N is taken as 256. The MSE lower than t_{lower} is considered as artifact-contaminated channel.

The threshold for kurtosis is calculated as:

$$t_{\text{upper}} = \bar{x} + \frac{s}{\sqrt{N}} \times t_{N-1} \tag{5}$$

The channels having kurtosis higher than the t_{upper} are artifact-contaminated channels.

2.4 Artifact Denoising

In the proposed method, artifact removal is done by discrete wavelet transform. The wavelet transform is applied only to the artifact-detected channels. Again, the wavelet transform is not applied to all the data samples of the artifact-detected channels. By doing this, the computation time and complexity drastically reduced. The artifact channel is visually inspected and artifact window is identified. The wavelet correction is applied to the identified artifact window. In the proposed algorithm, multiresolution-based discrete wavelet transform with biorthogonal wavelet as a mother wavelet is applied to the artifact window. The wavelet correction is based on a statistical threshold and the statistical threshold is calculated using Eq. (6).

$$t_w = \sqrt{2 \log N} \times \text{var} \tag{6}$$

where var is the variance of the wavelet coefficient and is calculated as:

$$\text{var}^2 = \text{median}(|w(j, k)|/0.6745) \tag{7}$$

where $|w(j, k)|$ = absolute value of wavelet coefficients. N = number of wavelet coefficients.

The wavelet coefficients, which are greater than the threshold, are substituted by zero. After substitution, inverse wavelet transform is applied for signal recovering the original signal.

3   Results and Discussion

Twenty-four channel EEG signal of 2560 samples is taken from CHB-MIT scalp EEG database [20]. Out of 24 channels, one channel is not used. So total 23 channel EEG signal is used in this proposed method for the processing. Out of 23 channels, 11 channels (1,2,5,8,9,10,13,14,16,19,20) are detected as artifact-contaminated channels. The detection of artifact channels is based on the threshold, which is calculated by considering kurtosis and multiscale sample entropy (MSE) of the EEG data samples. The threshold is calculated by using Eqs. (4) and (5). After the detection of artifact channels, a single channel (i.e., FP1) is taken for processing. Two windows, window 1 and window 2, where the signal attains its peak are considered as artifact position in the signal as shown in Fig. 2. The windows are chosen of the same sample size. Two artifact windows are shown in Figs. 3 and 4, respectively.

Instead of applying all data samples of channel 1 for wavelet correction, only window 1 and window 2 are applied for wavelet correction. This reduces execution time, thereby decreases the complexity.

The threshold is calculated from the wavelet coefficients using the Eq. (6). The wavelet coefficients which are greater than the threshold are substituted by zero.

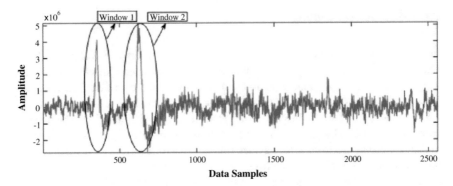

Fig. 2 Plot of signal of channel 1. Window 1 and window 2 which is indicated by ellipse are the artifact position

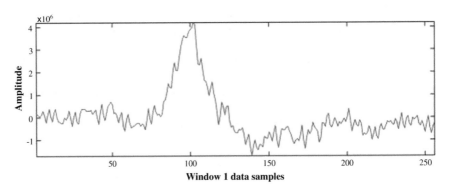

Fig. 3 Window 1 having 256 samples

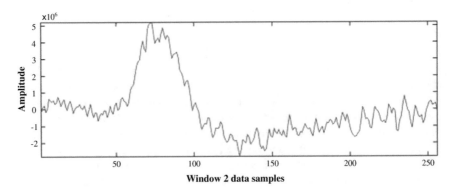

Fig. 4 Window 2 having 256 samples

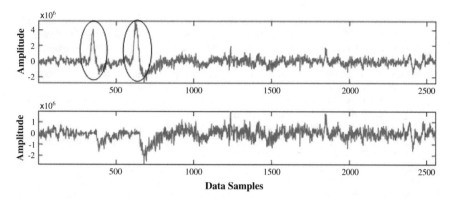

Fig. 5 Signal of channel 1 before and after removing the artifact

The inverse wavelet transform is applied for reconstruction. After reconstruction, the artifact-free windows are placed at their original position. The channel 1 signal before and after the proposed artifact-free method is shown in Fig. 5.

4 Conclusion

In this paper, the artifact removal method from the brain signal shows a good result. Because of the statistical threshold which is applied on kurtosis and multiscale entropy, the identification is excellent and helps for the removal of artifacts. The wavelet-based removal method which is applied to the artifact window in this work saves time and complexity. For future scope, the removal method may be optimized and will be considered for different types of artifact.

References

1. Mahajan, R., Morshed, B.I.: Unsupervised eye blink artifact denoising of EEG data with modified multiscale sample entropy, kurtosis, and Wavelet-ICA. IEEE J. Biomed. Health Inform. **19**(1), 158–165 (2015)
2. Aydemir, O., Pourzare, S., Kayikcioglu, T.: Classifying various EMG and EOG artifacts in EEG signals. Przegląd Elektrotechniczny **88**(11a), 218–220 (2012)
3. Tandle, A., Jog, N.: Classification of artefacts in EEG signal recordings and overview of removing techniques. Int. Conf. Comput. Technol. 46–50 (2015)
4. Dhiman, R., Saini, JS., Priyanka, AM.: Artifact removal from EEG recordings–an overview. Proc. NCCI. 1–6 (2010)
5. Woestenburg, J.C., Verbaten, M.N., Slangen, J.L.: The removal of the eye-movement artifact from the EEG by regression analysis in the frequency domain. Biol. Psychol. **16**(1–2), 127–147 (1983)

6. Shooshtari, P., Mohammadi, G., Molaee Ardekani, B., Shamsollahi, MB.: Removing ocular artifacts from EEG signals using adaptive filtering and ARMAX modeling. In: Proceeding of World Academy of Science, Engineering and Technology, Vol. 11. EPFL-CONF-153221, 277–280 (2006)
7. Mammone, N., La Foresta, F., Morabito, FC.: Automatic artifact rejection from multichannel scalp EEG by wavelet ICA. IEEE Sensors J. **12.3**, 533–42(2012)
8. Croft, RJ., Barry, RJ.: Removal of ocular artifact from the EEG: a review. Neurophysiol. Clin./Clin. Neurophysiology **30.1**, 5–19 (2000)
9. Teplan, M.: Fundamental of EEG measurement. Measurement science review (2002)
10. Li, M., Yang, L., Yang, J.: A fully automatic method of removing EOG artifacts from EEG recordings. Commun. Inf. Sci. Manag. Eng. 1–6 (2011)
11. Kumar, PS., Arumuganathan, R., Sivakumar, K. Vimal, C.: A wavelet based statistical method for de-noising of ocular artifacts in EEG signals. Int. Comput. Sci. Netw. Sec. **8.9**, 87–92 (2008)
12. Delorme, A., Sejnowski, T., Makeig, S.: Enhanced detection of artifacts in EEG data using higher-order statistics and independent component analysis. Neuroimage. **34.4**, 1443–9 (2007)
13. Akhtar, MT., Mitsuhashi, W., James, CJ.: Employing spatially constrained ICA and wavelet denoising, for automatic removal of artifacts from multichannel EEG data. Signal Process. **92.2**, 401–16 (2012)
14. Kannathal, N., Choo, ML., Acharya, UR., Sadasivan, PK.: Entropies for detection of epilepsy in EEG. Comput. Methods Programs Biomed. **80.3**, 187–94 (2005)
15. Wu, SD., Wu, CW., Lee, KY., Lin, SG.: Modified multiscale entropy for short-term time series analysis. Physica A Stat. Mech. Appl. **392**(23), 5865 (2013)
16. Hyvarinen, A., Oja, E.: Independent component analysis: algorithms and applications. Neural Netw. **13**(4–5), 411–430 (2000)
17. Ghandeharion, H., Erfanian, A.: A fully automatic ocular artifact suppression from EEG data using higher order statistics: Improved performance by wavelet analysis. Med. Eng. Phys. **32**(7), 720–9 (2010)
18. Shaw, L., Routray, A., Sanchay, S.: A robust motifs based artifacts removal technique from EEG. Biomed. Phys. Eng. Express 3(3), 035010 (2017)
19. Behera, S., & Mohanty, M.N.: A statistical approach for ocular artifact removal in brain signals. In: 2018 2nd International Conference on Data Science and Business Analytics (ICDSBA), pp. 500–503. IEEE (2018)
20. Goldberger, A.L., Amaral, L.A., Glass, L., Hausdorff, J.M., Ivanov, P.C., Mark, R.G. … Stanley, H.E.: PhysioBank, PhysioToolkit, and PhysioNet: components of a new research resource for complex physiologic signals. Circulation. **101**(23), e215–e220 (2000)

A Novel Approach to Heterogeneous Multi-class SVM Classification

Ashish Kumar Mourya, Harleen Kaur and Moin Uddin

Abstract The usability of multi-class support vector machine for processing and classifying heterogeneous data has defined in proposed study. Multi-class approaches are machine learning algorithms that construct a new subset of learning classifiers and then classify new models by adding the results of the base classifiers in various ways (characteristically by weighted or un-weighted voting). To construct the multi-class support vector machine classifier into subsets, first we have to decide different kernel tricks for getting better results. However, processing heterogeneous data by support vector machine based on different kernel functions have rarely used. The training time, cross-validation, classification accuracy of single class and multi-class are evaluated here. The sum rule for combining the subsets of multi-class classifiers has been used in this study. The experimental results analysis shows that our proposed multi-class approach achieves highest accuracy with heterogeneous data in comparison with other methods.

Keywords Support vector machine (SVM) · Multiple kernel learning (MKL) · Radial base function (RBF) · Classification · Decision boundary · Heterogeneity

1 Introduction

Machine learning classifiers have been used for classification tasks. Support vector machine had proposed during 1998, which has been elaborately defined [1, 2], as suggested and supported by the latter other researchers. The input vectors initially come in dual structure of dot products. This algorithm can exist comprehensive to non-linear cataloging; it can be accomplished by mapping to the inputs in a high-dimensional feature space via an Apriori algorithm, which is chosen as linear mapping function Φ. Therefore, by drawing a distinguishing hyperplane in the given

A. K. Mourya · H. Kaur (✉) · M. Uddin
Department of Computer Science & Engineering, School of Engineering Sciences and Technology, Jamia Hamdard, New Delhi, India
e-mail: harleen.unu@gmail.com

© Springer Nature Singapore Pte Ltd. 2020
S. Patnaik et al. (eds.), *New Paradigm in Decision Science and Management*,
Advances in Intelligent Systems and Computing 1030,
https://doi.org/10.1007/978-981-13-9330-3_5

vector space will lead to a new non-linear decision boundary in the n-dimensional vector. Classification problem diagnosis using support vector machine has two parts, i.e., training and prediction, respectively. During the phase of training, the support vector machine takes inputs as datasets with fixed vector length as an example. Furthermore, each example must have an associated binary classifier tag. Let set positive and negative labels to denote the '+1' class and '−1' class, respectively. To separate positive (+1) from the negative (−1) class, the support vector machine algorithms find the hyperplane in the input space.

High-dimensional classifier [3–5] is a modern learning paradigm best fitted for classification and vector regression. In classification learning problem, input datasets are defined as n times of observations

$$(x_1, y_1), (x_2, y_2), \ldots, (x_n, y_n) \in \mathbb{X} \times \mathbb{Y} \tag{1}$$

where set of X characterizes the input values unfolding the patterns and Y set includes the resultant class labels. In an easiest lemma, output class contains two elements, foremost to binary arrangement, where the classes are denoted via.

$$\mathbb{Y} = \{+1, -1\} \tag{2}$$

For non-linearly separable class data, support vector fits a non-linear method, maximizing the edge between the separable classes in order to furnish the finest generalization presentation. Nevertheless, as if features are habitually non-linearly separable, it is the core of learning classifiers that two annotations "being near in input space" should have a related cost. Consequently, support vectors fit in the kernel tricks.

$$\mathcal{K} = \mathbb{X} \times \mathbb{X} \Rightarrow \mathbb{R} \tag{3}$$

Signify the comparison of two different observations. The kernel trick (k) needs to be adequate several conditions, e.g., regularity, and positive semi-definiteness. The trick itself can be construed as a $(\cdot)$ product in a high-dimensional boundary [6–8]. It improves the learning classifier algorithm by an implied mapping of the input data into a vector attribute space, where a learning classifier is fitting.

Multi-class or ensemble the classifier is one of the best techniques currently used with support vector machine. Its principle relies on the divide and conquer method. Ensemble method works on the approach of dividing the original input boundary keen on a small group of several subspaces to increase classification accuracy and performance [9]. In this paper, support vector machine has used bagging and boosting algorithms and compared with the heterogeneous data. In this ensemble approach, the following facts of evidence have occurred: Combining more than one classifier can direct to an ensemble learner more accurate than each liner support vector machine on entire datasets and powerful than multi-class from a one-kernel type unaccompanied. On the other hand, boosting is only prolific if single support vector sufficiently weak;

otherwise, classifier may be overfitted. The ensemble method of support vector has attained higher accuracy in comparison with single one.

2   Literature Review

Support vector machines (SVMs) have shown their efficiency in processing and classifying image classification. [10] introduces a more competent algorithm that can gain knowledge of whole afforest with similar set of generalized assessment trees under unlike fall values concurrently without replicating any node that is general across several trees and provided an intricacy investigation of the classifier. Both filter and wrapper advantages have applied in the hybrid model by using the support vector machine. Researchers of [10] have used the C-ascending technique to modify the support vector machine to predict time series forecasting in financial data. They achieved accuracy by transforming the regularized the risk function.

The research scientist of this research demonstrates that the proposed study attempt can be beneficial to the soft computing and artificial intelligence community. It is also acknowledgeable that semantic consolidation lies in the core of many data warehouses and artificial intelligence problems will necessitate the solutions that blend database and artificial intelligence techniques. Mounting such solutions may be facilitating with even more efficient association between the different communities in the upcoming decade. [11] Developed PerCon library platform for large heterogeneous datasets to support access and analysis. It had been started with the goal to develop an environment that will integrate much if the data ingestion, management, and analysis activities for research. PerCon have three-layer architecture that consists of application layer, a middleware layer, and resource layer.

Privacy-preserving data mining has developed confident data cleaning and feature extracting computations whereas providing few safeguards for the primary data. Lots of work also has been going on seclusion preserving data cleaning strategies and classifiers with association rules, clustering, and statistical analysis. [12] has also been worked on privacy-preserving data mining by using cryptographic methods for the decision tree to provide strong privacy guaranteed. Nevertheless, randomized methods of [13] are considerably more proficient than cryptographic methods with minimum privacy guarantees.

Cooperative Bayesian machine learning classifier in a distributed environment for homogeneous datasets had introduced by [14]. In the proposed work, multiple Bayesian agents approximation the parameters (prior, posterior) of the objective distribution, and inhabitants classifiers join the Bayesian model results. [15] presents learning Bayesian network model from dispersed non-homogeneous datasets. All disseminated site gain knowledge of their local Bayesian network here. Indeed, after this all sites identify a few records to the main site. A Bayesian network has structure and associated probabilities.

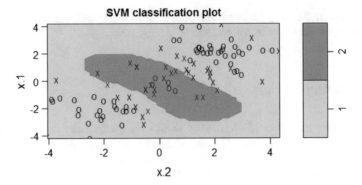

Fig. 1 SVM Classification using RBF kernel

3 Classification Problem

In machine learning, other than regression tasks, support vector classier is an effective method. The optimal separate hyperplane can be constructed in a feature space. Consider a simple example with two classes $\mathbf{y} = \mathbf{1}, -\mathbf{1}$, then a hyperplane that separates all data $\mathbf{w}^T\mathbf{X} + \mathbf{B} = \mathbf{0}_{+1}^{-1}$ as given in Fig. 1. By a function (higher dimensional) Φ, we map an input vector y into a feature space. If we have $\mathbf{g}$ samples $\{(x_i, y_i)\}_i^g = 1$, support vector machine searches for a non-linear resolution function $[\mathbf{f}(x) = \text{sgn}(\mathbf{w}^T x + \mathbf{b}]$, x: test data) with a maximum margin between other classes in the feature space. For solving the below quadratic equation problem (4), the values of W and b are taken [9, 16].

$$\min_{w, b, \xi} \frac{1}{2} \| W^T \|^2 + C \sum_{i=1}^{n} \xi \tag{4}$$

Already given w is an infinite variable, ξ used as slack variable, and for constant variable (0 or 1), we used C which penalizes for training error.

For obtaining the finest resulting function in the attribute space, the multi-class support vector maps a training dataset into an attribute space $\Phi(x_i) = z_i$. Non-linear version of the high-dimensional vector resolves the quadratic problem define below:

$$\min_{w, b} \frac{1}{2} \| w, w \| + C \sum_{i=1}^{n} \xi_i \tag{5}$$

Fulfill as: $y_i\big((w, z_i) + b\big) \geq 1 - \xi_i$ where $\xi_i > 0$.

Now, we demonstrate feature space Φ where data is not linear or multilayered. Φ function maps the key pattern into a higher feature space. Let's take an example of

$$\text{input vector space } [X] = \Phi[X] \tag{6}$$

$$[\![X_1, X_2]\!] = \left[\!\left[X_1^2, X_2^2, \sqrt{2X_1 \times X_2} \right]\!\right] \tag{7}$$

where X_1 and X_2 represents two-dimensional input vector space and X_1, X_2 $\sqrt{2X_1 * X_2}$ denotes to three-dimensional feature vector space.

Now applying (6) among (7) we get

$$\Phi : x \rightarrow X = \Phi\langle x \rangle \tag{8}$$

Now applying polynym trick as

$$\mathbb{K} [\![X_1, X_2]\!] = \langle X_1, X_2 \rangle^2 \tag{9}$$

Now applying feature space Φ, we get as given below:

$$\Phi(x_1) = \left\| x_{11}^2, x_{12}^2, \sqrt{2_{x_{11}x_{12}}} \right\| \tag{10}$$

$$\Phi(x_2) = \left\| x_{12}^2, x_{22}^2, \sqrt{2_{x_{21}x_{22}}} \right\| \tag{11}$$

After solving (10) and (11) we get

$$K[\![x_1 x_2]\!] = \langle \Phi x_1, \Phi x_2 \rangle \tag{12}$$

We have another kernel trick to transform non-linear classifiers text data mining and another dataset called Gaussian kernel function.

$$\mathcal{F}(\chi) = \sum_i^t \alpha_i \psi_i k\{\chi_i \chi\} + \mathscr{b} \tag{13}$$

where training datasets size is t and x_i is support vectors

$$k\langle x, x^2 \rangle = (\exp -\frac{\left\| x, x^2 \right\|^2}{2\sigma^2}) \tag{14}$$

The ethical meaning to represent the kernel methods has drawn from assorted data type (Fig. 2). The kernel function has explained by the similarity between any given sets of complex data objects. For example, an assortment of various objects may represent by $m \times n$ matrix of pair-wise similarity function values. The heterogeneous dataset of applied support vector machine contains distinct kernels that may use for all the data types, even matrices have to combine numerically. This approach can combine heterogeneous gene (micro-array gene synthesis) data and phylogenetic description in an un-weighted mode, afterward ignoring the correlations between

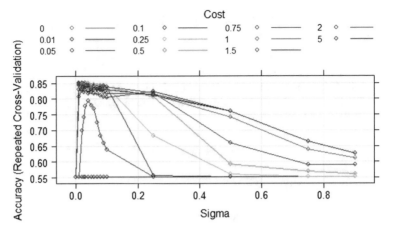

Fig. 2 Accuracy of single support vector

various datasets and applying numerical operators to the center of the support vector machine and within-dataset correlations surrounded by features [11, 17–20]. An unweighted addition of similarity function has been used successfully in the forecasting of protein relations. A few years ago, researchers proposed a method to combine different kernels within the SVM algorithm that is multiple kernel learning (MKL) methods [21–24].

4   Proposed Novel Approach for Heterogeneous Data

Combining more than one learner to accept different features as inputs is a significant topic in data mining. The input data is randomly partitioned into subsets of similar size for training purpose. We used the sum rule for combining subsets of odds results of multi-class support vector classifiers. Therefore, a two-layer classifier of support vector network for each one-to-all classification task has constructed (Fig. 3).

The sum rule function is the simple and competent method for adding the subsets results. This study used sum rule in ensemble architecture. When the learning algorithms have minimum faults and work in autonomous attribute space or domain, it is very knowledgeable to join their results. Therefore, the sum rule function has applied here to decide the final decision of the multi-class approach.

The resultant outputs $O_j^k(x^k)$ of classifier j of η diverse classifier added via the sum function as:

$$O_j(x^1, ..., x^n) = \sum_{\mathcal{K}=1}^{n} O_j^k(x^k) \tag{15}$$

Fig. 3 Proposed model for multi-class support vector

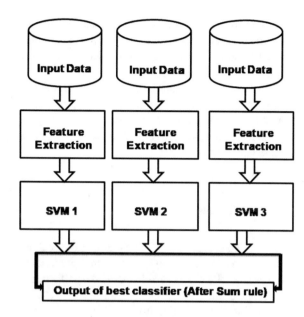

Table 1 Accuracy comparison of multi-class learner

Classifier	Accuracy of training data	Accuracy after cross-validation
Single	87.06	85.04
AdaBoosting	88.03	87.01
Bagging	86.04	83.06
Our method	92.25	90.09

where x^k is the nominal representation of Kth classifier. The maximum probability class sum is considering as the resultant feature label related with input feature.

5 Experimental Results

Within proposed experiment, we have used two input sets from UCI data warehouse. First inputs are relying on heart disease that contains 303 records. Among 303 records, six records have missing values. These missing values have removed by feature extraction. Another dataset used in this study is breast cancer [25] that has 699 records with 16 incomplete records. An experimental approach has been evaluated in the R programming software. All above-said datasets have partitioned keen on training data moreover testing data. The training accuracy and testing accuracy are defined in Table 1.

In this study, testing accuracy is lower in comparison with training. In boosting technique, AdaBoosting is most popular technique where each sub-classifier has trained using a different training set. However, the kth classifiers have been trained sequentially not independently.

The radial base function has used less training time in comparison with other kernel functions such as linear and polynomial. The AdaBoosting approaches use maximum training time in comparison with bagging approaches. We found that ensemble approaches provide better results on heterogeneous data in comparison with single-class classifier.

6 Conclusion

The non-linear approach of machine learning classifier for heterogeneous data has evaluated in proposed study. Distributing the unique input space into various input subsets generally works for advancing the performance. In this paper, performance of single- and multi-class support vector machine over heterogeneous data is evaluated. The training and testing accuracy of heterogeneous datasets has also compared here. This study combined the ensemble support vector with sum rule for combining subsets. We have assessed the proposed approach over heterogeneous data. Our tentative results indicate that a multi-class support vector with the sum rule handles more than one feature competently and achieves significantly good results in comparison with linear support vector machine. In future, we will work on the extension of multi-class approach with the multiple kernel learning.

References

1. Breiman, L.: Bagging predictors. Mach. Learn. **24**(2), 123–140 (1996)
2. Caputo, B., Sim, K., Furesjo, F., Smola, A.: Appearance-based object recognition using SVMs: which kernel should I use? In: Proceeding of NIPS workshop on Statistical methods for computational experiments in visual processing and computer vision, Whistler (2002)
3. Chang, C., Lin, C.: LIBSVM: a library for support vector machines. ACM Trans. Intell. Syst. Technol. (TIST) **2**(3), 27 (2011)
4. Chang, E.Y., Zhu, K., Wang, H., Bai, H., Li, J., Qiu, Z., Cui, H.: PSVM: Parallelizing support vector machines on distributed computers. Adv. Neural. Inf. Process. Syst. **20**, 16 (2007)
5. Cortes, C., Vapnik, V.: Support vector machine. Mach. Learn. **20**(3), 273–297 (1995)
6. Freund, Y., Schapire, R.E.: A decision-theoretic generalization of on-line learning and an application to boosting. In: Proceedings of the Second European Conference on Computational Learning Theory, EuroCOLT '95, pp. 23–37, London, UK (1995)
7. Mercer, J.: Functions of positive and negative type, and their connection with the theory of integral equations. Philosophical Trans. Royal Soc. London **209**, 415–446 (1909)
8. Meyer, O., Bischl, B., Weihs, C.: Support vector machines on large data sets: Simple parallel approaches. In: Spiliopoulou M., et al. (eds.) Data Analysis, Machine Learning and Knowledge Discovery. Springer (2013)

9. Opitz, D., Maxlin, R.: Popular ensemble methods: an empirical study. J. Artif. Intell. Res. **11**, 169–198 (1999)
10. Zhao, H., Sinha, A.P.: An efficient algorithm for generating generalized decision forests. IEEE Trans. Syst. Man Cybernetics Part A **35**(5), 754–762 (2005)
11. Wickramaratna, J., Holden, S.B., Buxton, B.: Performance degradation in boosting. In: Kittler, J., Roli, F. (eds.) Proceedings of the 2nd International Workshop on Multiple Classifier Systems, number 2096 in LNCS, pp. 11–21. Cambridge, UK (2001)
12. Dittrich, P.Z.: Three decades of data Integration. In: 18th IFIP World Computer Congress (2004)
13. Agrawal, D., Aggarwal, C.C.: On the design and quantification of privacy preserving data mining algorithms. In: Proceedings of the Twentieth ACM SIGMOD-SIGACT-SIGART Symposium on Principles of Database Systems, pp. 247–255. ACM (2001)
14. Troyanskaya, O.G., Dolinski, K., Owen, A.B., Altman, R.B., Botstein, D.: A Bayesian framework for combining heterogeneous data sources for gene function prediction (in Saccharomyces cerevisiae). Proc. Natl. Acad. Sci. U.S.A. **100**, 8348–8353 (2003). https://doi.org/10.1073/pnas.0832373100
15. Chen, R., Sivakumar, K., Kargupta, H.: Collective mining of Bayesian networks from distributed heterogeneous data. Knowl. Inf. Syst. **6**(2), 164–187 (2004)
16. Crammer, K., Singer, Y.: On the algorithmic implementation of multiclass kernel-based vector machines. J. Mach. Learn. Res. **2**, 265–292 (2002)
17. Scholkopf, B., Smola, A.: Learning with Kernels: support Vector Machines, Regularization, Optimization and Beyond. MIT Press (2002)
18. Wang, S., Mathew, A., Chen, Y., Xi, L., Ma, L., Lee, J.: Empirical analysis of support vector machine ensemble classifiers. Expert Syst. Appl. **36**(3), 6466–6476 (2009)
19. Weston, J., Watkins, C.: Support vector machines for multi-class pattern recognition. In: Proceedings of the 7th European Symposium on Artificial Neural Networks (ESANN), vol. 99, pp. 61–72 (1999)
20. Kaur, H., Alam, M.A., Jameel, R., Mourya, A.K., Chang, V.: A proposed solution and future direction for blockchain-based heterogeneous medicare data in cloud environment. J. Med. Syst. **42**(8), (2018)
21. Cristianini, N., Shawe-Taylor, J.: Support Vector Machines (2000)
22. Dimitriadou, E., Hornik, K., Leisch, F., Meyer, D., Weingessel, A.: Misc functions of the department of statistics (e1071), TU Wien. R Package. 1–5 (2008)
23. Kaur, H., Lechman, E., Marszk, A.: Catalyzing Development Through ICT Adoption: The Developing World Experience, pp. 288. Springer Publishers, Switzerland (2017)
24. Baboota, R., Kaur, H.: Predictive analysis and modelling football results using machine learning approach for english premier league. Int. J. Forecast. **35**(2), 741–755 (2019)
25. Mangasarian, O.L., Wolberg, W.H.: Cancer diagnosis via linear programming. SIAM News **23**(5), 1–18 (1990)

An Empirical Study of Big Data: Opportunities, Challenges and Technologies

Shafqat Ul Ahsaan, Harleen Kaur and Sameena Naaz

Abstract Nowadays, Big Data is considered to be the emerging field of research. However, it has gained momentum a decade ago and is still in its infancy stage. Big Data is a huge volume of data that is produced from various sources and thus traditional database systems are incapable of processing such a voluminous data. This is the need of hour to use advanced tools and methods to get value from Big Data. However, the pace of data generation is greater than ever which leads to numerous challenges such as data inconsistency, security, timeliness and scalability. Here, in this paper, a brief introduction to the Big Data technology and its importance in various fields. This paper mainly focuses on characteristics of Big Data and gives insights of a brief clarification of challenges related to Big Data and some analytical methods such as Hadoop and MapReduce. The various tools and techniques that can be used in Big Data technology have been discussed further.

Keywords Big Data · Hadoop · MapReduce · Machine learning · Artificial intelligence · Support vector machine

1 Introduction

The sudden increase of information that is being generated online by means of social media, Internet and worldwide communications has increasingly rendered the data-driven learning. A new study revealed that over 4 million queries are being received by Google every minute, e-mails sent by users reach the limit of 200 million messages, 72 h of videos are uploaded by YouTube users, 2 million chunks of content are shared over Facebook, and 277,000 Tweets are generated every minute on Twitter,

S. Ul Ahsaan (✉) · H. Kaur · S. Naaz
Department of Computer Science & Engineering, School of Engineering Sciences and Technology, Jamia Hamdard, New Delhi, India
e-mail: mailforshafqat@gmail.com

H. Kaur
e-mail: harleen.unu@gmail.com

© Springer Nature Singapore Pte Ltd. 2020
S. Patnaik et al. (eds.), *New Paradigm in Decision Science and Management*,
Advances in Intelligent Systems and Computing 1030,
https://doi.org/10.1007/978-981-13-9330-3_6

WhatsApp users share 347,222 photos, Instagram users post 216,000 new photos every minute [1, 2]. We are living in an era where data is growing on a large scale than ever before. According to the Computer World, 70–80% of data is considered to be in unstructured form in organizations [3]. The data, which derives from social media, forms 80% of the data globally and reports for 90% of Big Data. As stated by the International Data Corporations (IDC) annual digital universe study [4], the data is being produced too rapidly and by the estimation of 2020 it would touch the range of 44 zettabytes which would be ten times larger than it was in 2013 [5].

With the amount of data growing on a large and rapid scale, there may arise a situation when conventional analytical methods lack the ability to process such a voluminous data and therefore we require advanced algorithms and techniques in order to extract out information that best aligns with the user interests, which finally became the key to introduce a new technology to the world called Big Data. Hence, more powerful tools and technologies are compulsory to process the data and in fact, the generation of data on a large scale is the main foundation for the emergence and intensification of Big Data.

Big Data is gigantic and multifarious data that is out of the capability of traditional data warehousing tools to process. As the technology and services seemed to have progressed at a pace, it leads to the generation and extraction of such giant sum of data from several sources that can be heterogeneous [6]. The need for Big Data emerges from major companies like Google and Facebook. The data that is generated while using Facebook or Google is mostly of unstructured form, and it seems laborious to process a data that contains billions of records of millions of people. Therefore, Big Data can be defined as the quantity of data that is far-fetching from the potential of technology to pile up, handle and process in the most powerful and substantial way [7, 8].

Big Data can be elucidated on the basis of "3 V" model proposed by Gartner as: Big Data is a compilation of voluminous, frequently generated and multiple forms of data that requires cost-effective, advanced tools in order to process the data for better understanding and decision making [9].

Big Data consists of both unstructured and structured data. Big Data is a complex, unstructured data that requires a series of advanced and unique techniques for its storage, processing and analysis in order to transform it (data) into value.

This research has been arranged in seven sections. Introduction about Big Data is covered in this section. Section 2 highlights the importance of Big Data. Section 3 focuses on the characteristics of Big Data. Section 4 features the Hadoop architecture. Further, we have discussed Big Data challenges in depth in Sect. 5. In Sect. 6, we have looked and interpreted on various Big Data techniques, and lastly in Sect. 7 it emphasis on multiple Big Data tools that can be used to obtain the value from Big Data. Henceforth, the conclusion has been drawn in the last section.

2   Importance of Big Data

According to the survey in August 2010, it was derived out that the White House, Office of Management and Budget and Office of Science and Technology Policy announced that is a national challenge [10]. In order to unfold advanced Big Data tools and techniques, the National Institute of Health, the National Science Foundation, the U.S. Geological Survey, the Departments of Defense and Energy and the Defense Advanced Research Projects Agency declared a joint R&D proposal in March 2012 that will spend more than $200 million. The main objective and purpose of this conference are to make awareness and assimilate the technologies that are required to control and extract knowledge, also to reflect that awareness to other scientific fields [10, 11]. The focus of government is entirely confined on how to mine or extract value or knowledge from Big Data both within and across different disciplines and domains. Nowadays, Big Data seems to be employed in every field of life because of its crucial value. At present, Big Data is playing a very important role in our lives by changing the overall strategy of working, living and thinking. The significance of Big Data with distinguished perspective is discussed in this section.

2.1   Importance to National Development

As we can witness, the world from top to bottom has plunged into the ocean of data. The extreme access to Internet, cloud computing, IoT and other promising IT technologies has put a variety of data sources in a pace therefore making the data excessively complex. The study and deployment of Big Data in profundity will play a effective and crucial role in promoting unremitting financially viable development of countries and hence the companies across the world have competition among each other. Those days are not far away when Big Data will emerge as a turning point of fiscal increase; companies will promote and convert to the mode of Analysis as a Service (AaaS). With the rise of Big Data, the renowned companies of IT industry like IBM, Google and Microsoft have initiated their technical development. In the coming years, the strength of a country is measured and evaluated in terms of how its intensity to uphold the vast amount of data, its utilization and processing [5].

2.2   Importance to Industrialized Upgrades

At present, handling Big Data has become a general issue for IT industries and it comes with a series of striking challenges that these industries have to encounter. Industries have to make advancement in their core technologies in order to exploit the intricacy caused by unstructured data and to master ambiguity that occurs due to redundancy and shortage of data. The IT industries are in a race to mine demand-driven information and knowledge and finally taking advantage of value of Big Data. It is possible only when the IT companies upgrade their core technology and focus on how to extract more and more value from Big Data as possible [5].

2.3 Importance to Technical Research

The revolution of Big Data has forced the technical community to reassess its slant and has set off a swift change in scientific methods. The surfacing of Big Data has totally changed the scenario of research and gave birth to a new concept of research where the researchers have to mine only the information, knowledge and intelligence that is need of the hour [12].

2.4 Importance to Interdisciplinary Research

Big Data is emerging as a foundation of a new interdisciplinary discipline called Data Science. The discipline of Data Science accepts Big Data as its object of study and mainly focuses on the mining of facts from data. It covers a number of disciplines like information science, mathematics, network science, psychology and economics [13]. It uses different assumptions and approaches from different fields like machine learning, statistical learning, data warehousing, computer programming and pattern recognition [5].

2.5 Importance to Better Predict the Future

Multidimensional Data technology plays a key role in predicting future events by analyzing heterogeneous data efficiently and accurately that is being generated from multiple sources. Big Data analysis makes it possible to nourish viable development of society and also come through to new data service industries [14]. The predictive analysis that is carried out by virtue of Big Data addresses issues related to our society like hygiene and sustainable economic growth [5]. Ginsberg et al. found that if the number of queries that have been submitted to Google with keywords like "flu symptom" and "flu treatment" hikes in a particular province then subsequent to a few days/weeks the frequency of patients suffering from influenza will increase in hospitals in the respective areas [15]. After the analysis of such type of flu data, the data engineers will come to the light and get enlightened about the occurrence of influenza and deploy countermeasures in advance.

3 Big Data Characteristics

One view accepted by Gartner's Doug Laney defined Big Data as including three dimensions: volume, variety, and velocity. Thus, IDC defined it: "Big data technologies describe a new generation of technologies and architectures designed to economically extract value from very large volumes of a wide variety of data, by enabling high-velocity capture, discovery, and/or analysis" [16]. This paper defines Big Data based on 7 Vs'.

3.1 Volume

It is the infinite masses of data that are introduced each second.

3.2 Velocity

How frequently the data is being generated by means of different sources.

3.3 Variety

The multiple forms of data like music, pictures, text, financial transactions, videos, tweets, Facebook data, sensor data, etc.

3.4 Veracity

The intricacy of data which points to the certainty or quality of data. It means meaningfulness of data.

3.5 Validity

Validity and Veracity are not the same but have similar concept. Validity means the accuracy of data for the intended usage. Veracity leads to validity if the data is properly understood, it means that we have to check properly and appropriately whether the dataset is valid for a particular application or not.

3.6 Volatility

Volatility refers to the period for which we have to store the data. If volatility is not in place then a lot of storage space is wasted in storing data that is no more required, for instance a commerce company keeps the purchase history of a customer for 1 year only as after 1 year the warranty on the purchased item expires so there is no reason to store such data.

3.7 Value

It focuses on analytics and statistical methods, knowledge extraction and decision making.

4 Hadoop: Way to Process Big Data

4.1 Hadoop Distributed File System

With the rise of Big Data, accessing Big Data was a big challenge for the researchers that are stored on a cluster in a distributed manner; the common solution that has figured out to access Big Data is a cluster file system. The cluster file system provides access to data files that can be stored anywhere on a cluster. One of the most popular cluster file systems is Hadoop Distributed File System (HDFS). Hadoop Distributed File System as is designed to provide the facility to store voluminous data in a large-scale cluster [17], and it works on master–slave architecture as depicted in Fig. 1. It logically divides the file system into metadata and application data. In master–slave architecture, there are two types of computers called NameNode or master node and DataNode or slave node. The NameNode stores the metadata and another type of node called DataNode stores the application data. When a large data file is about to be stored, it splits into multiple blocks and these blocks are copied and stored among several DataNodes. In a Hadoop cluster, all the nodes are full-fledged connected in order to communicate with each other and the communication among different nodes in a cluster occurs with the help of TCP-based protocols. The namespace of the file system is maintained by NameNode and the mapping of file blocks is maintained by DataNodes. When a file has to be read, NameNode is communicated to get the location where the data blocks are stored and then data blocks are read and viewed from the closest DataNode [18].

HDFS file system offers two striking features as compared to traditional file systems; it is highly fault-tolerant and can store data in petabytes. It supports high bandwidth and scalability.

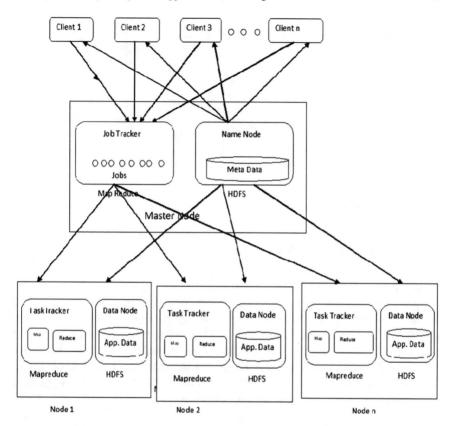

Fig. 1 Architecture of hadoop distributed file system (HDFS)

4.2 MapReduce

MapReduce is a programming model as shown in Fig. 2, used to refine for massive data files with the implementation of coordinated and disbursed algorithms on a cluster. MapReduce programming structure is sparked by the Map ( ) and the Reduce ( ) function. In Map ( ) step, the master node or the NameNode accepts the input file and partitions it into minor sub-problems, these sub-problems are then assigned to slave nodes or DataNodes. The slave nodes may further divide the problem into sub-problems. The slave node then handles these smaller problems and responds to the master node to which it is connected. In the Reduce ( ) step, the master node receives the result and combines them together to turn out the final result to the original problem that it has to solve [19].

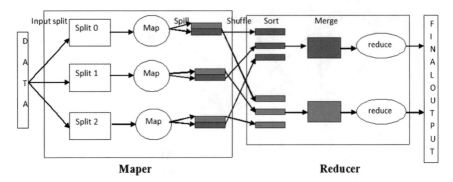

Fig. 2 MapReduce framework

5 Challenges of Big Data

As we know that Big Data is an emerging field of research and is still in underdeveloped stage. It comes with a set of characteristics that make Big Data easy and simple to understand but unfortunately these characteristics have become a challenging factor [20]. These characteristics make it a hindrance and a challenge to get useful information from. Researchers, data scientists and organizations have to put every effort in order to cope up with these challenges and make Big Data a revolution to the field of science. Big Data exists with big challenges that are mainly categorized as data challenges, processing challenges and management challenges.

5.1 Data Challenges

5.1.1 Volume

The massive quantity of data that is derived every second constitutes volume of Big Data [21]. There are multiple numbers of sources that play a key role in producing this vast portion of data like social media, surveillance cameras, sensor data, weather data, phone records, online transactions, etc. We are living in the age where data is generated in petabytes and zettabytes. This sudden boom in the production of data that is too large to store and analyze requires advanced tools and techniques that open the way for Big Data. To handle such a voluminous data is really a big challenge for the data scientists.

5.1.2 Velocity

Velocity is defined by how rapidly the new data is being generated. As we see how messages on social media go viral within no time, millions of photos are being uploaded by Facebook users each and every second, it takes milliseconds for the business systems to analyze social networking Web sites to gather message that set off the verdict to purchase or sell shares [22]. Big Data streaming processing method makes it possible to examine the data while it is emanated, in need of ever storing it into database.

5.1.3 Variety

Variety focuses on different forms of data like music, pictures, text, e-mails, medical records and images, weather records, log files, etc. generated from multiple sources. This means that the data produced belongs to different categories consisting of raw, unstructured, structured and semi-structured data which looks very difficult to deal with.

5.1.4 Veracity

It denotes the meaningfulness or value of data.

5.1.5 Value

It focuses on the analytics and statistical methods, knowledge extraction and decision making [21].

The data that is generated and it is not analyzed and processed then it is nothing other than garbage.

5.2 Processing Challenges

Data processing is common in every section of organization. As we know, today data is derived through various sources like Facebook, Twitter, WhatsApp, mobile phones, sensors, surveillance cameras, etc. The more data we collect, the more accurate result will be available for optimization. To process this data generated on a large scale, new technologies, advanced algorithms and a variety of different approaches are required [10, 21]. Processing of data comprises data gathering, fixing resemblance found in distinct sources, refinement of data to a type adequate for the scrutiny and output characterization. Let's have a look on various processing challenges of Big Data.

5.2.1 Heterogeneity

With the perspective of Big Data, heterogeneity refers to the variety of data. There are unlimited different sources that generate data to Big Data. The data from different origin consists of dissimilar forms. Heterogeneity in Big Data is a requirement to tackle and deal with multiple forms of data concurrently [23].

5.2.2 Timeliness

Timeliness refers to how swiftly the data is being handled and analyzed. As the magnitude of the data that is set to be refined, get larger, undoubtedly will get extra time to evaluate. However in some cases, outcome of the evaluation is needed without delay. For example, if we have a transaction and this transaction is made by a fraudulent, it must be identified before the transaction is over by blocking it from occurring. So here we have to build up unfinished consequences in advance so that a minute number of additional calculations with new data can be used to resolve the issue immediately [1, 2].

5.2.3 Complexity

Complexity refers to the rigidness of data [4]. It mainly focuses on unstructured data. According to Zikopoulos and Eaton [24], Multidimensional data can be classified into three types, namely, semi-structured, structured and unstructured. Structured data is homogeneous and lengths of data are defined in advance and are produced by automatic generators like computers or sensors without user communication. Unstructured data is complex such as client reviews, photos and other multimedia. The unpredictable growth of the Internet depicts that the size and range of Big Data keep on growing.

5.2.4 Scalability

As the data is generated at a large scale and is growing rapidly day-by-day. This rapid increase in the data rate can be mitigated by improving the processor speed. However, the quantity of data expands at a faster rate than computing resources and CPU speeds [21]. Taking all these parameters into consideration, we have to build such a scalable system with a persistent storage and high-processing speeds where a single node can share many hardware resources. In broader sense, we can say scalability provides fault tolerance to the Big Data platform.

5.2.5 Accuracy

Accuracy refers that data is accurate and is obtained from limited sources. When we gather data from authentic sources and are extracted out from such data analysis, the results are more accurate [23].

5.3 Management Challenges

The existing technologies of data management systems are incapable to gratify the requirements of Big Data, and the pace at which data storage space is increasing is far less than that of data generated, thus a reconstruction of information framework is need of an hour. Besides, the existing algorithms do not fulfill the need to store the data efficiently that is directly acquired from different sources because of the heterogeneity of the data. We have pointed out few important management challenges concerned with Big Data as below.

5.3.1 Security

As Big Data exceeds the data sources it can use, there is a need to verify the trust and worthiness of each source and therefore approaches should be used to find malevolent inserted data [25]. Information security is fetching to be a critical Big Data issue where huge quantity of data will be linked scrutinized and extracted for significant patterns [22].

5.3.2 Data Sharing

We exist in the world of Big Data. Everything is being shared on social networking Web sites. Information about almost everything and anything can be collected by means of a single click on "Google." Each and every individual and corporation has a immense amount of information at their disposal that can be used for their purpose and requirement. Every kind of content is available and is one click away only when everyone shares it. But there is a differentiation among what is private and what can be shared [2, 26].

5.3.3 Failure Handling

To build a 100% reliable system is no way a straightforward task. Systems can be put up to such a degree that the possibility of breakdown must fall in the acceptable threshold. When a process is started, its processing involves multiple setups of nodes and the total working out process becomes unwieldy. Maintaining control points and setting up the threshold level for process resume subject to failure is a big challenge. Therefore, in Big Data this is called fault tolerance [27].

6 Big Data Techniques

As we know that Big Data in addition to vast dimensions of data, it is also varied and growing increasingly. Big Data requires unique and advanced techniques or to upgrade the existing techniques as well in order to proficiently process outsized volume of data within a predefined period of time. Big Data techniques cover up a number of disciplines like statics, data mining, neural networks, machine learning, social network analysis, pattern recognition, etc.

Statics is the science which deals with the compilation, management and elucidation of data. The key point of statistical techniques is to find the co-relationships and casual relationships among different objectives. However, the basic statistical techniques did not come up with Big Data. So we need to focus on parallel statics and different algorithms regarding parallel statics. We should try our hands on large-scale and parallel application of statistical algorithms.

Data mining is a vast field that mainly focuses on the information retrieval (patterns) from data: Combines a series of techniques like association rule learning, clustering, classification and regression. Data mining is a combination of statics and machine learning. It makes use of both statics and machine learning models to get value from data. However, traditional data mining algorithms did not handle Big Data very well. Big Data mining is more challenging and requires new and advanced techniques such as models and algorithms to process data and get value from it. Now, let's take an example of clusterings such as hierarchical clustering, k-means clustering and fuzzy C-mean clustering; we need to extend the capabilities of these existing clustering techniques in order to switch to Big Data [28].

Machine learning is a sub-field of artificial intelligence, mainly emphasis on layout of algorithms. It is the power of machine learning through which systems are capable to learn from present or past and make fruitful predictions for future events automatically [4]. As far as the Big Data is concerned; we have to focus on both supervised and unsupervised machine learning algorithms [15]. Nowadays, deep machine learning has turned out to be a new and trendy research area of artificial intelligence [29]. Supervised learning is used in classification whereas clustering falls in unsupervised learning. Support vector machine (SVM) employed in regression and classification suffers from both usage of memory and computation time. However to lessen memory and time utilization, parallel support vector machines (PSVM) are being used [30].

Artificial neural network (ANN) is another machine learning technique that has a broad application area. The applications' areas of ANN include pattern recognition, image analysis, adaptive control, etc. The working of most of the ANNs is rest on statistical estimation, classification optimization and control theory [31]. ANN is mainly composed of three layers; input layer, hidden layer and output layer. The complexity of ANN can be influenced by the number of hidden layers: The more is the number of hidden layers the more complex the neural network is and the complexity also increases the learning time. However, it is proved and there are evidences that the neural network with additional hidden layers and nodes produce high-level of accuracy and authenticity. The employment of ANNs over Big Data is strictly time and memory consuming and often produces extremely huge networks. There are two major challenges regarding the use of ANNs over Big Data: Firstly, its poor performance of training algorithms and the second is the increasing training time and memory limitations [32]. This situation can be handled in two ways; one is to lessen the data size without changing the configuration of the neural network and the second one is to expand neural networks in parallel and distributed ways [33].

Social network analysis (SNA) consists of nodes and ties, is widely known as the main approach in recent sociology, evaluates social interaction regarding network theory and is nowadays regularly used as a client tool [34].

7 Big Data Hadoop-Based Tools

Tools are used to get value from Big Data. There are a series of tools that are used to process Big Data. Some important tools are described below.

7.1 Apache Hadoop

Apache Hadoop is considered as one of the powerful batch process-based Big Data tools. It acts as a platform for other Big Data applications. Hadoop is open-source software typed in Java that offers scattered processing of huge datasets over clusters of relatively inexpensive and widely available hardware in a fault-tolerant manner. Its key components are Hadoop kernel, Hadoop Distributed File System and MapReduce.

HDFS is a distributed file system that is used to store particularly large data files and provides high throughput of application data. It is highly fault-tolerant and scalable [35]. It follows a master/slave architecture, in which we have multiple devices known as slave devices or nodes and another single device called master device or node. The slave nodes are called DataNodes and the master node which controls all the DataNodes is called NameNode. The data file is split into small blocks of data and each block of data is copied over DataNodes [4]. The number of replicated copies per data block is three by default. DataNodes are controlled by NameNode and after every 3 s NameNode receives a heartbeat to check whether the DataNode is functional or not.

MapReduce is a designing framework used for processing of gigantic data sets [19]. MapReduce works on the strategy of divide and conquer, as it breaks down a compound problem into multiple sub-problems. The sub-problems are then assigned and distributed to DataNodes in parallel and are united to give result to the original problem.

7.2 Apache Mahout

Mahout is a machine learning tool of Hadoop ecosystem which provides scalable techniques for extensive data analysis applications [7, 36]. Mahout has been deployed in popular companies like Google, Amazon, Yahoo and Facebook [31]. Mahout includes a number of algorithms among these are clustering, classification, regression, pattern mining and evolutionary algorithms [37, 38].

7.3 Storm

Storm is an open-source distributed and fault-tolerant real-time computation system. Storm performs the real-time processing of unbounded streams of data easily and reliably [4, 39]. The working of storm solely depends on two types of working nodes, i.e., one master node and several working nodes and they work in a similar fashion as JobTracker and also TaskTracker in case of MapReduce.

7.4 S4

S4 was released by Yahoo in 2010 and begin to be an Apache project since 2011. It is a general-purpose, scalable, fault-tolerant and distributed platform that makes it possible to process limitless streams of data [40].

7.5 Apache Kafka

Kafka is a messaging system that provides high degree of efficiency, developed at Linkedin. It provides a platform for managing streaming and operational data for real-time decision-making process. It provides high throughput and promotes distributed processing and further loading parallel data into Hadoop.

8 Conclusion

In this paper, we have addressed and examined the number of challenges that comes in the way of Big Data. As we know that presently the data is generating at a very fast pace and will still continue to grow with every passing second. On the basis of research in the field of Big Data, the data will get doubled by the year 2020. To deal with such a gigantic data, there is a need to rectify the challenges and hindrances that arise as a result of gigantic quantity data. The various challenges concerned with Big Data are briefly discussed in this paper. In addition, the researchers must focus on Big Data analysis that will boost the business economy in the near future. Advanced tools and techniques are required to get value from Big Data such as Hadoop and MapReduce technologies. The Big Data V's model that defines the characteristics of Big Data is actually the challenges we need to tackle and to resolve. To harvest the reimbursement of Big Data, we have to convert the challenges into power.

References

1. Jaseena, K.U., David, J.M.: Issues, challenges, and solutions: big data mining. Comput. Sci. Inf. Technol. (CS&IT) 131–140 (2014)
2. Lawal, Z.K., Zakari, R.Y., Shuaibu, M.Z., Bala, A.: A review: issues and challenges in big data from analytic and storage perspectives. Int. J. Eng. Comput. Sci. 5(3), 15947–15961 (2016)
3. Holzinger, A., Stocker, C., Ofner, B., Prohaska, G., Brabenetz, A., Hofmann-Wellenhof, R.: Combining HCI, natural language processing, and knowledge discovery-potential of IBM content analytics as an assistive technology in the biomedical field. In: Human-Computer Interaction and Knowledge Discovery in Complex, Unstructured, Big Data (pp. 13–24). Springer, Berlin (2013)
4. Landset, S., Khoshgoftaar, T.M., Richter, A.N., Hasanin, T.: A survey of open source tools for machine learning with big data in the Hadoop ecosystem. J. Big Data 2(1), 24 (2015)
5. Jin, X., Wah, B.W., Cheng, X., Wang, Y.: Significance and challenges of big data research. Big Data Res. 2(2), 59–64 (2015)
6. Kaur, H., Alam, M.A., Jameel, R., Mourya, A., Chang, V.A.: Proposed solution and future direction for blockchain-based heterogeneous medicare data in cloud environment. J. Med. Sys. 42(8) (2018)
7. Anuradha, J.: A brief introduction on Big Data 5Vs characteristics and Hadoop technology. Procedia Comput. Sci. 48, 319–324 (2015)
8. Kaur, H., Lechman, E., Marszk, A.: Catalyzing Development through ICT Adoption: The Developing World Experience, pp. 288. Springer Publishers, Switzerland (2017)
9. Beyer, M.A., Laney, D.: The Importance of 'Big Data': A Definition, pp. 2014–2018. Gartner, Stamford (2012)
10. Kaisler, S., Armour, F., Espinosa, J.A., Money, W.: Big data: issues and challenges moving forward. In: 2013 46th Hawaii International Conference on System Sciences (HICSS) (pp. 995–1004). IEEE (2013)
11. Mervis, J. (2012). Agencies rally to tackle big data

12. Hey, T., Tansley, S., Tolle, K.M.: The Fourth Paradigm: Data-Intensive Scientific Discovery, vol. 1. Microsoft Research, Redmond (2009)
13. O'Neil, C., Schutt, R.: Doing Data Science: Straight Talk from the Frontline. O'Reilly Media, Inc. (2013)
14. Chen, H., Chiang, R.H., Storey, V.C.: Business intelligence and analytics: from big data to big impact. MIS Q. 1165–1188 (2012)
15. Ginsberg, J., Mohebbi, M.H., Patel, R.S., Brammer, L., Smolinski, M.S., Brilliant, L.: Detecting influenza epidemics using search engine query data. Nature **457**(7232), 1012 (2009)
16. Gantz, J., Reinsel, D.: Extracting value from chaos. IDC iview **1142**(2011), 1–12 (2011)
17. Shvachko, K., Kuang, H., Radia, S., Chansler, R.: The hadoop distributed file system. In: 2010 IEEE 26th Symposium on Mass Storage Systems and Technologies (MSST) (pp. 1–10). Ieee (2010)
18. Ghemawat, S., Gobioff, H., Leung, S.T.: The Google File System, **37**(5), 29–43. ACM (2003)
19. Dean, J., Ghemawat, S.: MapReduce: simplified data processing on large clusters. Commun. ACM **51**(1), 107–113 (2008)
20. Ahrens, J., Hendrickson, B., Long, G., Miller, S., Ross, R., Williams, D.: Data-intensive science in the US DOE: case studies and future challenges. Comput. Sci. Eng. **13**(6), 14–24 (2011)
21. Sivarajah, U., Kamal, M.M., Irani, Z., Weerakkody, V.: Critical analysis of Big Data challenges and analytical methods. J. Bus. Res. **70**, 263–286 (2017)
22. Lu, R., Zhu, H., Liu, X., Liu, J.K., Shao, J.: Toward efficient and privacy-preserving computing in big data era. IEEE Netw. **28**(4), 46–50 (2014)
23. Khan, N., Yaqoob, I., Hashem, I.A.T., Inayat, Z., Ali, M., Kamaleldin, W., Alam, M., Shiraz, M., Gani, A.: Big data: survey, technologies, opportunities, and challenges. Sci. World J. (2014)
24. Zikopoulos, P., Eaton, C.: Understanding Big Data: Analytics for Enterprise Class Hadoop and Streaming Data. McGraw-Hill Osborne Media (2011)
25. Demchenko, Y., Grosso, P., De Laat, C., Membrey, P.: Addressing big data issues in scientific data infrastructure. In: 2013 International Conference on Collaboration Technologies and Systems (CTS) (pp. 48–55). IEEE (2013)
26. Yi, X., Liu, F., Liu, J., Jin, H.: Building a network highway for big data: architecture and challenges. IEEE Netw. **28**(4), 5–13 (2014)
27. Samuel, S.J., RVP, K., Sashidhar, K., Bharathi, C.R.: A survey on big data and its research challenges. ARPN J. Eng. Appl. Sci. **10**(8), 3343–3347 (2015)
28. Bezdek, J.C.: Objective function clustering. In: Pattern Recognition with Fuzzy Objective Function Algorithms (pp. 43–93). Springer, Boston (1981)
29. Arel, I., Rose, D.C., Karnowski, T.P.: Deep machine learning-a new frontier in artificial intelligence research. IEEE Comput. Intell. Mag. **5**(4), 13–18 (2010)
30. Mitra, P., Murthy, C.A., Pal, S.K.: A probabilistic active support vector learning algorithm. IEEE Trans. Pattern Anal. Mach. Intell. **26**(3), 413–418 (2004)
31. Mansour, Y., Chang, A.Y., Tamby, J., Vaahedi, E., Corns, B.R., El-Sharkawi, M.A.: Large scale dynamic security screening and ranking using neural networks. IEEE Trans. Power Syst. **12**(2), 954–960 (1997)
32. Fujimoto, Y., Fukuda, N., Akabane, T.: Massively parallel architectures for large scale neural network simulations. IEEE Trans. Neural Netw. **3**(6), 876–888 (1992)
33. Ahn, J.B.: Neuron machine: parallel and pipelined digital neurocomputing architecture. In: 2012 IEEE International Conference on Computational Intelligence and Cybernetics (CyberneticsCom) (pp. 143–147). IEEE (2012)
34. Ma, H., King, I., Lyu, M.R.: Mining web graphs for recommendations. IEEE Trans. Knowl. Data Eng. **24**(6), 1051–1064 (2012)
35. White, T.: Hadoop: the definitive guide: the definitive guide. O'Reilly Media, Inc. (2009)

36. Ingersoll, G.: Introducing Apache Mahout. Scalable, Commercial Friendly Machine Learning for Building Intelligent Applications. IBM (2009)
37. Rong, C.: November. Using Mahout for clustering Wikipedia's latest articles: a comparison between k-means and fuzzy c-means in the cloud. In 2011 IEEE Third International Conference on Cloud Computing Technology and Science (CloudCom), pp. 565–569. IEEE (2011)
38. Chauhan, R., Kaur, H., Lechman, E., Marszk, A.: Big data analytics for ICT monitoring and development. In: Kaur et al. (eds.) Catalyzing Development through ICT Adoption: The Developing World Experience, pp. 25–36. Springer (2017)
39. Storm. http://storm-project.net (2012)
40. Chauhan, J., Chowdhury, S.A., Makaroff, D.: Performance evaluation of Yahoo! S4: a first look. In: 2012 Seventh International Conference on P2P, Parallel, Grid, Cloud and Internet Computing (pp. 58–65). IEEE (2012)

Forecasting World Petroleum Fuel Crisis by Nonlinear Autoregressive Network

Srikanta Kumar Mohapatra, Sushanta Kumar Kamilla, Tripti Swarnkar and Gyana Ranjan Patra

Abstract Petroleum is an essential commodity in today's world for the survival of mankind. The petroleum reserve in the world is limited and it is expected to get exhausted in the coming future, and the world is witnessing the crisis of this oil. The prediction of petroleum crisis in the world is a challenging problem to deal with. In this paper, we are using a novel and suitable time series prediction approach based on artificial neural network (ANN) for the forecast of future instances of a time series data. The prediction of petroleum crisis in the near future is obtained by a nonlinear and multistep method known as nonlinear autoregressive network (NARnet). The data set is obtained from different government sources of ten countries (like USA, Middle East countries, China and India, etc.) for over a period of more than 30 years and contains three features, viz. population, petroleum production, and petroleum consumption. The NARnet model requires known instances which are employed for the dynamic multistep prediction for the ahead time where one can predict as many numbers of future data instances as looked-for. The normalized mean square error (NMSE) and T-test of our prediction method, i.e., NARnet, have been verified by different standard predictive methods to handle the data set and found to be better in its performance. It has been forecasted that by 2050, hydrogen fuel (which can reduce pollution and boost the fuel efficiency at lower cost) could be thought of as an appropriate replacement for petroleum.

S. K. Mohapatra · S. K. Kamilla · T. Swarnkar · G. R. Patra (✉)
Institute of Technical Education & Research, Siksha 'O' Anusandhan (Deemed to be University), Bhubaneswar, Odisha, India
e-mail: gyanapatra@soa.ac.in

S. K. Mohapatra
e-mail: srikanta.2k7@gmail.com

S. K. Kamilla
e-mail: sushantakamilla@soa.ac.in

T. Swarnkar
e-mail: swarnkar.tripti@gmail.com

© Springer Nature Singapore Pte Ltd. 2020
S. Patnaik et al. (eds.), *New Paradigm in Decision Science and Management*,
Advances in Intelligent Systems and Computing 1030,
https://doi.org/10.1007/978-981-13-9330-3_7

Keywords Petroleum production · Hydrogen fuel · Petroleum consumption ·
Artificial neural network-based time series prediction · Nonlinear autoregressive
network (NARnet)

1 Introduction

Today's world is highly dependent on fossil fuel for industrial development. Nowadays, petroleum is used as the major source of running automobiles and the industries in the world [1]. In most of the countries, petroleum production cannot meet the consumption. According to the data book of transportation energy, manufacture of vehicles from 2000 to 2015 increased remarkably [2]. It is observed that vastly populated countries like China and India mostly depend upon the crude oil for their growth which is met by importing oil from others [3, 4]. The growing population is also one of the main factors for increasing oil price in the future [5, 6]. Lots of research is going on to deal with the problem of increasing oil prices [7, 8]. Researchers are trying to develop an alternative fuel, like liquid hydrogen as an alternate source of petroleum [9, 10].

This paper is an approach for prediction of petroleum crisis in the world. The data reference of the top ten countries in population, oil consumption and production has been used. The relative ranks of these countries are shown in Table 1. Most researchers predict that petroleum crisis in near future is must, but to forecast an accurate prediction of the same requires complex procedures where artificial neural networks (ANN) can achieve better accuracy [11, 12]. The prediction methodology is achieved by a novel ANN-based time series prediction approach known as nonlinear autoregressive network (NARnet) model. Forecasting of alternate types of fuel necessity for highly populated countries like China and India using NARnet has been reported earlier by this group in an earlier paper [13].

Table 1 Top ten countries w.r.t population, oil production and consumption (2016)

Rank	Population	Oil production	Oil consumption
1.	China	Russia	USA
2.	India	Saudi Arabia	China
3.	USA	USA	India
4.	Indonesia	Iraq	Japan
5.	Brazil	Iran	Saudi Arabia
6.	Pakistan	China	Brazil
7.	Nigeria	Canada	Russia
8.	Bangladesh	United Arab Emirates	South Korea
9.	Russia	Kuwait	Germany
10.	Japan	Brazil	Canada

2 Data Sources

The population data from 1985 to 2016 have been obtained from sources like the US Census Bureau for Population and World Bank.[1] Oil production and consumption data from 1985 to 2016 of various countries can be found in the US Energy Information Administration (USEIA)[2] and BP Statistical Review of World Energy.[3]

3 Methodology

3.1 Time Series Prediction Using Artificial Neural Network (ANN)

Prediction or forecasting is generally based upon the process of making predictions of future data by taking the current and past data set. Time series prediction of historical data is investigated for four known patterns of time decomposition, known as trends, seasonal patterns, cyclic patterns, and regularity. Marketing, finance, sales, and other institutions use some form of time series forecasting to evaluate feasible technical costs and customer demand. In most of the cases, it is too difficult to choose appropriate parameters for appropriate prediction. So, it is important to observe the minimum error value in the training and testing case. The primary benefit of ANN for prediction is its requirement of a smaller number of statistical training between dependent and non-dependent variables.

3.2 Comparisons of General Prediction Models

Various models are developed for the prediction of future data set by taking linear or nonlinear inputs from the past data set. The approach is different in different cases and is well developed in the literature. Most researchers used algorithms, networks for prediction of multiple types of commodities, and applications [5, 6, 14–26].

Vast majority of the forecast models are usually linear models or have the capability of single-step-ahead prediction. This paper considers a model based upon a nonlinear approach that is associated which can predict multiple steps in the future. Most of the prediction models predict future data by taking both actual and predicted data, but our model predicts future data by taking current and past actual data set. Our technique makes a demonstration that if the past data set $f(x) = t(1), t(2), \ldots t(n)$,

---

[1]Population of all the countries. Available at: www.data.worldbank.org.

[2]Crude Oil Production and Consumption by Year, **Source**: US Energy Information Administration. Available at: www.indexmundi.com.

[3]BP Statistical Review of World Energy. Available at www.bp.com.

then prediction value after successfully completion of training step, testing step, and validation is $t(2n)$.

3.3 Predictive Model Flow Control Diagram

The complete course of prediction can be summarized in four steps as shown in Fig. 1.

Step a: Take data from required sources from the various larger data sets.
Step b: Preprocess the data by keeping the vital data for the whole calculation.
Step c: Transform data to the neural network model.
Step d: Choice of neural network and training involves the following stages.

      i. First, feed forward inputs.
     ii. Determine the errors by back propagation algorithm.
    iii. Update weights for future predictions.

3.4 Predictive Model Structure

NARnet is a type of dynamic network in which employs two parallel looping architecture functions, viz. open and closed loops. Processing of the training data, validation, and testing are done by open-loop function and multistep-ahead prediction is carried out by closed loop. Figure 1 describes how the complete model performs from input to prediction by the execution of open-loop and closed-loop forms.

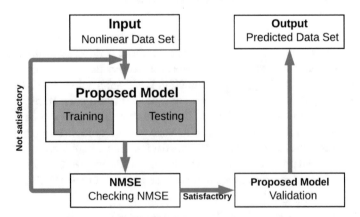

Fig. 1 Predictive model flow control diagram

4 Performance Measurement

Assessment of performance of the testing data of the NARnet is done by comparing with the actual different set of data. The training of the network is done by minimizing the error between actual and obtained data. Normalized mean square error (NMSE) is calculated, and the deviations are summed instead of differences [27]. NMSE is a very powerful tool for error calculation in comparison with other models.

Algorithm:

```
SETS:import data sets "Y" // importing the original data set in .mat
//function call of NARnet(InputDelays,HiddenSizes) with parameters Row vector of
//increasing 0 or positive delays (default = 1:2) hidden size is taken as null
narnet(1:2,[]);
view(net);        //view network
endof narnet();   // initiating function tonndata() for converting data to standard
                  // neural network cell array form with parameters 1) matrix 2) true
        if
                  // original samples are in columns, false if rows 3) timesteps are in
                  //rows if false otherwise in column for true
Target T:tonndata(Y,false,false);
function preparates(net,{},{},T); // prepare data set by network and get target output
                                  // train(net,shifted i/ps,shifted targets,initial
        i/p
                                  // delay states,initial layer delay staes)
train(net,Xs,Ts,Xi,Ai)
ts:cell2mat(shifted targets);    // conversion of cell to matrix
   rng('default');               // default number generator taken as default
   view(net);
Open-loop:        // openloop network is run to train the network and simulate the
                  // original data set
net = open loop (net) {
        net.outputs {i}.feedbackMode
        net.outputs{i}.feedbackInput }
        [net,xi,ai] = openloop(net,xi,ai)
           NMSEs=mse(Es)/var(ts,1) // error performance check of training set
endof open-loop;

Closed-loop:      // closedloop network is run to simulate the original data and
                  // predict target
net = closeloop(net) {
net.outputs{i}.feedbackMode
net.outputs{i}.feedbackInput }
   [net,xi,ai] = closeloop(net,xi,ai)
              NMSEs=mse(Ec)/var(ts,1)   // error performance check of target series
endof closed-loop;
```

5 Result and Analysis

In Fig. 2, it is seen that the production is remarkable increasing only in the case of countries of the Middle East like Saudi Arabia, Italy, Iran, and Kuwait, etc. and some extent Brazil. The future for all these countries except the Middle East, Brazil and others shows that these will not increase further. From Fig. 2 analysis, China is at the top by concerning population. But the prediction shows that around the year 2025, India will cross it and may become the world's most populated country. Population of almost all the countries are increasing but not like as rapid manner in these two countries. In Fig. 4, the predictive curves show that the situation may not in control after 30 years. Consumption of oil is highly increasing in almost all the countries except Russia. Transportation energy data book suggests that the production of cars

Fig. 2 Prediction of oil
production of referred
countries

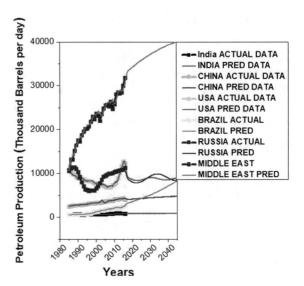

and trucks between 2000 and 2013 has managed to grow at a positive rate [2]. But
the scenario of the future oil consumption is revealing that consumption is increasing
highly in case of India and China, while the USA is managing it to a downward. To
reduce petroleum consumption, the developed and advanced country like the USA
has reduced the manufacturing of vehicles from 2010 onwards, which is shown in
Fig. 4. In Fig. 5, it concludes that in the year 2016, the petroleum production to
consumption is very low in case of India, i.e., petroleum production is very low
as compared to the population and oil consumption, but in Middle East countries,
Russia, Brazil, and the USA, it is quite high (Fig. 3).

The cost of petroleum production has an increasing trend due to the three reasons
cited here. (1) Growth of population gives rise to expensive procedures for petroleum
extraction. (2) Current techniques are not conducive for extraction of cheap fuel oil,
thereby increasing the cost of oil extraction. (3) Fuel sources with lower pollution
capability can be used to substitute for petroleum, adding to the expenses. This
increases the cost of production and usage, which has been discussed by Tverberg
et al. [28]. The demand for petroleum can be controlled by ensuring relatively higher
fuel prices [29]. In this regard, we can visualize **hydrogen fuel** that can act both
as an energy carrier and as an environmentally friendly alternative to fossil fuels.
Hydrogen fuel is an example of clean, renewable energy sources. This has led to the
development of hydrogen-based cars by several car manufacturers. Burning liquid
hydrogen directly in the engine produces hydrogen as residue gas instead of carbon
dioxide, and thus, it reduces global warming [30, 31]. It can be inferred that the usage
of hydrogen will be prevalent by 2040–2050.

Fig. 3 Prediction of change in population of referred countries with respect to time

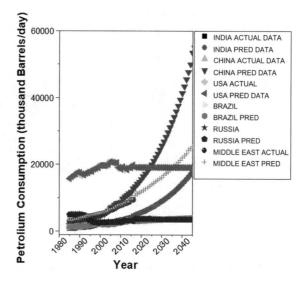

Fig. 4 Prediction of oil consumption of referred countries with respect to time

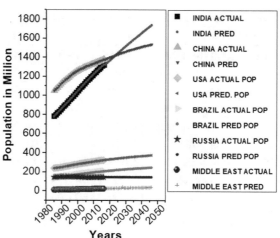

6 Comparison and Discussion

The comparison of our model with other models is demonstrated by taking three major countries like India, China, and the USA with three major factors, oil production, oil consumption, and population, given in Table 2. The NARnet model is found to be better in its performance.

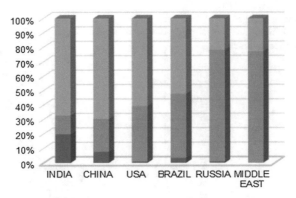

Fig. 5 Comparison of population, petroleum consumption, and production for the year 2016

Table 2 Comparison of NARnet result with other existing methods (MSE)

Country	Factor	GP	SC	ANFIS	GMDH	NARnet	P value
India	Oil production	1.2861	1.1643	0.6347	1.3732	0.1271	0.0383
	Oil consumption	1.6858	1.1493	1.6694	0.9625	0.3519	0.0411
	population	0.8750	0.8500	0.8389	2.3200	0.7320	0.0287
China	Oil production	1.6469	1.1403	1.2587	2.9903	1.0240	0.0484
	Oil consumption	1.7453	0.7087	1.0278	1.7782	0.6551	0.0324
	Population	2.5400	0.5700	0.1540	1.4500	0.1200	0.0235
USA	Oil production	2.7728	1.1607	1.7564	1.8131	1.0979	0.0367
	Oil consumption	2.1607	1.4170	1.8010	2.0759	1.4050	0.0185
	Population	1.7200	0.2300	0.1976	0.1400	0.0400	0.0254

7 Hypothesis Testing

The new algorithms and designs must be statistically significant before implementation [32, 33]. The statistical test must be done ensuring that the data set which is taken and the corresponding result which is achieved must be actual that means the result is not by hypothesis word "by chance." To ensure this, there are different statistical tools or the methods are present [34]. This has been demonstrated in the last column of Table 2.

8 Conclusion

It has been established on the basis of the various tests made on the results of the considered benchmark data set, it is confirmed that ANN with a multistep-ahead predictive model, i.e., the NARnet, has shown better performance in predicting population growth, production, and consumption of petroleum oil for six countries, viz. USA, Brazil, Russia, Middle East, China, and India, which are placed within the top ten. The result has shown better accuracy and lowers MSE. By performing T-test of our prediction method, i.e., NARnet has been verified by different standard predictive methods to handle the data set and found to be better in its performance. The fossil fuel production would tend to decline over time as production drains the existing reserve base. The whole scenario forecasts that nearly by the year 2050, the world has to rely on an alternative fuel like hydrogen fuel which reduces pollution and enhances fuel efficiency at a lower cost than petroleum. It can be foreseen that hydrogen could be a good replacement of fossil fuels in the future.

References

1. Mitchell, J., Marcel, V., Mitchell, B.: What Next for the Oil and Gas Industry? Chatham House (2012)
2. Davis, S.C., Diegel, S.W., Boundy, R.G.: Transportation Energy Data Book (2015)
3. Yuan, C., Liu, S., Fang, Z.: Comparison of China's primary energy consumption forecasting by using ARIMA (the autoregressive integrated moving average) model and GM (1,1) model. Energy **100**, 384–390 (2016)
4. Srinivasan, T.N.: China and India: economic performance, competition and cooperation: an update. J. Asian Econ. **15**(4), 613–636 (2004)
5. Hu, J.W.S., Hu, Y.C., Lin, R.R.W.: Applying neural networks to prices prediction of crude oil futures. Math. Probl. Eng. (2012)
6. Khazem, H., Mazouz, A.: Forecasting the price of crude oil using artificial neural networks. Int. J. Bus. Mark. Decis. Sci. **6**(1) (2013)
7. Bossel, U.: The physics of the hydrogen economy. Eur. Fuel Cell News **10**(2), 1–16 (2003)
8. Pillay, P.: Hydrogen economy and alternative fuels. IEEE Emerg. Technol. Portal (2006–2012)
9. Serrano, E., Rus, G., Garcia-Martinez, J.: Nanotechnology for sustainable energy. Renew. Sustain. Energy Rev. **13**(9), 2373–2384 (2009)
10. Sahaym, U., Norton, M.G.: Advances in the application of nanotechnology in enabling a 'hydrogen economy'. J. Mater. Sci. **43**(16), 5395–5429 (2008)
11. Armstrong, J.S.: Research needs in forecasting. Int. J. Forecast. **4**(3), 449–465 (1988)
12. Zhang, G., Patuwo, B.E., Hu, M.Y.: Forecasting with artificial neural networks: the state of the art. Int. J. Forecast. **14**(1), 35–62 (1998)
13. Mohapatra, S.K., Swarnkar, T., Kamilla, S.K., Mohapatra, S.K.: Forecasting hydrogen fuel requirement for highly populated countries using NARnet. Commun. Comput. Inf. Sci. **827**, 349–362 (2018)
14. Cui, X., Jiang, M.: Chaotic time series prediction based on binary particle swarm optimization. AASRI Procedia **1**, 377–383 (2012)
15. Gibson, D., Nur, D.: Threshold autoregressive models in finance: a comparative approach. In: Proceedings of the Fourth Annual ASEARC Conference, University of Western Sydney, Paramatta, Australia. http://ro.uow.edu.au/asearc/26 (2011)
16. Hansen, B.E.: Threshold autoregression in economics. Stat. Interface **4**(2), 123–127 (2011)

17. Kulkar, S., Haidar, I.: Forecasting model for crude oil price using artificial neural networks and commodity future prices. Int. J. Comput. Sci. Inf. Secur. **2**(1), 81–88 (2009)
18. Mohapatra, S.K., Kamilla, S.K., Mohapatra, S.K.: A pathway to hydrogen economy: artificial neural network an approach to prediction of population and number of registered vehicles in India. Adv. Sci. Lett. **22**(2), 359–362 (2016)
19. Zhang, H., Li, J.: Prediction of tourist quantity based on RBF neural network. JCP **7**(4), 965–970 (2012)
20. Azadeh, A., Sheikhalishahi, M., Shahmiri, S.: A hybrid neuro-fuzzy simulation approach for improvement of natural gas price forecasting in industrial sectors with vague indicators. Int. J. Adv. Manuf. Technol. **62**(1), 15–33 (2012)
21. Aksoy, F., Yabanova, I., Bayrakçeken, H.: Estimation of dynamic viscosities of vegetable oils using artificial neural networks. Indian J. Chem. Technol. **18**, 227–233 (2011)
22. Liu, J., Tang, Z.H., Zeng, F., Li, Z., Zhou, L.: Artificial neural network models for prediction of cardiovascular autonomic dysfunction in general Chinese population. BMC Med. Inform. Decis. Mak. **13**(1), 80 (2013)
23. Maliki, O.S., Agbo, A.O., Maliki, A.O., Ibeh, L.M., Agwu, C.O.: Comparison of regression model and artificial neural network model for the prediction of electrical power generated in Nigeria. Adv. Appl. Sci. Res. **2**(5), 329–339 (2011)
24. Tehrani, R., Khodayar, F.: A hybrid optimized artificial intelligent model to forecast crude oil using genetic algorithm. Afr. J. Bus. Manag. **5**(34), 13130 (2011)
25. Yadav, A.K., Chandel, S.S.: Artificial neural network-based prediction of solar radiation for Indian stations. Int. J. Comput. Appl. **50**(9) (2012)
26. Markopoulos, A.P., Georgiopoulos, S., Manolakos, D.E.: On the use of back propagation and radial basis function neural networks in surface roughness prediction. J. Ind. Eng. Int. **12**, 389–400 (2016)
27. Poli, A.A., Cirillo, M.C.: On the use of the normalized mean square error in evaluating dispersion model performance. Atmos. Environ. Part A Gen. Top. **27**(15), 2427–2434 (1993)
28. Tverberg, G.: Oil limits and the end of the debt super-cycle. Available online at: https://ourfiniteworld.com/2016/01/07/2016-oil-limits-and-the-end-of-the-debt-supercycle (2016)
29. Streifel, S.: Impact of China and India on global commodity markets: focus on metals and minerals and petroleum. Development Prospects Group/World Bank, UU World Investment Report (2006)
30. Offer, G.J., Howey, D., Contestabile, M., Clague, R., Brandon, N.P.: Comparative analysis of battery electric, hydrogen fuel cell and hybrid vehicles in a future sustainable road transport system. Energy policy **38**(1), 24–29 (2010)
31. Cheng, X., Shi, Z., Glass, N., Zhang, L., Zhang, J., Song, D., Liu, Z.S., Wang, H., Shen, J.: A review of PEM hydrogen fuel cell contamination: impacts, mechanisms, and mitigation. J. Power Sources **165**(2), 739–756 (2007)
32. Farlow, S.J.: Self-Organizing Methods in Modeling: GMDH Type Algorithms, vol. 54. CrC Press (1984)
33. Kuznetsova, A., Brockhoff, P.B., Christensen, R.H.B.: lmerTest package: tests in linear mixed effects models. J. Stat. Softw. **82**(13) (2017)
34. Preacher, K.J., Curran, P.J., Bauer, D.J.: Computational tools for probing interactions in multiple linear regression, multilevel modeling, and latent curve analysis. J. Educ. Behav. Stat. **31**(4), 437–448 (2006)

Stock Market Price Prediction Employing Artificial Neural Network Optimized by Gray Wolf Optimization

Sipra Sahoo and Mihir Narayan Mohanty

Abstract As the stock market is highly volatile and chaotic in nature, prediction about this is a highly challenging task. To achieve better prediction accuracy, this article presents a model that uses artificial neural network (ANN) optimized by gray wolf optimization (GWO) technique. The model is applied on Bombay stock exchange (BSE) data. The range of data selection was from 25 August 2004 to 24 October 2018. To evaluate the performance of the model, many evaluation metrics such as mean square error (MSE), root mean square error (RMSE), mean absolute error (MAE), and median average error (MedAE) are used. The end result shows that the proposed model outperforms ANN model.

Keywords ANN · GWO · Machine learning · Prediction · Pricing · Stock market

1 Introduction

Estimating or expectation of securities exchange is one of the most sizzling field of research of late for its business applications inferable from the high stakes and appealing advantages the field offers and above all it gives the understanding into nation's monetary development. Forecasting the pricing of stock market is a challenging task for all starting from common investors to industry personnel, from academicians to researchers and speculators. Stock market has many influential factors such as socioeconomic scenarios, political events, other stock's conditions, corporate policies, investor's psychological factors, and investor's expectations [1]. Prediction

S. Sahoo
Department of Computer Science and Engineering, ITER, Siksha 'O' Anusandhan (Deemed to be University), Bhubaneswar, Odisha, India
e-mail: siprasahoo@soa.ac.in

M. N. Mohanty (✉)
Department of Electronics and Communication Engineering, ITER, Siksha 'O' Anusandhan (Deemed to be University), Bhubaneswar, Odisha, India
e-mail: mihirmohanty@soa.ac.in

© Springer Nature Singapore Pte Ltd. 2020
S. Patnaik et al. (eds.), *New Paradigm in Decision Science and Management*,
Advances in Intelligent Systems and Computing 1030,
https://doi.org/10.1007/978-981-13-9330-3_8

model can be applied on the historical data to get future trend. Generally, data of stock market is highly dynamic, noisy, volatile, nonlinear, and nonparametric in nature. Hence, the data is not of fixed pattern. Beforehand factual methodologies, for example, moving normal, weighted moving normal, Kalman separating, exponential smoothing, relapse investigation, autoregressive moving normal (ARMA), autoregressive coordinated moving normal (ARIMA), and autoregressive moving normal with exogenous have been utilized for forecast errand [2]. But they were not good enough for the prediction task of stock market pricing. So machine learning algorithms have been developed to overcome the drawbacks of statistical techniques in recent years, and it is found that the performance of these is better than that of statistical techniques in predicting the stock market pricing. Machine learning is an emerging field of computer science; the machine learning approaches include genetic algorithm, decision tree learning, fuzzy system, neural network (NN), association rule learning, deep learning, artificial neural network (ANN), support vector machines (SVM); and many more and hybrid methods have been widely employed in forecasting stock prices. Supervised learning, unsupervised learning, and semi-supervised learning are the three categories of machine learning tasks, and the applications are regression, classification and clustering. Out of these above methodologies, strategies dependent on ANN have progressively picked up notoriety because of their natural abilities to rough the nonlinearity to a high level of exactness [3].

In this article, gray wolf optimization (GWO) is deployed to optimize the ANN for predicting the stock market price for one day, fifteen days, and thirty days ahead [4, 5]. The main objective of this article is to evaluate the performance of GWO in comparison with ANN. The obtained results are summarized and visualized with the respective graphical representations and tabulations.

The rest of the paper is organized as follows. Section 2 gives an idea about the related work of the problem. Some of the preliminary concepts like data normalization, ANN, and GWO are discussed in Sect. 3. Proposed model and result evaluation are done in Sects. 4 and 5, respectively. Finally, Sect. 6 concludes the article.

2 Related Work

Stock market price can be said as the highest amount someone can pay willingly or the lowest amount it can be bought for. For long been, it draws the attention of academicians, researchers, and also common men as the return is very lucrative. So prediction in stock market pricing is the focus point of many and it is highly challenging task. To fulfill the purpose of accurate prediction, a lot of effort have been made by the researchers and drawbacks of those models motivated researchers to go for optimization methods which are meta-heuristic and nature-inspired algorithms and pave the way to obtain global optimum solution with the integration of ANN model.

For modeling the volatility of stock market, IT2F-CE-EGARCH model has been developed which is the hybridization of time series model features, IT2FLS reasoning and learning component of ANN [6]. BSE Sensex and CNX Nifty Indices datasets are taken and differential harmony search is used for parameter estimation. Fluctuation behavior of stock market is addressed by proposing PCA-STNN which is used to predict the prices of SSE, HS 300, and S&P 500 [7]. In comparison with BPNN, STNN, and PCA-BPNN, the model performs better. As the stock market is very chaotic and noisy in nature, wavelet denoising-based back propagation (WDBP) neural network was developed to eradicate that component and was tested on the empirical dataset of Shanghai Composite Index (SCI) [8]. Using wavelet transform function, dataset can be divided into multiple layers having signal components of high frequency and low frequency where the low frequency signal helps to predict the future value. The model performs better in comparison with a single back propagation neural network. In stock market, genetic algorithms are generally used for generating optimal weights. In [9], a robust and novel hybrid model combining two linear models (exponential smoothing model, autoregressive moving average model) and one nonlinear model (recurrent neural network) optimized by genetic algorithm has been proposed which leads to improved accuracy of prediction of stock market. Mainly, data dependency and choosing regression order for ARMR model randomly are the limitations of this model. Although ANN is used widely in many stock market's modeling applications, it has the limitation of overfitting and under fitting problem. To address the problem, ANN is hybridized with meta-heuristic optimization methods such as harmonic search (HS) and genetic algorithm (GA) which also leads to improvement in prediction performance [10]. Turkish dataset was used for this purpose and drawbacks of this approach was number of hidden layers of ANN are fixed and quality of ANN model was greatly affected by combination of transfer function and training function.

As sentiments play a vital role in our daily lives, it has a great effect on stock market also. So sentiments are also modeled along with the stock market data using supervised machine learning algorithm which predicts the pricing on a daily basis and monthly basis [11]. Long-term prediction of stock market is as important as short-term prediction. So a new novel method called Bat-neural network multi-agent system has been proposed and applied on both technical and fundamental data of DAX stock price. The model is compared with GANN and GRNN giving the conclusion that stock price can be predicted suitably in a long-term period [12]. To address the problem volatility and inner dynamics of the stock market, a novel and hybrid model has been proposed which uses genetic fuzzy system and artificial neural network. As there are many influential factors in the stock market prediction, stepwise regression analysis is used to determine the most influencing factor on stock market. Self-organizing map neural network is also used on dataset of IBM and DELL corporations [13]. Another model named as SERNFIS has been proposed for tending to the issue of instability in the stock exchange [14]. Takagi Sugeno Kang (TSK) type fuzzy if then guidelines are utilized having two sorts of criticism circles, i.e., inner transient input circles and time postponed yield criticism circles which are in charge of improving the execution of forecast capacity of the model.

3   Preliminaries

Some of the preliminary concepts associated with the model are discussed below.

3.1   Dataset

For every research, dataset plays a vital role. In this article, dataset is taken from verifiable information from Bombay stock exchange as chronicled information for stocks and corporate information give standard items, i.e., without separating for client chose information focuses, or on altered premise. The dataset is isolated into two sections: for training and for testing. Whole depiction of dataset is depicted in Table 1.

3.2   Normalization

The primary objective of information standardization is to ensure the nature of the information before it is bolstered to any learning calculation. There are numerous sorts of information standardization. It tends to be utilized to scale the information in a similar scope of qualities for each info highlight with the end goal to limit predisposition inside the neural system for one component to another. Information standardization can likewise accelerate preparing time by beginning the prepara-tion procedure for each element inside a similar scale. It is particularly helpful for demonstrating application where the sources of info are for the most part on broadly extraordinary scales. Diverse methods can utilize distinctive standards, for example, min-max, Z-score, decimal scaling, median standardization, etc.

3.3   Artificial Neural Network

Neural networks are the data handling frameworks which are manufactured and actualized for displaying of the human mind as it is considered as generally keen. An artificial neural network is a productive data handling framework which takes

Table 1 Depiction of dataset

Dataset	Example	Data range	Training samples	Test samples
Stock market	*3518*	25 August 2004 to 24 October 2018	2518	1000

after with an organic neuron. ANNs comprise of profoundly interconnected preparing components called as hubs or neurons. The association connects between the neurons are portrayed by a few weights which demonstrates the data about the information flag which takes care of a specific issue. ANNs are having three layers, for example, input layer, hidden layer, and output layer fit for learn, review, and sum up preparing designs. For the most part, initiation capacities are utilized in shrouded layers and yield layers.

3.4 Gray Wolf Optimization

Gray wolf optimization works on the principle of leadership nature and hunting behavior of gray wolves. It is a powerful population-based stochastic algorithm having better convergence quality, near optimal solution in comparison with genetic algorithm (GA), differential evolutionary (DE) algorithm, gravitational search algorithm (GSA), ant colony optimization (ACO), and particle swarm optimization (PSO). Gradient information is not required for the optimized function, and it is a great advantage of this recently developed algorithm. In comparison with other optimization algorithms, it gains significant attention as it is easily implementable and able to optimize many practical optimization problems [4, 5].

As the algorithm is inspired by hunting and leadership behavior of wolves, it has three phases

- Social hierarchy of wolves for approaching the prey
- Encircling the prey
- Hunting the prey

For modeling, the first phase mathematically that is social hierarchy of gray wolves alpha (α) is taken as the best solution (the leader wolf who is the decision maker) is followed by beta (β) and delta (δ) as second and third best solutions (wolves assist the leader wolf) in decision making. All other solutions (Rest wolves) in the population are named as omega (ω) and those wolves follow alpha (α), beta (β), and delta (δ).

In second phase, for encircling the prey the mathematical model is stated as

$$\vec{w}(t + 1) = \vec{w}_P(t) - \vec{A} \cdot \left| \vec{C} \cdot \vec{w}_P(t) - \vec{w}(t) \right| \tag{1}$$

where position vector of a gray wolf is $\vec{w}$, 't' indicates the current iteration.
Position vector of the prey is indicated by $\vec{w}_p$.
$\vec{A}$ and $\vec{C}$ are coefficient vectors calculated as

$$\vec{A} = 2a \cdot \vec{r}_1 - \vec{a}$$
$$\vec{C} = 2 \cdot \vec{r}_2$$

where $\vec{r}_1, \vec{r}_2$ are random vectors and $\vec{a}$ is linearly decreasing over many number of iterations as follows

$$\vec{a}(t) = 2 - 2t/\text{Itr}_{\text{max}} \tag{2}$$

where current iteration is 't'.

Total no of iterations is Itr_{max}.

The positions of other wolves are updated according to the positions of $\alpha, \beta,$ and δ.

The last phase, i.e., 'hunting the prey,' can be mathematically simulated as follows: Positions of other wolves are updated according to the positions of $\alpha, \beta,$ and δ

$$\vec{w}_1 = \vec{w}_\alpha - \vec{A}_1 \cdot \left| \vec{C}_1 \cdot \vec{w}_\alpha - \vec{w} \right| \tag{3}$$

$$\vec{w}_2 = \vec{w}_\beta - \vec{A}_2 \cdot \left| \vec{C}_2 \cdot \vec{w}_\beta - \vec{w} \right| \tag{4}$$

$$\vec{w}_3 = \vec{w}_\delta - \vec{A}_3 \cdot \left| \vec{C}_3 \cdot \vec{w}_\delta - \vec{w} \right| \tag{5}$$

$$\vec{w}_{t+1} = \frac{\vec{w}_1(t) + \vec{w}_2(t) + \vec{w}_3(t)}{3} \tag{6}$$

$\vec{A}_1, \vec{A}_2, \vec{A}_3$ are same as $\vec{A}$ and $\vec{C}_1, \vec{C}_2, \vec{C}_3$ are same as $\vec{C}$.

The pseudo-code is as follows:

1. Generate initial population of hunting agents

$$w_i (i = 1, 2, \ldots, n)$$

2. Initialize the vector coefficient $\vec{a}$, $\vec{A}$ and $\vec{C}$
3. Estimate fitness value of each hunting agent

$$w_\alpha = \text{fittest hunting agent}$$
$$w_\beta = \text{second fittest hunting agent}$$
$$w_\delta = \text{third fittest hunting agent}$$

4. $\text{Itr} = 1$
5. Repeat
6. For $i = 1 : w_s$
   Update the position of current hunting agent using Eq. 6
   End for
7. Estimate the fitness value of all hunting agents
8. Update the values for $w_\alpha, w_\beta, w_\delta$

9. Update vector coefficient $\vec{a}$, $\vec{A}$ and $\vec{C}$
10. Itr = Itr + 1
11. Stop when Itr$_{\text{max}}$ reached
12. Output w_α
End

4 Proposed Approach

The proposed model takes the past data of BSE, which is collected from 25 August 2004 to 24 October 2018. Training data range is from 25 August 2004 to 8 October 14 (1000 samples), and testing data range is from 9 October 2014 to 24 October 2014 (2518 samples). Primarily, in this context data is normalized within the range 0–1. Min-Max normalization technique is applied on the above-said dataset, and then, the network is trained using ANN and ANN-GWO. Basic difference between ANN and ANN-GWO is ANN is optimized with the feedback of error to the network whereas ANN-GWO model is optimized by applying gray wolf optimization technique. In the next phase, after constructing the network, it is tested with the test data. The model predicts the pricing for one day ahead, fifteen days ahead, and one month ahead. The proposed model has been compared with ANN prediction model. The features of the historical dataset involve opening price, closing price, high price, low price. The layout of the proposed model is illustrated in Fig. 1.

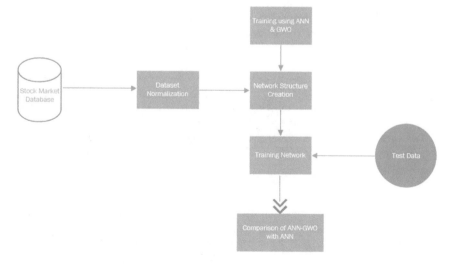

Fig. 1 Schematic layout of proposed model ANN-GWO

5 Experimental Setup and Simulation Result

For experimental simulation, MATLAB (version 6) was employed on PC with
2.13 GHz with i5 processor. The system is having RAM of 4.0 GB and x-64-based
processor which operates on Windows 8.1 Pro.

In this study, one standard benchmark dataset has been used for prediction of
stock market pricing from which 2518 are training samples and 1000 test samples. A
network structure is created using ANN which has been trained using gray wolf optimization (GWO) for prediction of the stock market pricing. The model is designed
using a suitable number of population size 50 and the number of nodes in input layer
is 4. The number of hidden layer in network is one having four number of nodes.
Execution is carried out for 50 iterations as shown in Table 2. The model predicts
for one day, 15 days, and 30 days in advance. The resultant weight from the trained
dataset has been tested using sample test data. The experimental results of GWO are
compared with simple ANN which predicts for one day, fifteen days, and thirty days.
From the below figures, it is observed GWO predicts more close to actual price and
it converges faster than ANN. By using GWO, ANN gives better result than simple
ANN. The performance is evaluated using MAE, MedAE, MSE, RMSE. Tables 3,
4, and 5 show the performance of the model for one day, fifteen days, and thirty days
in advance. Figures 2, 3 and 4 show the prediction of one day, fifteen days, and thirty
days in advance. Figure 5 shows the training of dataset for one day, fifteen days, and
thirty days.

6 Conclusion

In the era of economic globalization, everyone wants to make easy money and stock
market provides the platform to make it fruitful. However, on the other hand, there is

Table 2 Parameters of proposed model

ANN	ANN-GWO
No. of input nodes—4 No. of hidden layers—1	Population size—50
No. of nodes in the hidden layers—4	Iterations—50
Learning rate (alpha)—0.125	$a = 2$–0
No. of iteration—50	$A = [-1, 1]$
	$C = [0, 2]$

Table 3 One day ahead performance evaluation measure

Method	MAE	MedAE	MSE	RMSE
ANN	0.454300	0.454300	0.268888	0.518545
GWO	0.448400	0.448400	0.263563	0.513383

Table 4 15 days ahead performance evaluation measure

Method	MAE	MedAE	MSE	RMSE
ANN	0.473300	0.473300	0.286513	0.535269
GWO	0.470300	0.470300	0.283682	0.532618

Table 5 Thirty days ahead performance evaluation measure

Method	MAE	MedAE	MSE	RMSE
ANN	0.478100	0.478100	0.291080	0.539518
GWO	0.472500	0.472500	0.285756	0.534562

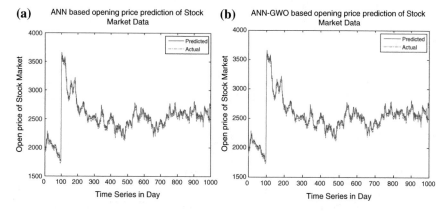

Fig. 2 One day in advance prediction **a** ANN, **b** ANN-GWO

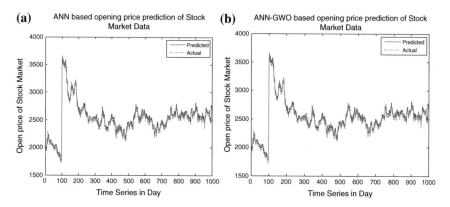

Fig. 3 15 days in advance prediction **a** ANN, **b** ANN-GWO

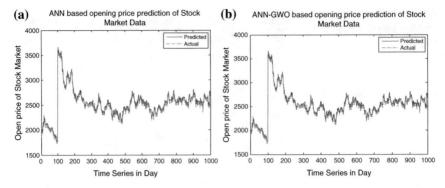

Fig. 4 30 days in advance prediction **a** ANN, **b** ANN-GWO

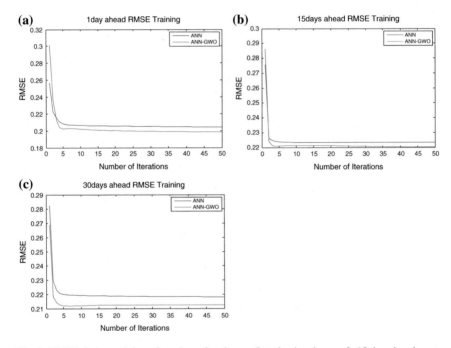

Fig. 5 RMSE during training of stock market data. **a** One day in advance, **b** 15 days in advance, **c** 30 days in advance

the high chance of losing money. So it is very necessary to know the future trend of the stock market. Many predictive models have been proposed to accurately predict the stock market pricing. In this experimental research, an effective hybrid optimization model is used for the prediction of stock market pricing 1 day ahead, 15 days ahead, and 30 days ahead. The model is the combination of artificial neural network and gray wolf optimization technique and is applied on Bombay stock exchange dataset.

References

1. Zhou, F., Zhou, H.M., Yang, Z., Yang, L.: EMD2FNN: a strategy combining empirical mode decomposition and factorization machine based neural network for stock market trend prediction. Expert Syst. Appl. **115**, 136–151 (2019)
2. Chen, Y., Hao, Y.: Integrating principle component analysis and weighted support vector machine for stock trading signals prediction. Neurocomputing **321**, 381–402 (2018)
3. Das, S.R., Mishra, D., Rout, M.: A hybridized ELM-Jaya forecasting model for currency exchange prediction. J. King Saud Univ.-Comput. Inf. Sci. (2017)
4. Mirjalili, S., Mirjalili, S.M., Lewis, A.: Grey wolf optimizer. Adv. Eng. Softw. **1**(69), 46–61 (2014)
5. Long, W., Jiao, J., Liang, X., Tang, M.: Inspired grey wolf optimizer for solving large-scale function optimization problems. Appl. Math. Model. **60**, 112–126 (2018)
6. Dash, R., Dash, P.K., Bisoi, R.: A differential harmony search based hybrid interval type 2 fuzzy EGARCH model for stock market volatility prediction. Int. J. Approximate Reasoning **59**, 81–104 (2015)
7. Wang, J., Wang, J.: Forecasting stock market indexes using principle component analysis and stochastic time effective neural networks. Neurocomputing **156**, 68–78 (2015)
8. Wang, J.Z., Wang, J.J., Zhang, Z.G., Guo, S.P.: Forecasting stock indices with back propagation neural network. Expert Syst. Appl. **38**(11), 14346–14355 (2011)
9. Rather, A.M., Agarwal, A., Sastry, V.N.: Recurrent neural network and a hybrid model for prediction of stock returns. Expert Syst. Appl. **42**(6), 3234–3241 (2015)
10. Göçken, M., Özçalıcı, M., Boru, A., Dosdoğru, A.T.: Integrating metaheuristics and artificial neural networks for improved stock price prediction. Expert Syst. Appl. **44**, 320–331 (2016)
11. Nayak, A., Pai, M.M., Pai, R.M.: Prediction models for indian stock market. Procedia Comput. Sci. **89**, 441–449 (2016)
12. Hafezi, R., Shahrabi, J., Hadavandi, E.: A bat-neural network multi-agent system (BNNMAS) for stock price prediction: case study of DAX stock price. Appl. Soft Comput. **29**, 196–210 (2015)
13. Hadavandi, E., Shavandi, H., Ghanbari, A.: Integration of genetic fuzzy systems and artificial neural networks for stock price forecasting. Knowl. Based Syst. **23**(8), 800–808 (2010)
14. Dash, R., Dash, P.: Efficient stock price prediction using a self evolving recurrent neuro-fuzzy inference system optimized through a modified differential harmony search technique. Expert Syst. Appl. **52**, 75–90 (2016)

Analysis of Techniques for Credit Card Fraud Detection: A Data Mining Perspective

Smita Prava Mishra and Priyanka Kumari

Abstract The expeditious development of World Wide Web tradition has derived the situation where online shopping and other payment services become most popular among people. They always buy goods and services online or off-line and use their credit or debit card for payment. With the swipe card employment, the fraud rate is also swelling day by day. Hence, the credit card has evolved as the standardized method of payment and the fraud associated with credit card is also expanding exponentially. As we know, many modern data mining techniques have been deployed for the detection of fraud in the domain of credit cards, such as Hidden Markov model, fuzzy logic, K-nearest neighbor, genetic algorithm, Bayesian network, artificial immune system, neural network, decision tree, support vector machine, hybridized method, and ensemble classification. The purpose addressed in this paper is to consolidate various data mining approaches used for finding credit card frauds by researchers to carry out research in the domain and has a state-of-the-art view of the financial domain.

Keywords Credit card fraud · Financial dataset · Fraud detection techniques · Performance metrics

1 Introduction

In the current date, most of the organizations and financial institutions have chosen electronics commerce and web-based services to heighten their turnover and to save effort and time. The usage of credit card for online as well as off-line payments is always accompanied with a risk of fraudulent transactions. The purchases based on

S. P. Mishra (✉)
Department of CS and IT, ITER, Siksha 'O' Anusandhan University, Bhubaneswar, India
e-mail: smitamishra@soa.ac.in

P. Kumari
Department of CSE, ITER, Siksha 'O' Anusandhan University, Bhubaneswar, India
e-mail: kumari.priyanka522@gmail.com

© Springer Nature Singapore Pte Ltd. 2020   89
S. Patnaik et al. (eds.), *New Paradigm in Decision Science and Management*,
Advances in Intelligent Systems and Computing 1030,
https://doi.org/10.1007/978-981-13-9330-3_9

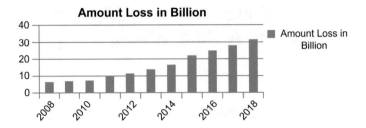

Fig. 1 Global losses annually caused by credit and debit card frauds

card may be categorized into (a) virtual card and (b) physical card [1]. In both cases, the fraudsters can perform fraud, if the card or card characteristics are abstracted or extracted somehow. The term credit card fraud refers as unauthorized use of other's card by fraudsters without the permission of authorized user [2].

Many financial institution and banking industries lose millions of dollars per annum resulted due to credit card fraud. Figure 1 shows the rate of global losses annually for years 2008–2018 caused by the credit or debit card frauds as per the Nilson Report, October 2018.

The remaining section of the paper is organized as follows: Sect. 2 presents the reviewed literature in light of datasets used, techniques adopted, and metrics for measuring performances. Section 3 discusses the challenges the researchers encounter for the said purpose and domain. Section 4 concludes the analysis for future researchers.

2 Literature Review

This review describes the different data mining techniques adopted by researchers to detect credit card fraud. Enlisted below are process parameters for devising a model for fraud detection.

2.1 Dataset Collection

In the credit card domain, the fraud detection process evidently requires a credit card dataset to test fraudulent data. In this field, lesser availability of datasets has been a restricting element for the detection of fiscal fraud [3]. Altogether, few works make their own dataset for detection [4, 5]. Table 1 shows some datasets used by researchers that we found by conducting a review on various research findings in the domain of credit card for detection of fraudulent transactions.

Table 1 Details of datasets used by researchers

Reference	Dataset	Collected from	Description of dataset
[6–8]	UCSD Dataset	UCSD-FICO Data Mining Contest 2009	100,000 instances with 20 attributes. It is imbalanced 98% of transactions are legal and 2% are fraud
[9, 4]	German Dataset	UCI Repository	1000 instances, 21 attributes including class label attribute
[10]	Australian Dataset	UCI Repository	690 instances, 15 attribute including class label attribute
[11]	Second Robotic & Artificial Intelligence Festival of Amirkabir University Dataset	From Authors [11]	It is an imbalanced dataset. It contains 96.25% legal transaction and 3.75% fraud

2.2 Various Methodology of Fraud Detection in Credit Card

After the literature review of popular methods used to identify credit card fraud, we found that each method has its own advantages and disadvantages. Here, we present a summary of all the effective methods used in several research literature for fraud detection in credit card domain.

Figure 2 shows the complete taxonomy of different classification techniques for spotting of credit card fraud which is found from the reviewed literature.

2.2.1 Hidden Markov Model (HMM)

HMM is a framework which can fix arbitrary method firmly and deeply with two distinct levels. HMM uses user's card spending behavior for fraud detection. If an inbound credit card processing request is not processable by the trained HMM, it is supposed to be fraudulent transactions. In 2008, Srivastava et al. [12] report the "Credit card fraud detection method by using Hidden Markov Model (HMM)." To identify credit card fraud, a HMM is made to learn for identifying the normal transactions of cardholder and encode that behavior in customer profiles [13]. HMM generates higher false positive rate [14] as reported by other researchers as well. Bhusari et al. [15] employed HMM for detection of credit card frauds with lesser false alarm rate. HMM processes transactions and identifies whether it is normal or fraudulent. If the model is sure about the suspiciousness of the transaction, then an alarm is raised and associated banks reject that transaction as per usability. Fraud detection rate is fast. But, model development cost is high whereas low accuracy is obtained.

Fig. 2 Taxonomy of overall classification techniques for fraud detection

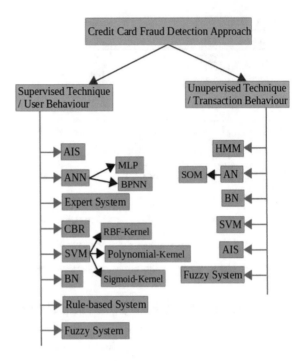

2.2.2 Genetic Algorithm (GA)

The concept of GA searches for an optimal solution from a population of candidates. The extension of GA is Genetic Programming (GP) [16]. In data mining techniques, the genetic algorithm is also used for feature selection. Duman et al. [4] proposed assignment of a cost-sensitive objective function to different misclassification errors such as false positive and false negative. The objectives of classifiers are to reduce the universal cost rather than the number of misclassified transactions. Rama Kalyani et al. [17] developed a model for fraud detection in credit card with genetic algorithm. Bentley et al. [18] developed a fuzzy system build on genetic programming to remove rules to classify data. GA behaves well with noisy data, it can enhance the performance of other algorithms, and also fraud detection rate is fast. However, it is complicated and difficult to interpret.

2.2.3 Bayesian Network (BN)

BN is a probabilistic directed acyclic graph-based framework or statistical model that represents group of random variables. The main principle of Bayesian networks is to find unknown parameter by given known parameter in the presence of uncertainty [19]. Bayesian networks are employed when some basic knowledge of the data is available and we have to classify new incoming data on the basis of known data. So King-Fung Pun et al. [20] have suggested two approaches for card fraud detection applying the above-said framework. In the first, the activity of fraudsters, while transacting with the card, and, in the second, the activity of normal user are modeled by the Bayesian network. This method is also adopted by other researchers [21]. Bayesian networks have also been used by researchers for pattern recognition [22], diagnostics [23, 24] also forecasting [25]. Beyond, the Bayesian networks are also in use for detection of anomalies in credit card transactions and even for anomaly detection in telecommunication networks [26, 27]. The probability of fraud for new transactions can be derived while applying Bayes rule [28]. Four stages of the Bayesian network was developed by Ezawa and Norton [29]. Processing speed is high and fraud detection rate is high, and accuracy of result is also high. At times, it seems to be expensive to build a model.

2.2.4 Artificial Neural Network (ANN)

ANN acts on the basis of the principle in which human brain works. In 1994, Ghosh and Reilly [30] presented a paper on the fraud detection of credit card with neural network. Again in 2011, Patidar and Sharma [11] tried to identify fraudulent transaction by using the neural network with the genetic algorithm. Some variants of ANN have also been used for fraud detection by other researchers [31]. Processing speed and detection rate are very high. Sometimes difficult to operate, in this we need to convert all nominal data into numerical data which is expensive.

2.2.5 Decision Tree (DT)

Decision trees can be used as the predictive model which can be used in statistic, data mining, and machine learning that indicate each attributes in a tree [32]. Dhanpal et al. [33] projected a credit card fraud detection model using decision tree algorithm. They used best split measure such as Gini, entropy, information gain ratio to choose the best classifier. The popular classifiers of the decision trees method are C5.0, CART, and CHAID. Then, they used CART classifier with decision tree which produced best results. In some other works [13], parallel granular neural networks (PGNNs) have been used to enhance the speed of detection process of credit card fraud. Implementation work is easy, and display of result is easy, easily understandable. In this, each condition should be checked one by one in case of fraud detection. Hence, complexity is more.

2.2.6 Support Vector Machine (SVM)

SVM is used to classify new incoming instances by making a hyper-plane between two classes under consideration. SVM can examine and identify patterns for classification and regression work [4, 34]. In 2012, Dheepa et al. [35] described "a fraud detection model using behavior-based credit card fraud detection and support vector machine." In 2016, Nancy Demla and Alankrita Aggarwal described "credit card fraud detection using SVM and Reduction of false alarms" [9], authors found SVM giving better result than other while using high dimensional dataset. Chen et al. [36] select support vectors by using genetic algorithm to develop a binary support vector system (BSVS). In [37], authors developed a decision tree and support vector machine-based classification model for credit card fraud detection. Chen et al. [38] proposed a questionnaire responder transaction (QRT) mechanism with SVM for detection of credit card fraud. Qibei et al. [39] developed an improvised model depending on class weighted SVM, which also applied principal component analysis for credit card fraud detection. It can easily handle high dimensional data. But processing speed is low and method is expensive.

2.2.7   Hybrid Method

In the domain of credit card fraud, using a hybrid approach gives better accuracy instead of single method. In the hybrid approach, the authors take two of the techniques individually and also implementing both together. Tripathi et al. [40] presented a fraud detection framework by using the hybrid method of Dempster–Shafer adder (DSA) and Bayesian algorithm. Here, they found DSA giving better performance, in terms of identifying true positives. A hybrid method for credit card fraud detection using rough set and decision tree technique [41], preprocessing is done by rough set and used decision tree-based J48 classifier for the classification. Tanmay Kumar et al. [42] developed a framework for credit card fraud identification by the use of hybrid approach with fuzzy clustering and neural network. They also could detect and minimize false alarm rate by it. Stephen Gbenga et al. [43] employed hybrid method by using K-means with HMM and MLP for fraud detection.

2.2.8   Ensemble Classification

Ensemble mechanism is the process of combining a series of models aimed at creating an improved model. The researchers have employed several well-known ensemble techniques like: bagging, voting, classification via regression, and random forest to analyze their ability to detect fraud. In the area of credit card fraud, detection ensemble techniques give more accuracy than other single classifiers. Masoumeh et al. [44] presented a fraud detection method using bagging ensemble classifier based on decision tree and found that bagging ensemble classifier outperformed and was more stable than individual classifiers. Seeja et al. [7] also used an ensemble fraud miner classifier and also compare the results with other four classifiers K-NN, NB, SVM, and RF and found ensemble works well for credit card fraud detection.

Table 2 Performance criteria for evaluation of credit card fraud detection

Measure	Formula	Description
Accuracy	TN + TP/TP + FP + FN + FP	Percentage of instances those are classified correctly
Precision/hit rate	TP/TP + FP	It is no. of classified fraud class that are actually legal class
Fraud catching rate	TP/P	It identifies the fraction of actual positive instances those are predicted as positive
False alarm rate	FP/N	It finds the part of actual negative which are predicted as positive
Balanced classification rate (BCR)	(TP/P) + (TN/N)	It is average of sensitivity and specificity
Matthews correlation coefficient (MCC)	(TP * TN) − (FP * FN)/($\sqrt{}$(TP + FP)(TP + FN)(TN + FP)(TN + FN))	It is used to measure the quality of binary classification
Cost	100 * FN + 10 * (FP + TP)	Cost of evaluation of the model

2.2.9 Performance Measurements

Various classification matrices are available in credit card fraud detection domain for performance measure [45], as described in Table 2.

3 Challenges in Credit Card Fraud Detection

There are several challenges encountered by researchers while trying to identify fraudulent transactions from legitimate ones. Some of the popular causes are enlisted as below:

- *Imbalanced data*: It means that, among the huge transaction of customers, very small percentages are fraudulent. This reasons the detection of fraud transactions more difficult.
- *Dynamics behavior of fraudsters*: In fraud detection task, it is difficult to identify fraud pattern based on previous historical data because fraudsters always change the way of committing fraud and use new ideas and techniques for the same.
- *Overlapping data*: Sometimes the normal transaction may be considered as fraudulent and vice versa, a fraud transaction may also look like an authorized one counting for false negative.

- *Lack of adaptability*: Detection of upcoming varieties of legitimate as well as fraudulent activity patterns is the main problem with the learning of classification algorithms.

4   Conclusion

The facts, we observed and examined in this review work that credit card has become the most important part of our today's developing economic environment. After that we observed that the area of credit card is always tracked by the fraudsters. Therefore, an efficient fraud detection system is much necessary in today's scenario. That's why researchers have developed many techniques for detection of credit card fraud and currently working on it. In this paper, we examined several effective techniques from existing research findings for fraud detection. All the techniques have their own strengths and weaknesses. Some of them give better performance than others based on their design criteria. Almost all the researchers suffer from the unavailability of real-world dataset problem. But still some of the techniques perform well with the available data as well. Performance of hybrid and ensemble classifiers opens a wider scope for future researchers for fraud detection in the fiscal domain.

References

1. Ravindra, P.S., Vijayalaxmi, K.: Survey on credit card fraud detection techniques. Int. J. Eng. Comput. Sci. **4**, 15010–15015 (2015)
2. Linda, D., Hussein, A., Pointon, J.: Credit card fraud and detection techniques: a review. Banks Bank Syst. **4**, 57–68 (2009)
3. Gadi, M.F.A., Wang, X., do Lago, A.P.: Credit card fraud detection with artificial immune system. In: Bentley, P.J., Lee, D., Jung, S. (eds.) Artificial Immune Systems. ICARIS 2008. Lecture Notes in Computer Science, vol. 5132, pp. 119–131. Springer, Berlin, Heidelberg (2008)
4. Duman, E., Ozcelik, M.H.: Detecting credit card fraud by genetic algorithm and scatter search. Expert. Syst. Appl. **38**, 13057–13063 (2011)
5. Guo, T., Yang, G.: Neural data mining for credit card fraud detection. In: Proceedings of the Seventh International Conference on Machine Learning and Cybernetics, Kunming, China, pp. 3630–3634 (2008)
6. Abdulla, N., Rakendu, R., Varghese, S.M.: A hybrid approach to detect credit card fraud. Int. J. Sci. Res. Publ. **5**, 304–314 (2015)
7. Seeja, K.R., Masoumeh, Z.: Fraud miner: a novel credit card fraud detection model based on frequent itemset mining. Sci. World J. **10**, 1–10 (2014)
8. Masoumeh, Z., Shamsolmoali, P.: Application of credit card fraud detection: based on bagging ensemble classifier. In: International Conference on Intelligent Computing, Communication and Convergence. Procedia Computer Science, vol. 48, pp. 679–685 (2015)
9. Demla, N., Aggarwal, A.: Credit card fraud detection using support vector machines and reduction of false alarms. Int. J. Innov. Eng. Technol. **7**, 176–182 (2016)
10. Ping, Y.: Credit scoring using ensemble machine learning. In: International Conference on Hybrid Intelligent Systems (HIS'09), Shenyang, China, vol. 3, pp. 244–246 (2009)

11. Patidar, R., Sharma, L.: Credit card fraud detection using neural network. Int. J. Soft Comput. Eng. **1**(NCAI2011), 32–38 (2011)
12. Srivastava, A., Kundu, A., Sural, S., Majumdar, K.A.: Credit card fraud detection using hidden Markov model. IEEE Trans. Dependable Secure Comput. **5**, 37–48 (2008)
13. Syeda, M., Zhang, Y.Q., Pan, Y.: Parallel granular neural networks for fast credit card fraud detection. In: Proceedings of IEEE International Conference on Fuzzy Systems, Honolulu, HI, USA, pp. 572–577 (2002)
14. Benson, S.E.R., Annie Portia, A.: Analysis on credit card fraud detection methods. In: IEEE-International Conference on Computer, Communication and Electrical Technology, Tamil Nadu, India, pp. 152–156 (2011)
15. Bhusari, V., Patil, S.: Application of hidden markov model in credit card fraud detection. Int. J. Distrib. Parallel Syst. (IJDPS) **2**(6), 1–9 (2011)
16. Hansena, J.V., Lowrya, P.B., Meservya, R.D., Mc Donald, D.: Genetic programming for prevention of cyber terrorism through dynamic and evolving intrusion detection. J Decis. Support Syst. **43**(4), 1362–1374 (2007)
17. Rama Kalyani, K., Uma Devi, D.: Fraud detection of credit card payment system by genetic algorithm. Int. J. Sci. Eng. Res. **3**(7), 1–6 (2012)
18. Bentley, P.J., Kim, J., Jung, G., Choi, J.: Fuzzy Darwinian detection of credit card fraud. In: Proceedings of 14th Annual Fall Symposium of the Korean Information Processing Society, pp. 1–4 (2000)
19. Monedero, I., Biscarri, F., Leon, C., Guerrero, J.I., Biscarri, J., Millan, R.: Detection of frauds and other non-technical losses in a power utility using Pearson coefficient, Bayesian networks and decision trees. J. Electr. Power Energy Syst. **34**, 90–98 (2012)
20. King-Fung Pun, J.: Improving credit card fraud detection using a meta-learning strategy. A Masters Thesis of Applied Science Graduate Department of Chemical Engineering and Applied Chemistry, University of Toronto (2011)
21. Lee, K.C., Jo, N.Y.: Bayesian network approach to predict mobile churn motivations: emphasis on general Bayesian network, Markov blanket, and what-if simulation. In: Kim, T., Lee, Y., Kang, B.H., Ślęzak, D. (eds.) Future Generation Information Technology. FGIT 2010. Lecture Notes in Computer Science, vol. 6485, pp. 304–313. Springer, Berlin, Heidelberg (2010)
22. Thornton, J., Gustafsson, T., Blumenstein, M., Hine, T.: Robust character recognition using a hierarchical Bayesian network. In: Sattar, A., Kang, B. (eds.) AI 2006: Advances in Artificial Intelligence. AI 2006. Lecture Notes in Computer Science, vol. 4304, pp. 1259–1266. Springer, Berlin, Heidelberg (2006)
23. Przytul, K.W., Dash, D., Thompson, D.: Evaluation of Bayesian networks used for diagnostics. In: Proceedings of IEEE Aerospace Conference, vol. 60, pp. 1–12 (2003)
24. Riascos, L.A.M., Simoes, M.G., Miyagi, P.E.: A Bayesian network fault diagnosis system for proton membrane exchange fuel cells. J. Power Sour. **165**, 267–278 (2007)
25. Aggarwal, S.K., Saini, L.M., Kumar, A.: Electricity price forecasting in deregulated markets: a review and evaluation. Int. J. Electr. Power Energy Syst. **31**, 13–22 (2009)
26. Kou, Y., Lu, C.-T., Sinvongwattana, S., Huang, Y.-P.: Survey of fraud detection techniques. In: IEEE International Conference on Network Sensor and Control, Taiwan, pp. 89–95 (2004)
27. Buschkes, R., Kesdogan, D., Reichl, P.: How to increase security in mobile networks by anomaly detection. In: Proceedings of 14th Computer Security Application Conference (ACSAC'98), Phoenix, AZ, USA, pp. 3–12 (1998)
28. Dheepa, V., Dhanapal, R.: Analysis of credit card fraud detection methods. Int. J. Recent Trends Eng. **2**(3), 126–128 (2009)
29. Ezawa, K., Norton, S.: Constructing Bayesian networks to predict uncollectible telecommunications accounts. IEEE Expert Intell. Syst. Appl. **11**(5), 45–51 (1996)
30. Ghosh, S., Reilly, D.L.: Credit card fraud detection with a neural-network. In: Proceedings of 27th Hawaii International Conference System Sciences: Information Systems: Decision Support and Knowledge Based Systems, vol. 3, pp. 621–630 (1994)
31. Quah, J.T.S., Sriganesh, M.: Real-time credit card fraud detection using computational intelligence. Expert Syst. Appl. **35**(4), 1721–1732 (2011)

32. Shen, A., Tong, R., Deng, Y.: Application of classification models on credit card fraud detection. In: 2007 International Conference on Service Systems and Service Management, Chengdu, China, pp. 1–4 (2007)
33. Dhanpal, R., Gayathiri, P.: Detecting credit card fraud using decision tree for tracing email and IP. Int. J. Comput. Sci. Issues **9**(5), No 2, 406–412 (2012)
34. Kamboj, M., Shankey, G.: Credit card fraud detection and false alarms reduction using support vector machines. Int. J. Adv. Res. Ideas Innov. Technol. **2**(4), 1–10 (2016)
35. Dheepa, V., Dhanpal, R.: Behaviour based credit card fraud detection using support vector machine. ICTACT J. Soft Comput. **2**, 391–397 (2012)
36. Chen, R.-C., Chen, T.-S., Lin, C.-C.: A new binary support vector system for increasing detection rate of credit card fraud. Int. J. Pattern Recognit. Artif. Intell. **20**(02), 227–239 (2006)
37. Sahin, Y., Duman, E.: Detecting credit card fraud by decision trees and support vector machines. In: Proceeding of International Multi-conference of Engineering and Computer Scientists, Hong Kong, vol. 1 (2011)
38. Chen, R., Chen, T., Chien, Y., Yang, Y.: Novel questionnaire responded transaction approach with SVM for credit card fraud detection. In: Lecture Notes in Computer Science, vol. 3497, pp. 916–921. Springer, Berlin, Heidelberg (2005)
39. Qibei, L., Ju, C.: Research on credit card fraud detection model based on class weighted support vector machine. J. Converg. Inf. Technol. **6**(1), 1–7 (2011)
40. Tripathi, K.K., Ragha, L.: Hybrid approach for credit card fraud detection. Int. J. Soft Comput. Eng. (IJSCE) **3**(4) (2013)
41. Jain, R., Gour, B., Dubey, S.: A hybrid approach for credit card fraud detection using rough set and decision tree technique. Int. J. Comput. Appl. **139**(10), 1–6 (2016). (Foundation of Computer Science (FCS), NY, USA)
42. Behera, T.K., Panigrahi, S.: Credit card fraud detection: a hybrid approach using fuzzy clustering & neural network. In: Second International Conference on Advances in Computing and Communication Engineering (ICACCE), pp. 494–499. IEEE (2015)
43. Fashoto, S.G., Owolabi, O., Adeleye, O., Wandera, J.: Hybrid methods for credit card fraud detection using K-means clustering with hidden markov model and multilayer perceptron algorithm. Br. J. Appl. Sci. Technol. **13**(5), 1–11 (2016)
44. Masoumeh, Z., Shamsolmoali, M.: Application of credit card fraud detection: based on bagging ensemble classifier. In: International Conference on Intelligent Computing, Communication and Convergence, vol. 48, pp. 679–685 (2015)
45. Ayano, O., Akinola, S.O.: A Multi-algorithm data mining classification approach for bank fraudulent transactions. Afr. J. Math. Comput. Sci. Res. **10**(1), 5–13 (2017)

Predictive Analytics for Weather Forecasting Using Back Propagation and Resilient Back Propagation Neural Networks

Bhavya Alankar, Nowsheena Yousf and Shafqat Ul Ahsaan

Abstract Weather can be elucidated as the appearance of atmosphere for a short span of time for a specific place. Weather an important factor in our day-to-day lives and also the one that controls how and what people should do. For researchers and scientists, weather has been the most demanding task. The practice of technology that has been practiced to presume the conditions of atmosphere for a given area at a specific time is weather forecasting. Rationally accurate forecasts are obtained because of modern weather forecasting methods that involve computer models, observations and information about current patterns, and trends. Due to the presence of nonlinearity in climatic data, neural networks have been the satisfactory for prediction of weather. An algorithm based on artificial neural networks is used for predicting the future weather as the artificial neural network (ANN) package supports a number of learning or training algorithms. The benefits of using artificial neural networks for predicting the weather have been presented in this paper. In this paper, weather predictions are made by building training and testing datasets with the usage of learning algorithms back propagation and resilient back propagation.

Keywords Artificial neural network (ANN) · Back propagation · Resilient back propagation · Weather forecasting · Feed forward network

1 Introduction

The condition of the atmosphere for a short period of time is weather. Weather involves short period changes like changes in minutes, hours, days, months, and even seasonal changes. The number of components that weather comprises of involves sunlight, rain, winds, cloud, freezing rain, thunder storms, excessive heat, heat waves, and many more. Exact weather forecasting is important for a number of reasons like

B. Alankar · N. Yousf · S. U. Ahsaan (✉)
Department of Computer Science and Engineering, School of Engineering Sciences and Technology, Jamia Hamdard, New Delhi, India
e-mail: mailforshafqat@gmail.com

© Springer Nature Singapore Pte Ltd. 2020
S. Patnaik et al. (eds.), *New Paradigm in Decision Science and Management*,
Advances in Intelligent Systems and Computing 1030,
https://doi.org/10.1007/978-981-13-9330-3_10

planning of day-to-day activities, agrarian activities, industrial activities, airlines for scheduling their flight, etc. [1–4]. Weather is the most significant circumstance for obtaining the results of agricultural enterprises. Weather also greatly contributes to the existing crop and animal diseases. Farmers decide their do's and don'ts completely on the basis of weather forecasts for their day-to-day lives; for example, prediction of a dry weather leads farmers to do works like drying hay, other crops, and even veggies. Dryness for a long period of time can prove harmful for a number of crops like cotton, wheat, corn crops, etc. Gas companies are also dependent on weather forecasts for anticipating their demands. To prevent and control harmful wildfires for wildlife in forests accurate weather forecasting of wind, precipitations, humidity, and pressure is necessary. Harmful insect's development conditions can also be obtained by accurate prediction of weather (http://en.wikipedia.org/wiki/Weather forecasting). Landing and takeoff of many airplanes or aircrafts can become less due to the presence of air traffic fog.

There are number of methods that have been used for weather forecasting. In ancient times, a number of people tried their hands upon forecasting weather. The forecasting was based on observing patterns of events. For example, if in the evening sky seems to be red, it was assumed that weather will be fair next day. If the sky seemed to be red and overcast, a stormy day might get predicted. However, all of these predictions do not prove to be reliable. To overcome with these problems a number of instruments were invented like hygrometer which is an instrument for measuring humidity of air. Then, thermometer was invented after that a barometer was invented for measuring atmospheric pressure eventually leading to the invention of number of instruments for future weather forecasting. Then, numerical weather predictions were used which is the prediction of weather using "models" of atmosphere and complex computations. operating on huge datasets and accomplishing complex calculations needed to do this on a resolution small enough to make it correct most powerful super computers are required [1].

In numerical weather prediction, current observations of weather are necessary for the numerical computer models to process data and provide outputs of temperature, precipitation, and other meteorological elements. Also time series analysis is used for predicting future weather. A time series comprises of series of data points registered in time manner that is allowing us to know what is effect on some variable Z of a change in variable Y over time. In time series forecasting, a model is used for analyzing data to extract meaningful statistics based on previously observed value of weather to predict future values of weather, hence proving an important tool for weather forecasting [5, 6].

Soft computing also proved useful in predicting weather. It involves techniques for learning of images representing the real data. They particularize data on their past records and history identified by the systems for upcoming weather events. Data explanations were done by soft computing techniques. This technique was based on the concept of adaptive forecasting model that describes a number of methods for finding likeliness of events that occur on the basis of statistical data and variable analysis [2, 3]. Weather forecasting using data mining classification and clustering techniques also helped in predicting future weather. In 1986, artificial neural network (ANN) came into real world and aroused as data modeling tool being able to catch and produce complicated input/output affiliations. The basic inclination actually arises because of the passion to establish a system that would be able to carry out "intelligent" works same to those accomplished by the human brain. ANNs have similarity with the brain in two subsequent ways. First is the way used by neural networks to obtain knowledge through learning. Second is the way how neural network store knowledge in the synaptic weights [7]. ANN was preferred technique as it fairly administers nonlinear difficulties than the traditional techniques. Section 1 of the research paper reflected the introduction part; Section 2 highlights the study of several researchers that used ANN techniques for the better prediction of weather; Section 3 focuses overview of supervised learning algorithms; Section 4 contains overview of ANN and learning algorithms used in this research work; and Sect. 5 explains results and discussions. Finally, the paper ends up with the conclusion.

2 Related Works

Literature is filled with disparate approaches for weather forecasting using ANN. In [8], a neural network model is represented to make assessment about the surface temp and humidity. ANN was built with a single hidden layer, and it was perceived that forecasting by this model came out as satisfactory for very small-scale and high-frequency weather systems.

In [9], the authors put forward an ANN model for the forecasting of weather for the city of Iran. The model was used for forecasting temperature for one day in advance. Four networks were presented for four different seasons that is spring, summer, autumn, and winter to achieve better accuracy for predictions. Error was calculated between exact and the predicted values which was very small and better performance and accuracy was achieved for each day temperature prediction. Within [10], the writer proposed an approach for weather prediction. ANN was used for the training of datasets. This trained network help in obtaining good prediction with lesser error. In this approach, N neural networks were taken and trained to obtain the desired results from each network [11]. The neural network may be able to provide high speed also. In this research, radial basis function and back propagation were selected as different architectures and were trained with the help of differential evolution algorithm.

In [12], the data used for temperature prediction was the daily weather data of the Ludhiana. Proposed time series based model was supposed to start by collecting the data related to weather, selection of appropriate weather parameters, finding the similarity between the weather parameters, forming datasets (comprising of inputs and outputs), and testing datasets. Within [13], the authors presented a model for rainfall prediction. ANN model is known to smartly analyze the trends or patterns from the already available voluminous historic sets of data. Like all other weather happenings rainfall is having its own importance in human life. Thus, the paper presents the study of monthly rainfall of Karnataka. Back propagation ANN has been used for the average summer monsoon rainfall prediction. In this research, three different algorithms were tested in multilayer architecture. The results showed that back propagation algorithm proved to be the best algorithm for the prediction of rainfall. Sawaitul et al. [14] presented a paper by providing a way for the prediction of weather by ANN. In this paper, weather forecasting kit was used for the transmission of data by usage of wireless medium. The weather parameters were recorded by the wireless kit. The wireless parameters can be shown on wireless display. In this rough step-by-step description, first collection of data is done. Then, the recorded data is placed in the datasheets. Now in data preprocessing step data is sent to statistical software as input [15]. Techniques like data mining were applied to transfer data. After that BP algorithm was used for training of the ANN that was achieved by creating some logic or program like it can be change on all other parameters keeping one parameter constant.

In [16], authors presented a paper on weather classification using back propagation. In that paper, a kit for forecasting was used to collect data about weather. Those parameters were given as input to artificial neural network for the training purpose. In this paper, the kit was used for classifying and displaying the weather for future which was also used for the display of information related to weather [17]. Within [18], authors presented weather predictions using artificial neural network. In this research paper, the different weather parameters of Rice Research Kaul station were utilized for effective weather forecasting using ANN. In this paper, forecasting was performed by collecting past plus present data of the atmosphere to train the neural network. After this, the performance plot was obtained that clearly showed that mean square error of the network starts getting decreased as it moves toward the end in contrast that was seen in the starting a huge mean squared error (MSE).

In [19], the authors developed model for weather forecasting based on temperature using neural networks and fuzzy logics. The authors proposed computer-based models for forecasting of weather based on the temp using the two different techniques ANN and fuzzy logic for prediction of daily temperatures. Different regions were taken for this forecasting purpose, and different models were formed using artificial neural networks and fuzzy logic. Two different regions taken as sampling sites involved the airport of Amman and Taipei from china. Back propagation (BP) was used to evaluate efficiencies for weather forecasting models by the usage of two measures one called as variance accounted for and other one is mean absolute error. Inside [20], there is presented a live weather forecasting. This paper conducted the review in order to obtain the best approach for forecasting of weather by comparing the number of techniques that perform different types of forecasting's. Among all of these ANN with back propagation happened to perform the prediction of weather with the minimal error.

Within [21], a model has been proposed neural network for ideal crop selection using weather predictions. This research composed of two main objectives: First one was prediction of weather conditions and second one was prediction of suitable crops based on that weather condition. The scope of this project lies in helping farmers to get knowledge about crops by seeing weather conditions. Weather was predicted using the parameters temperature, pressure, humidity, rainfall, and wind speed for the past year. Time series prediction with multilayer perceptron neural network was used in this proposed work. 80% accuracy was found in the predicted parameters which undoubtedly helped farmers in the productivity of their crops. Within [22], the authors presented an article on weather forecasting using hybrid neural model. Both MLP and RBF are notable types of ANN that work on the basis of exchanging information/messages using single hidden layer between inputs to outputs. Both multilayer perceptron and radial basis function use feed forward strategy but differ in the working of their activation functions. In this paper, it was shown that MLP forecasting came out to be as better in accuracy individually than RBF. But together the hybrid of both MLP and RBF showed better performance than the individual networks. This article showed that hybrid model has better generalization ability and also better learning ability.

3 Supervised Learning Algorithms

In the supervised type of learning, an algorithm is applied on the given input variables suppose (Y) and an output variable suppose (Z) in order to learn input to output mapping functions.

$$Z = f(Y)$$

The major aim of this type of learning is to approximate mapping function so well that whenever next time we have new input it should be able to provide us with correct output. The most prevailing network architecture today that is widely used is neural network architecture. In this type of network, biased weighted sum of the input units is performed and this activation level passes through transfer function in order to get output while units following the feed forward topology.

4 Artificial Neural Network

ANN is multilayered feed forward neural network comprises of subgroups or the layers of elements to be processed. In this type of network processing, elements perform on data independently and pass the obtained results to another layer. The other layers also perform independent computations on data and finally determine the output. Thus, the neural network has a simple type of interpretation in the form of input–output models along with the free parameters weights and biases (threshold). Thus, ANN is known to model the functions of arbitrary complexities even with many numbers of intermediate layers. Very important issue in neural networks involves the design specification that is number of hidden layers that should be present in ANN and number of units to be present in these layers, weights and biases. All these parameters must be set in such a manner that the prediction error gets minimized that will be made by the network. It has been seen more often that appropriate topology fails to give a better model unless and until training is performed by suitable learning algorithm. Less training time and better accuracy are the characteristics of good learning algorithms. Hence, training process of artificial neural networks is very important phase in which representative examples of knowledge are given to the network iteratively. A number of training algorithms are used to train ANNs. Some of the learning algorithms are:

4.1 Back Propagation Algorithm

Back propagation is considered to be the analytical method for training of artificial neural networks; clear and brief elucidation of back propagation algorithm was given by Rumelhart, Hinton, and Williams. It is a supervised form of learning. The error obtained at the output layer is back propagated to the earlier layers, and there occurs the phenomenon of weight updating. Difference between this algorithm and other lies in its process that is used for calculation of weights during learning of the network [23]. The network structure of back propagation neural network is shown in Fig. 1.

A meagre choice of the network structural design, i.e., the number of neurons in the hidden layer, will result in poor generalization even with best possible values of its weights after training. Back propagation is considered to be the analytical method for training of artificial neural networks; clear and brief elucidation of back propagation algorithm was given [24]. The error obtained at the output layer is back propagated to the earlier layers, and there occurs the phenomenon of weight updating. Difference between this algorithm and other lies in its process that is used for calculation of weights during learning of the network.

The pseudo-code of back propagation algorithm is given below:

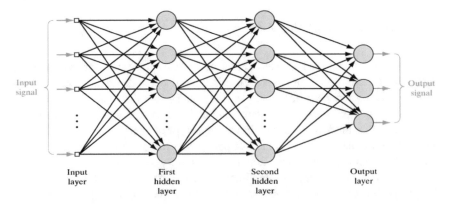

Fig. 1 Back propagation neural network architecture

All weights in network are haphazardly initialized.
Do
For every instance in training set
D = desired or target output
O = output of neural network
Calculate the error
Backward pass computes del_Whidden for all weights from hidden layer to output layer
Again during backward pass compute del_Winput for all weights from input layer to hidden layer
Revise weights till stopping criterion is meet
Return the network.

The mathematical derivation of back propagation algorithm is started first by the calculation of partial derivative of error. The equation representing the partial derivative of error (E) is given as:

$$E_n^p = \frac{1}{2} \sum \left(x_n^i - d_n^i \right) \tag{1}$$

E_n^p represents the error at last layer n.
d_n^i represents the desired output obtained from the neural network.
x_n^i represents the actual value of output at last layer of network.

Taking partial derivatives on both sides of the Eq. 1, we have

$$\partial E_n^p / \partial x_n^i = x_n^i - d_n^i \tag{2}$$

The above equation represents back propagation algorithms starting value. The numerical values for derivative are calculated by taking numeric values of right-hand side. The numeric value of this derivative helps in calculating changes in weights with the help of following equations.

$$\partial E_n^p / \partial y_n^i = G\left(x_n^i \right) \partial E_n^p / \partial x_n^i \tag{3}$$

Gx_n^i represents the derivative of the activation function.

$$\partial E_n^p / \partial w_n^{ij} = x_{n-1}^j \, \partial E_n^p / \partial y_n^i \tag{4}$$

Afterward, the Eqs. 2 and 3 are again used along with the equation below and are used to calculate error for previous layer

$$\partial E_n^p - 1 / \partial x_n^k - 1 = \sum i w_n^{ik} \partial E_n^p / \partial y_n^i \tag{5}$$

Finally, the weights are adjusted by the below given equation:

$$\left(w_n^{ij}\right)\text{new} = \left(w_n^{ij}\right)\text{old} + \text{eta}\left(\partial E_n^p / \partial w_n^{ij}\right) \tag{6}$$

where eta is a learning rate. It can also be used to check if the over learning of validation data is taking place. It is one of the free parameters that are very essential in case of back propagation learning algorithm. Learning rate should not be too large because it may miss local minima then and may not converge. If the learning rate used is too small, the learning of network may get affected as it would be very slow.

4.2 Resilient Back Propagation

Resilient back propagation (Rprop) is another learning algorithm that is used to train artificial neural networks (ANNs). Rprop, an iterative type of process, uses the same concept of gradient. It has got two important characteristics, it trains the neural network faster, and secondly, it does not require any free parameters like learning rate [25]. However, the major drawback of Rprop is that it is very complex algorithm to be implemented.

Rprop algorithm uses the sign of the gradient in order to determine weight delta. For each weight and bias, it also maintains separate weight deltas that are used during training. It first came on 1993 and then afterward a number of variations of Rprop were published. In case of Rprop, if the previous and the current weight or bias does not change or are same then a minimum of error is obtained and kept moving in the same direction. But if the previous and current weights or bias are different that means weight value has skipped the past value then we have to decrease the previously used delta and revert the weight to the previous value so that on next iteration it won't go so far. The size of partial derivative on weight step is the main idea that Rprop wants to remove. Thus, only sign of derivative is taken into consideration.

$$\Delta w_{jk}^t = \begin{cases} -\Delta_{jk}^t; & \text{if } \partial E^t / \partial w_{jk} > 0 \\ +\Delta_{jk}^t; & \text{if } \partial E^t / \partial w_{jk} > 0 \\ 0; & \text{else} \end{cases}$$

$\partial E^t / \partial w_{jk}$ represents the gradient information.

Pseudo-code of resilient back propagation

While epoch < maxEpochs loop
Calculate over all training items the Gradient
For every weight loop
If current and previous partial derivatives have same sign
Previously used delta is increased
Weights are updated with new delta
Else if current and previous partials derivatives have different signs

Previously used delta is decreased
Return weight to previous value
End if
Previous delta = new delta
Previous gradient = current gradient
End-for
Increase epoch
End-while
Return current and bias value.

5 Results and Discussion

The four important steps required for building of artificial neural network model involves:

Selecting or gathering of dataset for the supervised back propagation algorithm (BPA) learning and resilient back propagation (Rprop) learning;
Second important step involves the normalization of input and output data;
Third step involves the training of the normalized data;
Finally, in forth step de-scaling and testing are performed.

(a) **Data Collection**: In this research, the data for five years was taken from 2013 to 2017 with weather parameters as high temp, average temp, low temp, high-level dew point, average-level dew point, low-level dew point, high humid, avg. humid, less humid, high sea level pressure, average sea level pressure, low sea level pressure, high visibility, average visibility, low visibility, low wind speed, average wind speed, and precipitation of Mumbai and were used as the input. The data used for ANN training was collected from a known Web site www. wunderground.com.

(b) **Normalization of Data**: Preprocessing of data is very important as it plays very important role in order to get better accuracy. Normalization of data is very essential step before the training of neural network. Skipping of normalization process can lead to useless results or at many times algorithm may not converge before max number of iterations is allowed. A number of methods are used for data normalization; we have used min-max scale for the scaling or normalizing of data. It has been seen that min-max scale tends to provide better results.

(c) **Training of Input Data**: After the completion of scaling of data, the coming step comprises of input data training using RStudio BPA and Rprop. The training rope 60% of data from the available data. These algorithms usually select the training sample from given dataset, and the training data selection may be random or we can even specify the range of data that is to be used for training and testing. In case of random data selection, every time data gets trained it may provide different mean square error depending upon the 60% of data used for training. For the testing purpose, 40% data is used. In this experiment, rain is used as target variable.

(d) **Testing**: Testing is performed after data training is completed. Before testing, first de-scaling of data is performed and then in our experiment forty percent of the data is used for the testing purpose.

5.1 Backprop and Rprop Implemented in RStudio

Backprop and Rprop were implemented in RStudio. RStudio supports the neural net package. For this experiment, different weather parameters of Mumbai were taken for the purpose of training of ANN using learning algorithms Bprop and Rprop. The correlation matrix for the parameters is shown in Fig. 2. The parameters sea level pressure, high humidity, low humidity, average humidity, average dew point, temperature, dew point, visibility, and others are highly correlated parameters that have great influence on the target that is rain.

After providing input to the network normalization of data is performed using min-max scale. Performing normalization is very important in order to remove if there appeared to be any outlier or missing value or any kind of noise. Then, the training of data is performed by initializing weights in a random manner more often. In this research, there are 18 input parameters that are provided to the network and rain is taken as an output parameter. The training plots of both the algorithms are

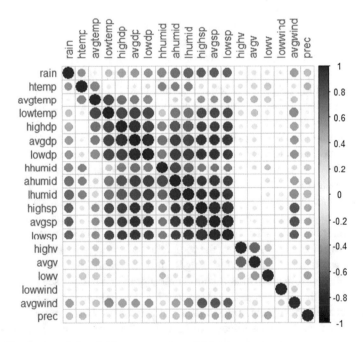

Fig. 2 Showing correlations between parameters of dataset

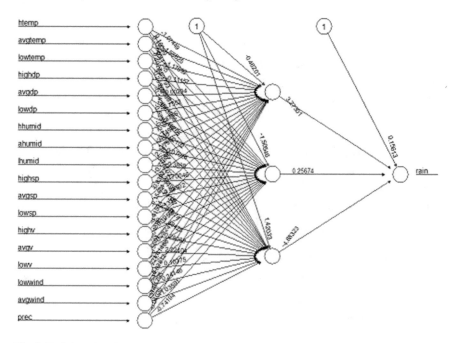

Fig. 3 Training plot of back propagation

Table 1 Algorithms performances

Learning algorithms	MSE	RMSE	MAE
Back propagation algorithm	0.12	0.33	0.24
Resilient back propagation	0.20	0.50	0.42

shown separately; in Fig. 3, the training plot for back propagation is shown whereas Fig. 4 gives training plot for resilient back propagation. Table 1 shows different performance parameters.

For the proposed approaches, performances are evaluated using the following parameters

Mean square error (MSE)
Root mean square error (RMSE)
Mean absolute error (MAE)

Table 2 provides the training and testing accuracies of the proposed back propagation and resilient back propagation algorithms. It is observed that the back propagation showed both better training and testing accuracies in comparison with Rprop which fails to show the better results for testing in case of difficult data.

Although it is known that Rprop provides better training and testing results than the regular back propagation for the simple datasets. But for a difficult dataset that increases in the complexity a standard back propagation shows better results which is true in our experimentation also.

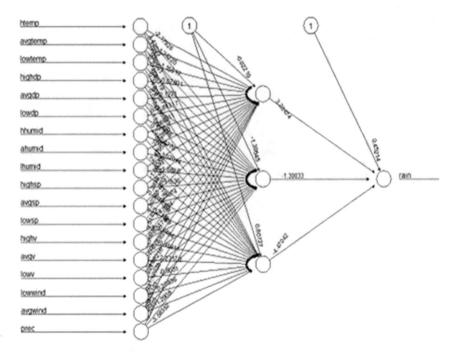

Fig. 4 Training plot of Rprop

Table 2 Showing the training and testing accuracies of the proposed approaches

Learning algorithms	Training accuracy (%)	Testing accuracy (%)
Back propagation algorithm	88	81
Resilient back propagation	80	67

5.2   Receiver Operating Characteristic Curve (ROC) and Area Under Curve (AUC)

In statistics, receiver operating characteristic curve (ROC curve) is a representational plot that shows the analytical capability of a binary classifier system as its bias threshold is varied. The ROC curve is produced by plotting the true positive rate (TPR) in opposition to the false positive rate (FPR) at different threshold settings. ROC is the most commonly used way to plot or show the performance of the classifiers. The performance of Bprop and Rprop is visualized by the below given ROC plots, respectively, in Figs. 5 and 6; also, the performance is summarized in a single number by specifying area under curve (AUC) shown in Fig. 7. In case of back propagation, the area under curve is 87.4% while as in case of Rprop area under curve is 73%. Hence, back propagation visualized better performance than Rprop.

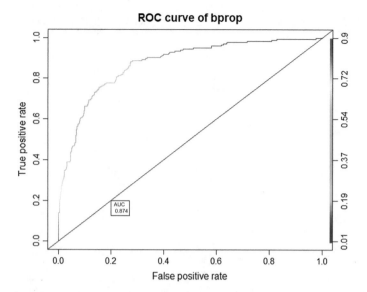

Fig. 5 ROC curve with AUC value of back propagation

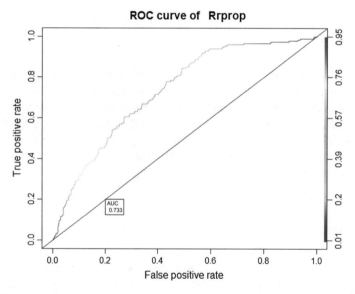

Fig. 6 ROC curve with AUC value of Rprop

Fig. 7 ROC curve for both
backprop and Rprop together

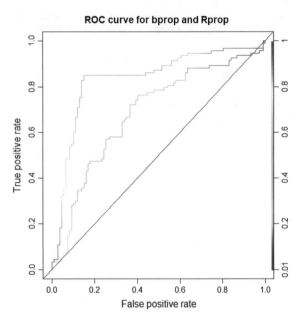

5.3 ggplot

It is another plotting system of R totally based on graphics. It works like by picking up the best parts of base and lattice and ignoring the bad parts. Grammar of graphics can be seen as a visualization technique for the data by which the graphs are broken up into scales and layers. With the help of ggplot2, we can create graphs that help us to visualize numerical, univariate, multivariate, and categorical data in a simple manner. The ggplots for the proposed approaches are provided below, Fig. 8 gives ggplot for back propagation, and ggplot for Rprop is shown in Fig. 9.

6 Conclusion

In ancient times, weather forecasting was done by only observations after that numerical methods came into existence. The satellite data was used for data retrieval for forecasting of weather. After that soft computing or data mining concepts were also used for forecasting purposes. But ANN nowadays is being put forward more and more for the purpose of weather forecasting because of its iterative nature that it uses for training purpose so that observed and target output are repeatedly compared to obtain the error which is readjusted by back propagating it to the neural network in order to get better accuracy. As weather forecasting is a huge challenge for prediction to get accuracy in its results because of its usage in a number of real time systems like

Fig. 8 ggplot of back
propagation

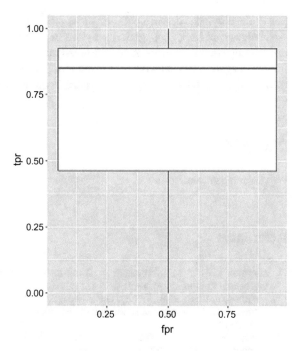

Fig. 9 ggplot of Rprop

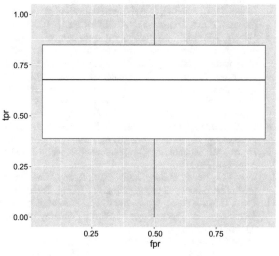

power development departments, airports, industries, etc. This paper presents that ANN algorithm back propagation provides better accuracy and yields good results and has appeared an alternative to traditional metrological approaches. Also resilient back propagation has shown better results for performance parameters but is less in comparison with back propagation because of the increase in complexity of data.

Appendix

www.wunderground.com.

References

1. Maqsood, I., Khan, M.R., Abraham, A.: An ensemble of neural networks for weather forecasting. Neural Comput. Appl. **13**(2), 112–122 (2004)
2. Maqsood, I., Khan, M.R., Huang, G.H., Abdalla, R.: Application of soft computing models to hourly weather analysis in southern Saskatchewan, Canada. Eng. Appl. Artif. Intell. **18**(1), 115–125 (2005)
3. Kumari, A., Alankar, B., Grover, J.: Feature level fusion of multispectral palmprint. Int. J. Comput. Appl. (0975-8887) **144**(3) (2016)
4. Mahdi, O.A., Alankar, B.: Wireless controlling of remote electrical device using android smartphone. IOSR J. Comput. Eng. (IOSR-JCE) **16**(3), 23–27 (2014). e-ISSN: 2278-0661
5. Donate, J.P., Li, X., Sánchez, G.G., de Miguel, A.S.: Time series forecasting by evolving artificial neural networks with genetic algorithms, differential evolution and estimation of distribution algorithm. Neural Comput. Appl. **22**(1), 11–20 (2013)
6. Mellit, A., Pavan, A.M., Benghanem, M.: Least squares support vector machine for short-term prediction of meteorological time series. Theoret. Appl. Climatol. **111**(1–2), 297–307 (2013)
7. Shank, D.B., Hoogenboom, G., McClendon, R.W.: Dewpoint temperature prediction using artificial neural networks. J. Appl. Meteorol. Climatol. **47**(6), 1757–1769 (2008)
8. Chaudhuri, S., Chattopadhyay, S.: Neuro-computing based short range prediction of some meteorological parameters during the pre-monsoon season. Soft. Comput. **9**(5), 349–354 (2005)
9. Hayati, M., Mohebi, Z.: Temperature forecasting based on neural network approach. World Appl. Sci. (2007)
10. Devi, C.J., Reddy, B.S.P., Kumar, K.V., Reddy, B.M., Nayak, N.R.: ANN approach for weather prediction using back propagation. Int. J. Eng. Trends Technol. **3**(1), 19–23 (2012)
11. Paras, S.M., Kumar, A., Chandra, M.: A feature based neural network model for weather forecasting. In: Proceedings of World Academy of Science, Engineering and Technology, pp. 66–74 (2007)
12. Singh, S., Bhambri, P., Gill, J.: Time series based temperature prediction using back propagation with genetic algorithm technique. Int. J. Comput. Sci Issues (IJCSI) **8**(5), 28 (2011)
13. Abhishek, K., Kumar, A., Ranjan, R., Kumar, S.: A rainfall prediction model using artificial neural network. In: Control and System Graduate Research Colloquium (ICSGRC), 2012 IEEE, pp. 82–87. IEEE (July, 2012)
14. Sawaitul, S.D., Wagh, K.P., Chatur, P.N.: Classification and prediction of future weather by using back propagation algorithm-an approach. Int. J. Emerg. Technol. Adv. Eng. **2**(1), 110–113 (2012)
15. Srinivasulu, S., Sakthivel, P.: Extracting spatial semantics in association rules for weather forecasting image. In: Trendz in Information Sciences & Computing (TISC), 2010, pp. 54–57. IEEE (December, 2010)

16. Naik, A.R., Pathan, S.K.: Weather classification and forecasting using back propagation feed-forward neural network. Int. J. Sci. Res. Publ. **2**(12), 1–3 (2012)
17. Geetha, G., Selvaraj, R.S.: Prediction of monthly rainfall in Chennai using back propagation neural network model. Int. J. Eng. Sci. Technol. **3**(1) (2011)
18. Malik, P., Singh, S., Arora, B.: An effective weather forecasting using neural network. Int. J. Emerg. Eng. Res. Technol. **2**(2), 209–212 (2014)
19. Al-Matarneh, L., Sheta, A., Bani-Ahmad, S., Alshaer, J., Al-Oqily, I.: Development of temperature-based weather forecasting models using neural networks and fuzzy logic. Int. J. Multimedia Ubiquit. Eng. **9**(12), 343–366 (2014)
20. Narvekar, M., Fargose, P.: Daily weather forecasting using artificial neural network. Int. J. Comput. Appl. **121**(22) (2015)
21. Shastri, M.S., Gunge, A.K.: Neural network based weather prediction model towards ideal crop selection. Int. J. Eng. Comput. Sci. **5**(9) (2016)
22. Saba, T., Rehman, A., AlGhamdi, J.S.: Weather forecasting based on hybrid neural model. Appl. Water Sci. **7**(7), 3869–3874 (2017)
23. Lee, J.H., Delbruck, T., Pfeiffer, M.: Training deep spiking neural networks using backpropagation. Front. Neurosci. **10**, 508 (2016)
24. Rumelhart, D.E., McClelland, J.L.: Parallel Distributed Processing: Explorations in the Microstructure of Cognition. Volume 1. Foundations (1986)
25. Günther, F., Fritsch, S.: neuralnet: training of neural networks. R J. **2**(1), 30–38 (2010)

Analysis of the Leadership of Undergraduate Freshmen in Military Schools from the Perspective of Space

Zelin Wu and Chaomin Ou

Abstract According to the university student leadership model, 1,550 undergraduate freshmen were tested, and SPSS17.0 and Lisrel8.72 were used for descriptive statistical analysis and system cluster analysis. The study starts from the perspective of the ability required by the leadership to construct the five structural dimensions of university student leadership, namely decision-making ability, stress-resistance ability, communication ability, unity and cooperation ability, and innovation ability. The survey results in the form of questionnaire survey. The results show that the leadership level shows the stepwise difference in the eastern, central, and western regions as a whole, and the geographical distribution of each capability dimension is consistent with the geographical division. It highlights the imbalance in the development of the dimensions of leadership in the first-tier cities.

Keywords Spatial perspective · Leadership of military undergraduate freshmen · Analysis

1 Introduction

At present, domestic scholars study the leadership of college students mainly from two dimensions: first, based on the leadership behavior dimension that leadership is a series of traits or abilities that college students already possess or need to master [1]. From the practical point of view, domestic scholars believe that the necessary characteristics of college students' leadership are: self-knowledge ability, ability to effectively deal with interpersonal relationships, flexible adaptability, creative thinking ability, ability to promise service, and ability to grasp public policy [2]. The second is based on the leadership process dimension. Leadership is seen as the process by which college students use their own resources and the resources around them to achieve self and group goals [3]. Leadership is seen as the process by which

Z. Wu (✉) · C. Ou

College of Systems Engineering, National University of Defense Technology, Changsha, China

e-mail: 781277764@qq.com

© Springer Nature Singapore Pte Ltd. 2020

S. Patnaik et al. (eds.), *New Paradigm in Decision Science and Management*,

Advances in Intelligent Systems and Computing 1030,

https://doi.org/10.1007/978-981-13-9330-3_11

college students use their own resources and the resources around them to achieve self and group goals [4]. Bian Huimin believes that university student leadership is composed of factors such as decision-making, communication, influence, coordination, resilience, anti-frustration, and innovation [5]. Zhang Huimin believes that youth leadership is the process by which young people can grasp their own abilities by fully utilizing their own resources and surrounding resources, and ultimately achieve self and group goals. Foreign scholars believe that university students' leadership includes self-knowledge, ability to deal effectively with interpersonal relationships, flexible adaptability, creative thinking, and ability to commit services [6].

In summary, most of the researches on the leadership of college students at home and abroad focus on the study of leadership factors and behavioral research, and there are few theoretical studies combining spatial geography and cultural background. The research object of college students' leadership is the undergraduate freshmen of military academy. The essence is to study the unique geographical characteristics of China in the leadership of college students.

2   Research Methods

The 2016 undergraduate freshmen of military academies were selected for questionnaire survey. A total of 1,553 questionnaires were distributed, and 1,553 were collected. After eliminating the invalid questionnaire, 1,550 valid questionnaires were finally obtained, and the effective recovery rate was 99.8%. This article uses this effective data for exploratory analysis.

2.1   Research Tools

Prepare a university student leadership questionnaire. The university students' leadership factors are divided into five parts: decision-making ability, anti-stress ability, communication ability, unity and cooperation ability, and innovation ability: 16 topics: (1) decision-making ability, including the ability of scientific decision-making, thinking ability; (2) compressive ability, including emotional anti-interference ability, emotional adjustment ability, risk-taking ability; (3) communication ability, including communication ability, communication comprehension, language expression ability, flexibility, and adaptability; (4) unity and cooperation ability, including organization and management competence, coordination, personality influence, fairness, affinity; (5) innovation ability, including learning ability, integration of resources, information collection and utilization, problem solving ability. The questionnaire asked the participants what level of leadership traits or abilities they have at the moment. They use the Likert 5-point scoring method. From 1 to 5, they are very weak, weak, general, strong, and very strong.

2.2   Research Procedures

First, determine the attitude questionnaire for the measurement and evaluation of university students' leadership; second, select college students to conduct questionnaire surveys; third, collect and select valid data, use EXCEL2010 for data entry; fourth, use SPSS20 to score data and data analysis. The main analytical methods include statistical analysis, correlation analysis, and system cluster analysis.

3   The Results Analysis

3.1   Reliability and Validity Test

Reliability test. Cronbach's alpha coefficient was used to test the homogeneity reliability of the questionnaire. The results showed that the total coefficient of the questionnaire was .658, indicating that the questionnaire had good homogeneity reliability (Tables 1 and 2).

Validity test. KMO statistic and Bartlett sphericity test were used to judge the applicability. The results showed a KMO value of .817 and a Bartlett sphericity test result of less than .01. The questionnaire is very effective.

3.2   Overall Analysis of Leadership

Based on the spatial characteristics of the space, the source areas are divided into eastern, central, and western regions. The use of average data summation way statistical data on military leadership undergraduate freshmen descriptive analysis. Preliminary judgment of the relationship between spatial region and leadership comprehensive score. The results are shown in Table 3.

Table 1 Reliability statistics

Cronbach's alpha	Number of items
.658	16

Table 2 Inspection of KMO and Bartlett

Sampling enough Kaiser–Meyer–Olkin metrics		.817
Bartlett's sphericity test	Approximate chi-square	2154.397
	df	120
	Sig.	.000

Table 3 Average score of leadership of military academy students in the eastern, central, and western regions

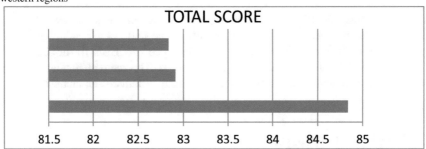

Table 3 shows that the leadership score for the eastern region is 84.83529. The leadership rating for the central region is 82.91091. The western region leadership score is 82.83657. It can be seen that the leadership score in the eastern region is higher than that in the central region and the western region. The central leadership score is close to the western leadership score. It can be seen that the eastern part of China has a great influence on accepting foreign multiculturalism and has a role in promoting the different needs of modern leadership. The cultures in the central and western regions are relatively single, and there are obvious shortcomings in leadership.

3.3 Analysis of the Correlation Between Leadership and Per Capita Income

To provide more data support for the regional characteristics of the above leadership ratings, we attempt to explore the correlation between leadership and per capita income. In order to obtain accurate data, we use the per capita GDP data to represent the per capita income, and use SPSS20 to analyze the bivariate correlation between leadership score and per capita GDP. The Pearson correlation coefficient was used to test the significance. The validation results are shown in Table 2.

Table 4 shows that the correlation between leadership score and GDP per capita is .217, and there is no significant correlation between the two. It can be seen that the relationship between leadership training and income is not significant. It should be a comprehensive embodiment of educational achievements. It cannot be directly measured by the level of economic development.

Table 4 Results of correlation test between leadership score and per capita GDP relevance

		Per capita GDP	Total score
Per capita GDP	Pearson correlation	1	.232
	Saliency (bilateral)		.217
	N	30	30
Total score	Pearson relevance	.232	1
	Saliency (bilateral)	.217	
	N	30	30

3.4 Analysis of the Correlation Between Leadership and Per Capita Income

Cluster analysis is a method of multivariate statistical analysis. It can automatically classify a batch of sample data according to their characteristics and the degree of affinity in nature. When we analyze each ability dimension, we use the method of cluster analysis to divide the scores of each province into three categories. The concrete results are as follows.

3.4.1 Decision-Making Ability

The results of cluster analysis of Cadets' decision-making ability are as follows:

Figures 1 and 2 show that the decision-making ability of 31 provinces can be divided into three categories by cluster analysis. Beijing is the first category alone; Inner Mongolia, Anhui, Hainan, Hebei, Guangdong, Hunan, Fujian, Sichuan, Guangxi, and Chongqing are the second; the remaining provinces are the third. The overall classification shows a continuous horizontal band distribution, indicating that there is a correlation between adjacent provinces.

3.4.2 Compression Capacity

A cluster analysis was conducted on the scores of military students' ability to withstand stress. The results are as follows:

Figures 3 and 4 show that 31 provinces can be divided into 3 categories by cluster analysis. Shanghai is the first category alone; Hunan, Guangdong, Fujian, Guangxi, Guizhou, Yunnan, Hainan, and Xinjiang are the second; the remaining provinces are the third. The overall distribution presents block characteristics, and most provinces

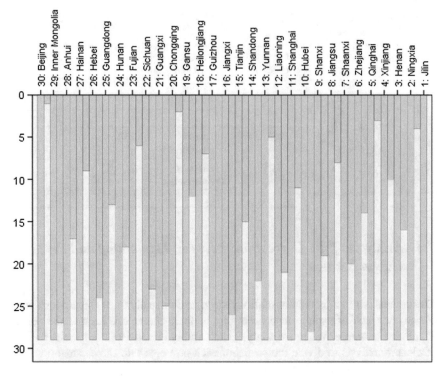

Fig. 1 Cluster diagram of decision-making ability of undergraduates in military academies

Fig. 2 Geographical distribution of decision-making ability of undergraduates in military academies

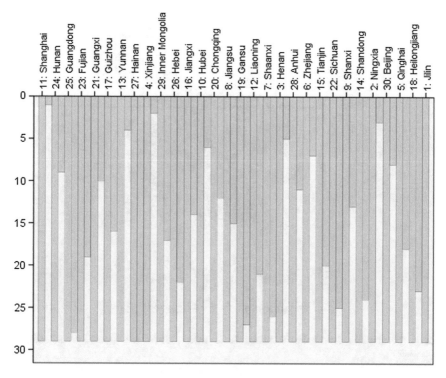

Fig. 3 Cluster chart of compressive capacity of undergraduates in military academies

are of the same type, reflecting that the compressive capacity in the overall scope has certain average characteristics.

3.4.3 Communication Skills

The cluster analysis of the communication ability of cadets was conducted. The results are as follows:

Figures 5 and 6 show that the communication ability of 31 provinces can be divided into 3 categories by cluster analysis. Beijing, Chongqing, Guangdong, Yunnan, Jiangxi, Shanxi, Hunan, Guangxi, Gansu, Zhejiang, Fujian, Sichuan, and Xinjiang are in the first category; Inner Mongolia, Hebei, Anhui, Jiangsu, Guizhou, Shandong, Henan, Shaanxi, Hubei, and Ningxia are in the second category; the remaining provinces are in the third category. The overall distribution conforms to the geographical distribution characteristics of northeast China, north China, south China, and northwest China, indicating that the communication ability based on language expression and understanding has certain spatial and geographical characteristics.

First kind   Third kind

Second kind   Unenrolled area

Fig. 4 Geographical distribution of undergraduate students' ability to withstand stress in military academies

3.4.4 Ability to Work Together

The cluster analysis of military cadets' solidarity and cooperation ability was carried out. The results are as follows:

Figures 7 and 8 show that 31 provinces can be divided into 3 categories by cluster analysis. Beijing is the first category alone; Gansu, Tianjin, Sichuan, Hunan, Jiangxi, Liaoning, Guangxi, Guangdong, Hubei, Hainan, Anhui, Hebei, Fujian, Shanxi, Inner Mongolia, Chongqing, Yunnan, Shandong, Qinghai, Jiangsu, and Henan are the second; the remaining provinces are the third. The overall distribution shows the distribution characteristics of the whole and fragments, which may be related to the differences in cultural characteristics.

3.4.5 Ability to Work Together

The ranking of military cadets' innovative ability was analyzed and ranked as follows (Table 5).

Table 4 shows that seven of the top 10 provinces are in the eastern coastal provinces of China. The other provinces are Henan, Qinghai, and Ningxia, respectively. This shows that coastal cities are more direct in accepting new things than inland cities, which is conducive to cultivating certain advantages in innovative ability.

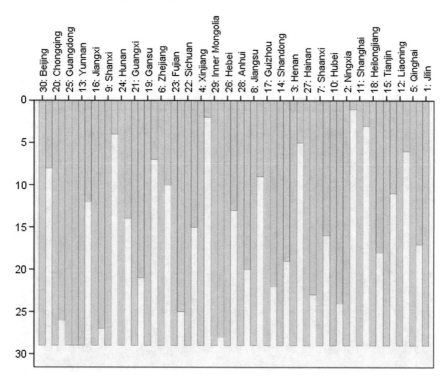

Fig. 5 Clustering chart of communication ability among undergraduates in military academy

Fig. 6 Geographical distribution map of communication ability among undergraduates in military academy

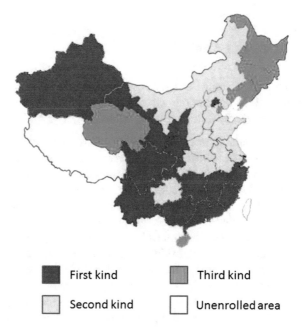

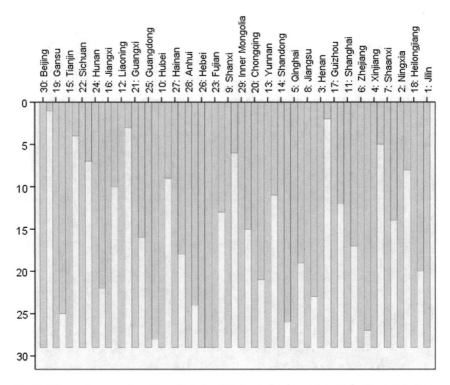

Fig. 7 Cluster diagram of undergraduates' solidarity and cooperation ability in each military academy

4 Result Analysis

Through literature analysis, questionnaire survey and data analysis, this paper tries to find the correlation between spatial and regional characteristics and leadership, and draws the following conclusions:

1. Leadership as a whole presents differences in ladder development. On the whole, the leadership of cadets in the eastern provinces is superior to the central and western provinces, and the central and western provinces are basically equal. Although leadership has overall trapezoid structure, the west has developed rapidly in recent years. Whether in economy or education, the differences between the central and western regions are gradually narrowing. Leading to the lack of regional advantages in central China. When analyzing the relationship between per capita income and leadership, we take GDP per capita as a measure of local economic income. Through the bivariate correlation analysis, we find that there is no obvious relationship between per capita income and leadership. It can be seen that income is not the key factor leading to differences. The score of leadership reflects more balanced and superior quality education in developed areas.

Fig. 8 Geographical
distribution map of
undergraduate students'
solidarity and cooperation
ability in each military
academy

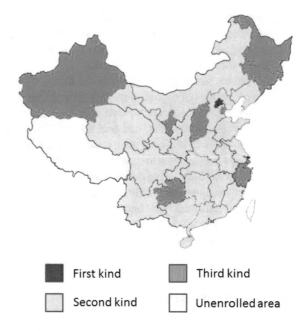

First kind   Third kind

Second kind   Unenrolled area

2. The dimensions of ability are related to the region. In the analysis of each ability
   dimension score, we found that the ability classification and geographical distribution have a significant correlation, excluding the heterogeneity of individual
   provinces, most of the provinces and geographical neighboring provinces are
   classified into one category, which shows that the development of leadership in
   the same region has a certain integrity. When analyzing communication skills, we
   also found strong geographical features. However, with the development of society, cultural boundaries gradually blurred, and beyond the geographical bound-

Table 5 Statistics of the top 10 provinces of the cadets' innovative ability

Province	Innovation capability
Shanghai	18.83333
Qinghai	18.83333
Heilongjiang	18.20155
Jiangsu	18.1
Liaoning	18.0226
Zhejiang	18.01149
Henan	17.97867
Shandong	17.9537
Jilin	17.90476
Ningxia	17.82051

aries, some of the dimensions of competence across the country also showed a balanced feature. From the ranking of innovation capability, we can see that the vast majority of the eastern coastal cities, while the central and western regions do not show significant differences.

3. The development of leadership in the first-tier cities is uneven. When studying the dimensions of capabilities, we find that Beijing and Shanghai show obvious heterogeneity, respectively. Beijing is classified as the first category in terms of unity and cooperation and decision-making ability. Shanghai is classified as the first class in terms of compression capability, and the Shanghai innovation capability rating is the first. It can be seen that Beijing and Shanghai, as the first-tier cities in China, have outstanding advantages in leadership training, but their abilities and qualities are not balanced. The other dimensions of the two cities are classified into third categories. It reflects that the education in the first-tier cities is moving toward modernization, while there are imbalances in the educational structure and the allocation of resources.

5   Concluding Remarks

As an exploratory research, this study has some limitations. First of all, the enrollment of military academies varies greatly from region to region, and their military training may have an impact on their leadership style. In the future, we can increase the diversity of samples and make research findings more universal. Secondly, the study has strong subjectivity in data generalization. In the future, we can conduct quantitative research on all abilities based on the concept of university students' leadership to verify the structural relationship and influence of each capability, to understand the actual impact of leadership ability among college students, and to reduce the subjectivity of researchers.

Acknowledgements Fund Project: This research was supported by National Natural Science Foundation of China under Grant Number 71774168. The key construction project of "Information System and Decision Technology" Graduate Innovation Research Base of National Defense University of Science and Technology.
Statement
This study has been agreed by all authors.

References

1. Adams, T.C., Keim, M.C.: Leadership practices and effectiveness among Greek student leaders. College Student J. **34**(2), 259–270 (2000)
2. Couzin, I.D., Krause, J., Franks, N.R., et al.: Effective leadership and decision-making in animal groups on the move. Nature **433**(7025), 513–6 (2005)
3. Judge, T.A., Bono, J.E., Ilies, R., et al.: Personality and leadership: a qualitative and quantitative review. J. Appl. Psychol. 87(4), 765–80 (2002)

4. Nonaka, I., Toyama, R., Konno, N.: SECI, Ba and leadership: a unified model of dynamic knowledge creation. Long Range Plann. **33**(1), 5–34 (2001)
5. Uhlbien, M.: Relationship-based approach to leadership: development of leader-member exchange (LMX) theory of leadership over 25 years: applying a multi-level multi-domain perspective. Leadersh. Quart. **6**(2), 219–247 (1995)
6. Spillane, J.P., Halverson, R., Diamond, J.B.: Towards a theory of leadership practice: a distributed perspective. J. Curriculum Stud. **36**(1), 3–34 (2004)

Correlation Analysis for CSI300 Index Return and Realized Volatility

Zhenzhen Yue

Abstract The return and volatility of financial assets are important determinants in making risk management band investment. At the present stage, the realized volatility is not only dependent on the return series, continuous-time volatility and jump volatility which have different statistical characteristics are used to model separately. Considering there do exist statistically significant correlations among innovation sequences of its return and volatility. Therefore, I use vine structure to study the dynamic correlation for the innovations of continuous-time volatility, jump volatility and the return series; it is found that significant dynamic correlation is found between two pairs of innovation sequences.

Keywords Realized volatility · Continuous-time volatility · Jump volatility · Vine · High-frequency data · Time-varying copula

1 Introduction

The yield and volatility of financial assets are important factors in designing portfolios and risk management. Research on yield and volatility has attracted a large number of researchers. With the development of computers, researchers' research has not only focused on low-frequency data in daily units, but high-frequency data in minutes and seconds has become an important means for researchers. The realized volatility based on high-frequency data transforms the volatility from unobservable hidden variables to explicit variables. The observability of realized volatility allows scholars to model the rate of return model separately. Andersen and Bollerslev [1–3]

Z. Yue (✉)

College of Mathematics and System Sciences, Shandong University of Science and Technology, Qingdao 266590, China
e-mail: 906494391@qq.com

© Springer Nature Singapore Pte Ltd. 2020
S. Patnaik et al. (eds.), *New Paradigm in Decision Science and Management*,
Advances in Intelligent Systems and Computing 1030,
https://doi.org/10.1007/978-981-13-9330-3_12

proposed to estimate the true volatility by the square of the intraday high-frequency yield, and model and predict it; Andersen proves that the price is subject to the assumption of no jump and half-turn process, and the realized volatility is a consistent unbiased estimate of the true volatility. Another scholar has found that the yield of financial assets may fluctuate greatly under the condition of continuous intraday, so the realized volatility is divided into jumping volatility and continuous volatility according to the different statistical characteristics. Bollerslev [4–6] presents the commonly used realized estimators for continuous volatility, including second power variances and more widely realized multiple power variances. As for the selection of the estimation model, the heterogeneous autoregression model (HAR) proposed by Corsi [7] is based on heterogeneous market hypothesis. By superimposing the realized autoregressive processes at three different timescales, namely day, week and month, it forms a model that can better describe volatility. Hillebrand [8] introduced the virtual variable Sunday effect in the log-realized volatility model (HAR), which improved the predictive ability of the model; Liu Xiaoqian [9] used the model to predict the CSI300 index and made the most comprehensive information available; it is found that the prediction effect of this model is better. Zhang bo and Zhong Yujie [10] found that the model of HAR-RV is more effective than FARIMA in describing the long memory of Shanghai index volatility. Moreover, the prediction effect of HAR-RV is much better than that of FARIMA; at the same time, because the HAR-RV comprehensively considers realized volatility at different time levels, it verifies the heterogeneity of Chinese stock market and the leverage effect of volatility in depth.

The modeling of the above-realized estimators does not take into account the possible correlation between the rate of return, continuous volatility and jump volatility. Zhai Hui et al.[11] combined modeling of yield rate, continuous volatility and jump volatility, and the correlation between the new information of time series model is described. The maximum likelihood estimation method of the joint model is given, and it is found that there is a significant correlation between the rate of return, continuous volatility and new interest with jump volatility.

Vine-copula has become mature in depicting the correlation between financial markets. Fischer et al. [12] studied the vine-copula model, and the analysis showed that the model can well fit the multivariate financial data. Holfmann and Czada [13] used D-vine-copula to study portfolio risk management and compared with the traditional multi-variable copula. It is proved that D-vine-copula is more flexible and the fitting effect is more remarkable.

Zhai Hui et al.[11] jointly model the correlation between stock return rate and realized volatility. It improved the validity of parameter estimation to further explore the correlation between yields and realized volatility. Based on the modeling of high-frequency price yield and realized volatility of stock index futures in Shanghai and Shenzhen, Vine-copula is used to describe the dynamic correlation.

2 Model Construction

2.1 Model Realized Estimate

Suppose the price of financial asset on the t day is $p(t)$, logarithmic price is $y(t)$, $y(t) = \ln p(t)$, and $y(t)$ satisfies with continuous-time jumping-diffusion semi-martingale model:

$$dy(t) = \mu(t)dt + \sigma(t)d\omega(t) + k(t)dq(t), \quad 0 \le t \le T \qquad (1)$$

In the form, $\mu(t)$ is the mean, $\sigma(t)$ is the instant volatility rate, $\omega(t)$ is the standard Brownian motion; $k(t)$ is the discrete jump magnitude of the logarithmic price in random process, $k(t) \equiv y(t) - y(t-1)$, $q(t)$ is the counting process whose intensity $r(t)$ meets $P\{dq(t) = 1\} = \lambda(t)dt$. Besides, the logarithmic return rate of this financial asset on the t day is $r(t)$, $r(t) = y(t) - y(t-1)$. Its quadratic variation is:

$$QV(t) = \int_1^t \sigma^2(s)ds + \sum_{t-1 < s \le t} k^2(s) \equiv IV(t) + \sum_{t-1 < s \le t} k^2(s) \qquad (2)$$

In the form, s corresponds to any moment on the t day, $QV(t)$ is the quadratic variation, it describes price volatilities of all assets on t day; $IV(t)$ is the integral square deviation, it describes the continuous part of volatility, also called as the continuous volatility; $\sum_{t-1 < s \le t} k^2(s)$, the difference between $QV(t)$ and $IV(t)$, describes the discrete part of volatility, called as jump volatility. In order to reduce the impact of microscopic noise, mode of staggered sampling is applied to calculate double power variation $BPV(t)$, and the calculated result is taken as the estimation of continuous volatility of the financial asset on the t day, that is:

$$BPV(t) = \frac{\pi}{2}\left(\frac{M}{M-2}\right)\sum_{j=1}^M |r_{t,j}| \cdot |r_{t,j+2}| \longrightarrow pIV(t, 1) \qquad (3)$$

In the form, M is the number of intervals of staggered sampling, its interval is Δ. $r_{t,j}$ is the logarithmic return rate of length of Δ of the j on the t day, $r_{t,j} = y_{t,j} - y_{t,j}$. And the realized volatility of the financial asset on the t day is $RV(t)$, $RV(t) \equiv \sum_j^M r_{t,j}^2$. With $\Delta \to 0$, $RV(t) \longrightarrow pQV(t)$, that is, the realized volatility is the consistent estimation of quadratic variation; correspondingly, $\ln RV(t) - \ln BPV(t)$ is the consistent estimation of jump volatility on the t day. A lot of demonstrations also have shown that $RJ(t)$ is a more stable estimation; $RJ(t) \equiv \ln RV(t) - \ln BPV(t)$, which is used to describe the characteristics of jump volatility. In order to prevent from information loss, no pretreatment is applied to the part of less than zero.

2.2   Edge Distribution Model Construction

2.2.1   Model Realized Estimate

For the logarithmic day return rate $r(t)$ of high-frequency data of CSI300 index construct ARMA model:

$$r(t) = \varphi_0 + \sum_{i=1}^{p} \varphi_i r_{t-i} + \varepsilon(t) + \sum_{j=1}^{q} \eta_j \varepsilon_{t-j} \tag{4}$$

In the form, p is the autoregressive order, q is the smoothness index determined by AIC criterion and ε_t is the innovation process.

2.2.2   Continuous Volatility Model

HAR model to model construction of continuous volatility model is applied:

$$\ln BPV(t) = \alpha_0 + \alpha_d \ln BPV\ (t-1) + \alpha_w \ln BPV\ (t-5:t-1)$$
$$+ \alpha_m \ln BPV\ (t-22:t-1) + \alpha_1 |z(t-1)| + \alpha_2 I[z(t-1) < 0]$$
$$+ \alpha_3 z(t-1) + \sum_{j=1}^{4} \alpha_{4,j} D(t)^j + \mu(t) \tag{5}$$

In the form, $\ln BPV(t-k:t-1) = \frac{1}{k}\sum_{j=1}^{k} \ln BPV(t-j)$ is the average of continuous volatility during the last k days; μ_t is the innovation process; $D(t)^j$ represents the effect of Sunday, $t = 1$ is the effect of Monday; α_d, α_w and α_m are used, respectively, to measure the impact of the average of logarithmic continuous volatility of Sunday, week and month to present situation; α_2 and α_3 are used to measure the effect of lever; α_1 measures the effect of scale; and $\alpha_{4,j}$ measures the effect of Sunday.

2.2.3   Jump Volatility Model

Demonstration shows that jump volatility is of lagged correlation. To set up autoregression model:

$$RJ(t) = \beta_0 + \beta_1 |z(t-1)| + \beta_2 I[z(t-1) < 0]$$
$$+ \beta_3 z(t-1) + \sum_{j=1}^{4} \beta_{4,j} D(t)^j + \sum_{i=1}^{q} \beta_{5,i} RJ(t-i)$$
$$+ \nu(t) \tag{6}$$

In the form, i is the order determined by BIC criterion; $v(t)$ is the innovation process; $\beta_{5,j}$ is applied to measure the impact of the lagged term to volatility; $\beta_{4,j}$ measures the effect of Sunday; β_2 and β_3 measure the effect of lever; and β_1 measures the effect of scale.

2.3 Construction of Time Change D-Vine Copula

2.3.1 Construction of D-Vine

For the model of logarithmic return rate, the model of logarithmic continuous volatility and the model of jump volatility extract the residual error sequence, respectively, and record them as $\varepsilon(t)$, $\mu(t)$ and $v(t)$ The next is to construct the residual error sequence Vine-copula. First, residual error needs to be probabilistic integral transformation to meet the condition of model construction and the steps are as follows:

(1) Calculate the rank correlation coefficient Kendall between two of the three variables. The rank correlation coefficient Kendall between variables i and j is recorded as $\tau_{i,j}$ and calculate $S_i = \sum_{j=1}^{3} |\tau_{i,j}|$.

(2) Select the structure of the first layer of trees so as for S_i the maximum.

(3) According to the selected structure of the first layer of trees, select the category of the function of pair-copula between variables and estimate its parameter.

(4) As per the function Pair-Copula determined by step (3), calculate h function and the variable needed to construct the second layer of trees.

(5) As per step (2), calculate the variable resulted from step (4) to obtain the structure of the second layer of trees and in turn perform step (3) and step (4).

(6) Repeat performing from step (2) to step (5) till the third layer of trees is obtained.

2.3.2 Construction of Time-Varying Copula

Time-varying copula may describe better the dynamic dependent relation among variables, while the core to construct time change Copula model is just how to obtain the parameter which changes with time in an evolution equation.

(1) The binary distribution function t-copula is:

$$C(u, v; \rho, \upsilon) = \int\limits_{-\infty}^{T_{\upsilon}^{-1}(u)} \int\limits_{-\infty}^{T_{\upsilon}^{-1}(v)} \frac{1}{2\pi\sqrt{1-\rho^2}}\left[1 + \frac{s^2 + t^2 - 2\rho st}{\upsilon(1-\rho)^2}\right]^{-\frac{\upsilon+2}{2}} ds dt \quad (7)$$

In the form, $\rho \in (-1, 1)$ is the linear correlation coefficient, $T_{\upsilon}^{-1}(\cdot)$ is the inverse function of unitary t distribution function $T_{\upsilon}(\cdot)$ with degree of freedom υ. The evolution equation of time change correlation parameter ρ_t of binary $t -$ Copula is:

Table 1 Descriptive statistics

Sequence	Mean	Standard	Median	Skewness	Kurtosis	J–B statistics
ln RV(t)	−9.123	0.997	−9.189	0.507	3.906	151.774***
ln BPV(t)	−9.363	0.988	−9.421	0.484	3.738	121.738***
RJ(t)	0.240	0.336	0.163	2.884	17.954	2187.15***
$r(t)$	0.028	0.015	0.0003	−0.730	8.0136	2238.25***
$z(t)$	−0.026	1.189	0.030	−0.114	3.956	38.305

Note *** is significant at the 1% level, ** is significant at the 5% level, the same below

$$\rho_t = \tilde{\Lambda}\left(\omega_T + \beta_T \rho_{t-1} + \alpha_T \times \frac{1}{10}\sum_{i=1}^{10} T^{-1}(u_{t-i}, \upsilon_x)T^{-1}(\upsilon_{t-i}, \upsilon_y)\right) \tag{8}$$

(2) The binary distribution function of Clayton Copula is:

$$C(u, v; \theta) = \left(u^{-\theta} + v^{-\theta} - 1\right)^{-1/\theta} \tag{9}$$

In the form, the relationship between rank correlation coefficient τ and correlation parameter θ of Kendall is $\tau = \theta/(\theta + 2)$, its evolution equation of time change correlation parameter is:

$$\tau_t = \Lambda(\omega + \beta\tau_{t-1} + \alpha \cdot |u_{t-1} - v_{t-1}|) \tag{10}$$

3 Empirical Analysis

3.1 Characteristic Analysis

Select high-frequency data every 5 min for the Shanghai and Shenzhen 300 Index from January 4, 2010, to February 9, 2018, excluding weekends and holidays, and eliminate outliers. The data come from Wind, using E views and R for analysis (Fig. 1).

Table 1 gives descriptive statistics for each time series and logarithmic rate of return $r(t)$ and normalized logarithmic rate of return $z(t)$, among them, $z(t) = \frac{r(t)}{\sqrt{RV(t)}}$. It can be seen from the JB-statistic that ln RV(t) and ln BPV(t) approximate a normal distribution, $z(t)$ obeys normal distribution. Zhai Hui et al. have shown that the leverage effect exists in the realized volatility, so the leverage effect is considered in the modeling of continuous volatility and jump volatility.

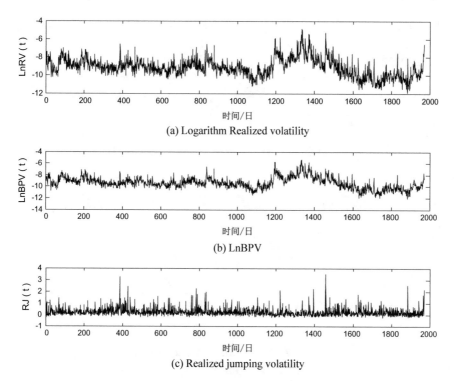

(a) Logarithm Realized volatility

(b) LnBPV

(c) Realized jumping volatility

Fig. 1 Time series plots

3.2 Parameter Estimation

3.2.1 Parameter Estimation of Log Daily Result Model

See Table 2.

Table 2 Parameter estimation results for the log daily result model

Coefficient	p_1	p_2	p_3	q_1	q_2	φ_0
Estimated value	0.2212	−0.9771	0.0731	−0.1725	0.9412	0.0017
Standard	0.0287	0.0201	0.0233	0.0184	0.0302	0.0549
t-statistics	6.343***	3.672**	3.539**	2.563***	3.864**	2.712***
$R^2 = 0.4102$ log likelihood $= -1917.73$, AIC $= 3849.45$						

Note ** is significant at the 5% level

Table 3 Parameter estimation results for the continuous-time volatility and the jump volatility models

Continuous-time volatility				Jump volatility			
Coefficient	Estimated	Standard	t-statistics	Coefficient	Estimated	Standard	t-statistics
Value				Value			
α_0	−0.659	0.129	−5.118 ***	β_0	0.170	0.014	11.750 ***
α_d	0.401	0.026	15.472***	β_1	0.028	0.009	3.002 **
α_w	0.443	0.041	10.772***	β_3	−0.030	0.006	−4.806***
α_m	0.109	0.033	3.237 **	$\beta_{5,2}$	0.080	0.023	3.556 ***
α_1	0.254	0.014	17.560 **	$\beta_{5,3}$	0.046	0.023	2.043*
α_3	−0.072	0.010	−7.591***	$\beta_{5,5}$	0.069	0.023	3.021**
$R^2 = 0.7493$				$R^2 = 0.0297$			

Note ***1% is significant; ** 5% is significant; *10% is significant

3.3 Parameter Estimation for Continuous-Time Volatility and the Jump Volatility Models

Table 2 gives the parameter estimation results of the rate of return model. It can be seen from the results that the variables of the model are significant, so the model of the rate model is established. Table 3 gives the results of the continuous-time volatility and the jump volatility models after the significant amount is removed. It can be seen from the estimation results of the continuous-time volatility model in Table 3 that continuous volatility has significant daily, weekly and monthly effects, indicating that the average value of continuous volatility in the previous trading day, the previous trading week and the previous trading month has significantly affected, and the estimated values are 0.401, 0.443 and 0.109, respectively. It is explained that the volatility is continuous, and continuous-time volatility has significant scale effects and leverage.

The estimation results of the jump volatility model in Table 3 show that the jump volatility also has a scale effect, but it does not have a leverage effect and the average value of the log jump volatility of the previous trading day, the previous trading week and the previous trading month. The logarithmic jump volatility of the day also had a significant impact.

3.4 Correlation Analysis

3.4.1 Dynamic Correlation Analysis

The structure D-vine is established by the Kendall correlation coefficient of Table 4, and the vine structure diagram is shown in Fig. 2. Table 5 gives the estimation results of the dynamic correlation of the three interest rate processes of return rate, continuity-time volatility and jump volatility. The results show that there is a dynamic correlation between the rate of return, continuous volatility and the dynamic process of jump volatility.

Figure 3 shows the dynamic correlation coefficient between the three rates of interest rate, continuous volatility and jump volatility. It can be seen from the figure that the correlation coefficient constant value of the continuous volatility and jump volatility information process is −0.372, the range of volatility is concentrated between − 0.43 and −0.31. The constant value of the correlation coefficient between the con-

Table 4 Correlation coefficient and its absolute value of Kendall

	$\varepsilon(t)$	$\mu(t)$	$\nu(t)$	S
$\varepsilon(t)$	1	−0.321	−0.372	1.693
$\mu(t)$	−0.321	1	−0.224	1.545
$\nu(t)$	−0.372	−0.224	1	1.596

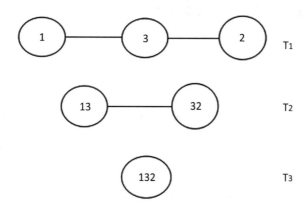

Fig. 2 D-vine of residual error sequence

Table 5 Results of D-vine model

		Copula type	Copula	Par1	Par2	Par3	AIC
T_1	c_{13}	Time-varying	Clayton	−0.124	−0.643	0.612	−383
	c_{32}	Time-varying	t	−0.014	0.543	0.347	−327
T_2	$c_{13\|32}$	Time-varying	Clayton	0.016	0.174	−0.073	−459

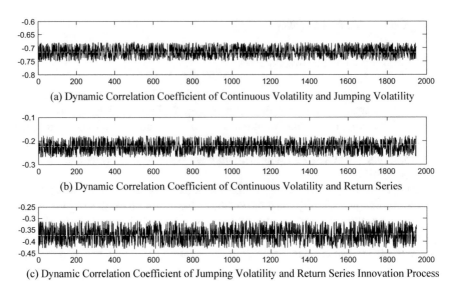

(a) Dynamic Correlation Coefficient of Continuous Volatility and Jumping Volatility

(b) Dynamic Correlation Coefficient of Continuous Volatility and Return Series

(c) Dynamic Correlation Coefficient of Jumping Volatility and Return Series Innovation Process

Fig. 3 Plot of dynamic correlation coefficient

tinuous volatility and the yield improvement process is -0.224. The volatility range is between -0.27 and -0.18. The constant value of the dynamic correlation coefficient between the jump volatility and the yield innovation process is -0.321, and the volatility range is between -0.37 and -0.28.

4 Conclusion

This paper uses the high-frequency data of the Shanghai and Shenzhen 300 Index every 5 min from January 4, 2010, to February 9, 2018, to model the yield, continuity-time volatility and jump volatility, respectively, then to extract the innovation process separately, in order to discuss the correlation further between the innovation process; and to adopt the time-varying analysis of the dynamic correlation of the innovation process. Finally, it is found that the innovation process of the rate of return, continuous volatility and jump volatility has dynamic correlation and is certain volatility within the scope.

References

1. Andersen, T.G., Bollerslev, T.: Answering the skeptics: yes, standard volatility models do provide accurate forecasts. Int. Econ. Rev. **39**(4), 885–905 (1998)
2. Andersen, T.G., Bollerslev, T., Diebold, F.X., et al.: The distribution of realized stock return volatility. J. Financ. Econ. **61**(1), 43–76.3 (2001)

3. Barndorff-Nielsen, O.E., Shephard, N.: Estimating quadratic variation using realized variance. J. Appl. Econometrics **17**(5), 21 (2002)
4. Andersen, T.G., Bollerslev, T., Diebold, F.X., et al.: Modeling and forecasting realized volatility. Soc. Sci. Electron. Publishing (2002)
5. Barndorff-Nielsen, O.E.: Power and bipower variation with stochastic volatility and jumps. J. Financ. Econometrics **2**(1), 1–37 (2004)
6. Barndorff-Nielsen, O.E., Shephard, N., Winkel, M.: Limit theorems for multipower variation in the presence of jumps. Stochast. Process. Appl. **116**(5), 796–806 (2006)
7. Corsi, F.: A simple long memory model of realized volatility. J. Financ. Econometrics **7**(2), 174–196 (2004)
8. Hillebrand, E., Medeiros, M.C.: The benefits of bagging for forecast models of realized volatility. Econometric Rev. **29**(5–6), 571–593 (2010)
9. Liu, X., Wang, J.: A predictive study of the CSI 300 index based on the high-frequency data HAR–CVX. Stat. Inf. BBS (06) (2017)
10. Zhang, B.. Zhong, Y.: Analysis of long memory driving factors of Shanghai stock index fluctuation based on high frequency data. Stat. Inf. BBS (06) (2009)
11. Qu, H., Liu, Y.: Study on the yield rate of CSI 300 index and the joint modeling of volatility. Manage. Sci. (06) (2012)
12. Fischer, M., Christian, K., Stephan, S., et al.: An empirical analysis of multivariate copula models. Quant. Finance (2009)
13. Hofmann, M., Czado, C.: Assessing the VaR of a portfolio using D-vine copula based multivariate GARCH models. Lehrstuhl für Mathematische Statistik, Germany (2010)

Prescription Basket Analysis: Identifying Association Rule Among Drugs in Prescription of Dentist in Bhubaneswar City

Manoranjan Dash, Preeti Y. Shadangi, Sunil Kar and Sroojani Mohanty

Abstract Amount of clinical and diagnosis data produced within the healthcare industry has grown to be vast, and thus, it provides a prospective for data mining research. Data mining is the process of pattern discovery and extraction which originally used in the field of marketing and now been used effectively in other fields such as bioinformatics, immunology, and nuclear science. Discovery of knowledge has potentially possibilities and thus will enhance the quality of health care to patients. Drug manufacturer needs to use the information efficiently and to discover patterns from large data and build predictive models. The objective of prescription basket analysis is to identify association rules between groups of drugs prescribed by dentists in their prescription. The data have been collected from the prescription of the dentists from different dental colleges as well as the dental clinics. Apriori algorithm applied to data collected from the prescriptions by dentists, and various association rules were found from the prescription data. It can provide information to pharmacy store to understand the prescription behavior of dentists which can help the pharmacy store in correct decision making in keeping their inventory. The research tried to discover the prescription behavior and association between drugs. Based on the result, the manufacturers can redesign their marketing strategy in the intention of improving patient care and thus decreasing healthcare cost. The results will reveal valuable insights into the association between drugs prescribed by dentists.

Keywords Data mining · Association rules · Apriori method · Knowledge discovery

M. Dash (✉) · P. Y. Shadangi
Faculty of Management Sciences, Siksha O Anusandhan (Deemed to Be University), Bhubaneswar, India
e-mail: manoranjanibcs@gmail.com

S. Kar
Hi-Tech Dental College & Hospital, Bhubaneswar, India

S. Mohanty
Interscience Institute of Management & Technology, Bhubaneswar, India

© Springer Nature Singapore Pte Ltd. 2020
S. Patnaik et al. (eds.), *New Paradigm in Decision Science and Management*,
Advances in Intelligent Systems and Computing 1030,
https://doi.org/10.1007/978-981-13-9330-3_13

1 Introduction

In recent years, data mining has drawn the great attention among the researchers to gain valuable and deep insights from large biomedical data sets. For better decision making in clinical and hospital administrative functioning, data mining is being used in the healthcare sector and has great potentiality in areas of quality as well as cost reduction. Researchers suggest using data mining in health care can reduce their overall healthcare costs by 30%. But due to the complexity of healthcare technology adoption still, data mining has been not been implemented successfully [1, 2]. Patients with chronic drug needs can be identified through Market basket Analysis (MBA). MBAs are performed using association rules. Data mining is used to discover relevant and meaningful information by analyzing data from different dimensions, categorize that information, and summarize the relationships identified in the database. Data mining is a good decision-making tool, for taking a medical decision for doctors and staff [3–6]. The process involves mathematical analysis to find the patterns or trends in the data by which the prediction can be made. In the areas like preventive medicine, CRM (Customer relationship management), healthcare management, data mining has proven to be effective and by which dramatic improvement has been done in the critical areas of decision making [7, 8]. Classification approach divides data into sample into classes, and it predicts the samples. It is used in the estimation of cost of treatment, diagnosis of patients, etc. K-nearest neighbor uses for discovering unknowing points from know points. Different data mining techniques are being used in the health sectors like support vector machine (SVM), artificial network neural network (ANN), Bayesian classification, regression, association for identifying different diseases, classification, and finding out the relationship and factors between them [9, 10]. Data mining techniques improve the efficiency of healthcare processes. Data mining is also referred as knowledge discovery which is an interactive and iterative process which consists of selection, creation, preprocessing, and data transformation. Apriori algorithm is used for frequent item set mining and also for the association rule learning. It proceeds with the identification of frequent items in the database [11, 12]. The frequent sets of items are determined by the Apriori which can be used to determine the association rule which usually highlights the general trends in the database. Health industry generates gigantic amount of data of patients, medical devices, etc. Larger amounts of data are the resource which is to be processed and analyzed for knowledge extraction which enables for better decision making and cost savings. Association rule includes determining patterns, or associations, between elements in data sets (Table 1).

2 Research Methodology

The database was prepared from the prescription prescribed by the dentists in order to find the association between the drugs. The prescriptions were collected from the

Table 1 Application of data mining algorithm in different research in medical fields

Research in medical field	Methods of data mining	Data mining algorithm	Description
Prediction of diseases	Clustering rules	K-Means Apriori	Determining factors affecting cancer types
Prediction ion emergency patients	Clustering	K-Means NN	Timely decision and reduce cost
Best type of treatment	Clustering	K-Means Apriori	Best type of cancer treatment

dentists working in dental clinics as well as dental hospitals. Data were standardized into a format. The association rules were extracted from the data set. The Apriori algorithm was used for finding the association rules and applied to the research.

Apriori Algorithm: It is an algorithm useful in mining frequent item sets and association rules. It has three significant components, i.e., support, confidence, lift. The algorithm works as follows

(a) Discover the item sets.
(b) Designing the association rule based on the sets.
(c) Hypotheses formulated based on iterative and non-iterative.

Association rules discover how different items are associated with each other. How particular item set is being measured by the proportion of transactions in which the item set appears.

Support = Number of transaction containing both M and N/Total Number of Transactions. It refers to the probability of both antecedent and subsequent in a transaction.

Confidence = Number of transaction containing Both M and N/Number of Transactions in M. It refers to the accuracy or truth of rule.

3   Data Analysis and Results

Data were taken from the prescription of the drugs prescribed by dentist working in different dental hospitals and clinics. The data taken were stored in a text file and the text file was given as input to the SPSS Clementine software for analysis. Ten rules were found from the data by using the Apriori algorithm, and the result is given given in Table 2.

Rule-1 Amoxicillin is combined with clavulanic acid to destroy the β lactamase of the bacteria sas they release an enzyme β lactamase. Hence act as a broad spectrum antibiotic. Combination of Aceclofenac With Paracetamol i.e Zerodal SP is given to reduce pain and swelling of the infected area.

Table 2 Output of Apriori algorithm

S. No.	Support	Confidence	Antecedent	Consequent
1	4.2	34.4	Amoxicillin	Aceclofenac + Paracetamol
2	4.3	31.8	Metrogyl 400 MG Augmentin 625 mg	Ketorol DT
3	4.7	29.6	Amoxicillin	Lactobacillus
4	4.0	26.0	Aceclofenac + Diclofenac	Rab D Pan 40
5	4.0	26.0	Augmentin *625 mg*	Metrogyl 400 *MG*
6	4.0	26.0	Tramadol	Domperidone
7	5.9	24.0	Tegretol	Lycopene
8	4.7	23.0	Vitamin B complex	Hexigel Ointment
9	6.8	21.0	Acyclovir	Hydrocortisone
10	6.8	21.0	Thiocolchicoside	Mefenamic Acid + Paracetamol

Rule-2 Metrogyl 400 combined with Augmentin 625 given for acute chronic dental infection involving gram-negative bacteria, with that Ketorol DT is given to relieve pain.

Rule-3 Amoxicillin may cause gastrointestinal disorder by destroying the bacteria flora. Hence, Lactobacillus Caps area added to provide health GI Flora and to avoid GastroIntestinal disorder.

Rule-4 Aceclofenac +Diclofenac may cause GI tract inflammation leading to gastric ulcers to avoid that RAB D/PAN 40 containing Pantoprazole can be given to reduce GI intolerance, erosion in the lining of stomach, gastroesophageal reflux diseases (GERD) or gastritis.

Rule-5 Augmentin 625 combined with Chymoral forte and Metrogyl 400 to reduce infection and selling because of odema and abscess.

Rule-6 Tramadol combined with Domperidone reduces dizziness and use of vomiting as Tramadol activates the chemoreceptor trigger zone which induces vomiting. Tramadol is a potential pain killer for Nephronic particular.

Rule-7 Tegretol, i.e., Carbamazepine reduces neurologic pain with it Lycopene is combined to repair cellular damage because of its antioxidant effect.

Rule-8 Vitamin B Complex combined with Hexigel ointment to repair or heal oral ulcer. Vitamin B Complex is prescribed and Hexigel, i.e., Chlorhexidine Gluconate reduces the chance of tissue infection and enhances healing.

Rule-9 Acyclovir (Zovirax) with Hydrocortisone 1% used for oral viral infection like Herpes Simplex with it Hydrocortisone 1% is combined to reduce inflammation and healing.

Rule-10 Thiocolchicoside with Mefenamic acid and paracetamol used for spasm, muscle relaxant, and jaw muscle spasm.

4 Conclusion

Data mining has played a very vital role in the healthcare sector and nowadays extensively used in decision making for better efficiency and productivity. Association among the drugs in the prescription of the dentist was the main objective of this paper. Association among the drugs in the prescription was found together. Association rules, i.e., Apriori algorithm was applied and found Rule 1 is the most important drug interrelated and is applied beside each other in the prescription. Second, Rule-2, similarly, the other association rules were found. The findings are beneficial for both manufacturer and stockist to know which type of dental disease that happens in that locality while patients purchasing their medicine regarding dental problems. Further, the manufacturer can understand the prescription prescribed by the doctors. The stockist will be benefited by reducing its inventory cost as well as manufacturer redesign their marketing strategy in promoting their drugs with the similar items prescribed by the dentist in their prescription.

Acknowledgements This research was supported and funded by Sai Shradha Dental Clinic, Bhubaneswar, Odisha, India. We thank to the dentist in the dental clinic who provided insight and expertise that greatly assisted the research, although they may not agree with all of the conclusions in this paper. Necessary permission has been taken from the doctors who are working in different dental hospitals in Odisha. We would like to show our gratitude to all dental professionals belonging to different dental clinic and dental hospitals for sharing their data and expertise with us during the research.

References

1. Kharya, S., Sing, U.: Data mining—techniques for diagnosis and prognosis. Int. J. Comput. Sci. Eng. Inf. Technol. (IJCSEIT) **2**(2) (2012)
2. Bodenreider, O.: Ontologies for mining biomedical data. In: IEEE International Conference on Bioinformatics and Biomedicine, Philadelphia, Pennsylvania (2008)
3. Chen, H., Fuller, S., Friedman, C., Hersh, W. (eds.). Medical Informatics: Knowledge Management and Data Mining in Biomedicine. Springer, New York (2005)
4. Ramageri Bharati, M.: Data mining techniques and applications. Indian J. Comput. Sci. Eng. **1**(4), 301–305
5. Sharma, A., Guhan, P.C.: Number of blood donors through their age and blood group by using data mining techniques. Int. J. Commun. Comput. Technol. **1**(6), 6–10 (2012)
6. Humayun, A., Waqar, A.: A comparative study on usage of data mining techniques in healthcare sector. Int. J. Comput. Appl. **162**(6), 13–15 (2017)
7. Tomar, D., Agarwal, S.: A survey on data mining approaches for healthcare. Int. J. BioSci. Bio-Technol. **5**, 241–266 (2013)
8. Cios, K.J., Moore, G.W.: Uniqueness of medical data mining. Artif. Intell. Med. **26**, 1–24 (2002)
9. Verma, S., Bhatnagar, S.: An effective dynamic unsupervised clustering algorithmic approach for market basket analysis. Int. J. Enterp. Comput. Bus. Syst. **4**(2) (2014)
10. Saurkar Anand, V., Bhujade, V., Bhagat, P., Khaparde, A.: A review paper on various data mining techniques. Int. J. Adv. Res. Comput. Sci. Softw. Eng. **4**(4), 98–101 (2014)

11. Mannila, H., Toivonen, H., Verkamo, A.I.: Efficient algorithms for discovering association rules. In: Proceedings of the AAAI Workshop on Knowledge Discovery in Databases (KDD-94), pp. 181–192 (1994)
12. Raorane, A.A., Kulkarni, R.V., Jitkar, B.D.: Association rule—extracting knowledge using market basket analysis. Res. J. Recent Sci. **1**(2), 19–27 (2012)

Decision Science and Management

Impact of Decision Science on e-Governance: A Study on Odisha Land Records System

Pabitrananda Patnaik and Subhashree Pattnaik

Abstract The objective of e-governance applications is to provide hassle-free governance to the public. The measures are taken to provide services at the doorstep of the citizen with low cost. Basically, the four types of e-governance services such as government to government (G2G), government to citizens (G2C), government to business (G2B) and government to employees (G2E) are emphasized. Most importantly, G2C and G2G services are having direct impact for the benefits of the society. The role of decision science which is a collaborative approach is more effective to G2G applications to strengthen the government to take right decision. On the other hand, it provides strong G2C service to the public with right decision made at right time by the government functionaries. The use of decision science will surely help to handle risk and uncertainty in decision-making. It will also help to use statistical analysis, operation research tools and behavioural decision-making to analyse the decisions made by government and their impact on the society. In this paper, the role of decision science on Land Records project, one of the important e-governance applications of Government of Odisha, is studied and the benefit of providing effective services are analysed. This study will provide a useful decision-making system with the use of decision science in e-governance.

Keywords Data analytics · Decision-making · Decision science · e-Governance · Land records · Risk management

P. Patnaik (✉)
National Informatics Centre, Unit—4, Bhubaneswar, Odisha, India
e-mail: p.patnaik@nic.in

S. Pattnaik
Capital Institute of Management & Science, 1309, Panchagaon, Jatni Road, Bhubaneswar, Odisha, India
e-mail: spattnaik12@gmail.com

© Springer Nature Singapore Pte Ltd. 2020
S. Patnaik et al. (eds.), *New Paradigm in Decision Science and Management*,
Advances in Intelligent Systems and Computing 1030,
https://doi.org/10.1007/978-981-13-9330-3_14

151

1 Introduction

All leading organizations including government sector are adopting data-driven cultures. This culture is certainly impacting the decision-making process of an organization. The big data analytics, machine learning, computational intelligence, artificial intelligence, etc., are the base for these data analysis systems. The decision-makers enhance their skill with the use of these tools for better decision-making. Their managerial instincts are improved. In government, the authorities take right decisions through e-governance [1].

1.1 e-Governance

The electronic governance is aiming at provisioning good services to the citizens. The traditional governance is upgraded to ICT-based governance to reach the people. The communication and information technology (ICT) plays a major role in today's governance. The hassle-free services at low cost are the objective. The faster, transparent and error-free communication creates confidence in implementing government schemes for the beneficiaries. In India, since 1976, the electronic services started. Gradually, the refinement of these services leads to better services. The establishment of National Informatics Centre (NIC) led the e-governance applications in the country. The wide area network (NICNET) connecting the districts, states and centre gave infrastructural set-up. Then, other organizations and networks were developed for more facilities at grassroot levels [2, 3].

Now, Digital India is more focusing on better provisioning of IT services to the citizens. The agenda is more emphasising on information technology utilization in the public domain. The right decision at right time for the citizens is essential. Thus, decision-making at government level is to be more accurate and timely one. With the use of technology, the execution of public services is easier and faster. Different sectors of government are automated, and Web-based e-governance applications are implemented to fetch the data from anywhere in the globe. Mobile apps are also built up for various citizen-centric applications. One such area is Revenue Administration, which deals with Land Records system. The landed properties are safeguarded by government. The record of rights (RoRs) is prepared and distributed to the tenants. The hierarchy of Revenue Administration is as follows:

In Fig. 1, the hierarchy of Revenue Administration and the status of revenue set-up in Odisha are shown. To provide the core services in time, Government of Odisha defined various services under Odisha Right To Public Services (ORTPS) Act, 2012. In this act, the services are to be delivered within stipulated time period. Otherwise, penalty is imposed on the authority [4–6].

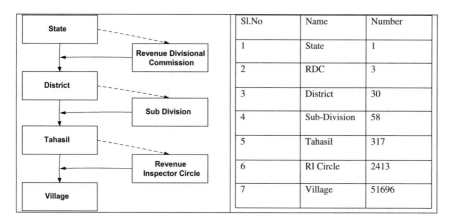

Sl.No	Name	Number
1	State	1
2	RDC	3
3	District	30
4	Sub-Division	58
5	Tahasil	317
6	RI Circle	2413
7	Village	51696

Fig. 1 Hierarchy of Revenue Administration

1.2 Odisha Right to Public Services Act

In Odisha, around 341 services have been identified. The delivery timelines are defined for each service. The penal action against the officials failing to provide the services in due time is also provisioned. This act enables the public to demand their rights from the government for the basic services they ought to get [6, 7].

The Revenue and Disaster Management Department covers 31 such services. These services strictly monitored by the top authority and public are made aware of these acts as well. In this paper, the provisioning of these services with the use of decision science for taking best decision is studied.

2 Literature Review

In today's age, each organization, government or private, is having a lot of data in electronic formats. The analytics on these data are used for taking many strategic decisions. In government sectors, a huge amount of data are available. Particularly, e-governance applications generate a lot of data. These data are in structured format making data analytics easy. The application of decision science on these data and data analytics is simplified. Different techniques and approaches in decision science are available, and few of them useful for governance are described below.

2.1 Brainstorming

The brainstorming sessions are conducted under the chairmanship of top authorities. In government, these authorities are exercised by ministers or bureaucrats. The problem is discussed in a group of top-level bureaucrats, technocrats, academicians, etc. The idea and solution are discussed threadbare. Its pros and cons are thoroughly analysed by the experts. The decisions are recorded, and minutes are prepared. Finally, the conclusion is derived for the problem solution [1, 8].

2.2 Committee Meetings

The decision of any individual might be a biased one. Thus, in government, a committee of competent officers from the same or different departments is formed. The committee is headed by one senior most officer known as chairman. The committee sits together on specified date and venue and analyses the problem for finalising solutions. The convenor of the committee calls the meeting presided by the chairman. The output of the committee meeting is recorded. The members analyse the problems and suggest for action to be taken. In these meetings, voting may be done if required. The approved committee decisions are communicated to all for implementation.

2.3 Delphi

In this technique, the facilitator allows the team members to express their ideas individually. The idea given by one is not known to another one. Then, the facilitator collects all the inputs and circulates them among others for studying and improvements. This process is continued until a final decision is made. The use of this technique is very rare in government functions [1, 9].

2.4 Electronic Meeting

To avoid the journey and saving time, video conferences are conducted to include the officers or officials in remote locations. The officers in the distant and remote locations participate through two-way audio and video communication. This helps a lot to include more members with less expenditure, movements and time consumption. Day-to-day usage of video conferencing in government departments is increasing. The proceeding of these conferences is recorded, and decisions are taken [10].

2.5 Survey

For data collection and decision-making, surveys are conducted through filled-up questionnaire. Then, the door-to-door survey is conducted. The filled-up formats are fed into the computer system for compilation. Then, analytics are applied to it for deriving inferences. This is a very traditional system in government sectors. Now, the paper formats are converted to Web-based applications and responses are received electronically. This makes the compilation work ease.

2.6 Decision Tree and Decision Tables

The conditions, actions and information flow are put in the structure of decision table or in decision trees. This helps to find the combination of condition—action set and the path to traverse from root to each node and back.

2.7 Game Theory

Game theory is applicable in the situations for complex strategic decisions, where the likely responses or outcomes are taken into account. The resulting impact on the competitors or stakeholders is considered. Minimising the maximum loss and maximising the minimum gain are the two concepts of this theory. In government, the win–loss methodology is rarely used. It mostly uses the win–win situation for its actions and the result at the field levels [11].

2.8 Simulation

In this technique, a real model of the existing system is built up. Then, it is put in use and observed for a time period. Its behaviour, outcome and result are observed, and decisions are taken. The computerized programs help a lot for running simulation models. The different variables, the correlation among them and other characteristics are recorded and analysed. It helps in analysing various alternatives and selecting the best one [12].

All the above-mentioned techniques are applied for decision-making in an easy and effective manner. The application of these techniques in government sector is highly required in today's age. In this paper, the Land Records project of Odisha has been taken for the study and how decision is useful for making project successful is also studied.

3   Decision Science for e-Governance

The e-governance applications are implemented for the service of the public. The key services are covered under ORTPS. These services need to be closely monitored by the higher authority ensuring delivery in time. There are 31 services rendered by Revenue Administration under ORTPS. To make result-oriented decisions, Revenue and Disaster Management Department, Government of Odisha, decided to publish a daily bulletin on ORTPS for the 31 services provided by them. Accordingly, the software was developed and bulletin was published in public Website. It created an ecosystem for all the 317 tahasils to reduce the percentage of cases beyond ORTPS timeline. The regular review by government created a pressure as well as a healthy competitive environment among the tahasildars. In this paper, out of 31 services, the mutation cases under ORTPS are covered for the study [7, 13, 14].

The highest number of mutation cases pending beyond ORTPS timeline is 35%, while the lowest is 4%. The average for the state is 18.41%.

The disposal of mutation cases from April to October is analysed and seen that the percentage of case disposal is increasing after the decision of government to publish the daily bulletin of ORTPS. This bulletin is in public view. Everyday people may see the bulletin status from the Website. The Revenue Authorities at different levels of government see the status and take a review on it. The periodic reviews such as weekly, fortnightly and monthly are taken at administration levels. Review meeting, Steering committee meeting and Advisory committee meeting are conducted to see the progress, discuss the issues and deciding the steps for removal of bottlenecks in the system [6, 7, 15].

The software outputs from dashboard, MIS reports and daily bulletin are seen online during review of the system.

Particularly, the ORTPS issues are discussed minutely for each case. The results are compared among tahasils and districts.

The comparison of case disposal status from April to October 2018 is shown in the figure below.

From the graph in Fig. 2, it is observed that the case disposal status is steadily increasing every month. Evidently, October 2018 shows the highest percentage of case disposal. This approach shows that the systematic use of technology and transparency of disclosing the result to public has created a healthy competitive environment in Revenue Administration. The significant improvement in public services is directly felt by all.

Besides ORTPS, which is the key area of this study, the use of decision science techniques for other important areas of e-governance applications is also described below [7].

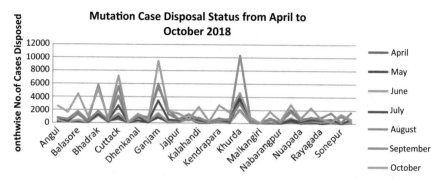

Fig. 2 Case disposal status from April to October 2018

3.1 Risk Analysis

The risk likelihood and risk impact are always associated with e-governance projects. The major risk areas relating to e-governance projects are (Fig. 3):

- *Technological*: The risk factors relating to technical activities are coming in this category. They are generally related to hardware, system software, application software, communication and other such areas. The rapid changes in information technology also impact the projects. Some of the technical risk areas are mentioned below.

Fig. 3 Risk factors of e-governance projects

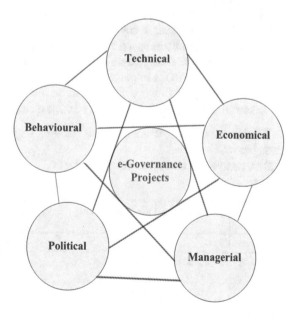

- Improper hardware, software and ICT architecture.
- Wrong or poor design of application software.
- Security problem of application and data.
- Poor design of data management architecture.
- Poor understanding of user requirements.
- Improper skill set of developers.
- Frequent changes in process flow and user requirements.
- Lack of user participation.

- *Managerial*: Management of limited resources is always important. Failure of managing the project in time causes heavy loss and resource consumption. Few areas of managerial risk in e-governance are:

  - The risk factors related to poor managerial skill of project leader, project manager and administrators.
  - Poor time management in project development and implementation.
  - Not using the proper project management tools.
  - The leadership quality of the team leaders, project leaders and the administrators.

- *Economical*: The e-governance projects need to be developed on a long-term basis. The cost–benefit analysis is required before execution of the project. As these projects are high fund projects, the risk is also high. Many a times, the benefits of e-governance projects are drawn in long run. Inaccurate cost–benefit analysis and project management may cause a heavy financial loss.
- *Political*: The political impact in governance is always there. The changes in political scenario, leaders and bureaucrats put a great impact in project output. The stakeholders such as ministers, bureaucrats, consultants and many others are directly or indirectly involved in an e-governance projects. Thus, the unhealthy interpersonal relations would badly impact the e-governance projects. The healthy interpersonal relation among the stakeholders leads the project in a positive direction [16, 17].
- *Behavioural*: One of the major areas is the mindset of stakeholders. Their attitude, behaviour and mentality is a major input for making the e-governance application successful. Finally, the behaviour of citizens also have a lot of impact [17].

The impact of risk factors on Land Records project is furnished in the matrix given below.

Risk analysis of Land Records project

Sl. no	Risk areas	Risk parameters		Risk impact	
			Low	Moderate	High
1	Technical risk	Hardware and connectivity			X
		Poor software design			X
		Data security			X

(continued)

(continued)

Sl. no	Risk areas	Risk parameters		Risk impact	
		Frequent requirement changes		X	
		Poor user participation		X	
		Look of the pages not good	X		
2	Managerial	Poor management skill of administrator			X
		Poor managerial skill project leader			X
		Wrong time scheduling		X	
3	Economic	Improper cost–benefit analysis		X	
		Mismanagement of project fund			X
4	Political	Conflict among stakeholders			X
		Conflict within development team		X	
		Changes in government functionaries	X		
5	Behavioural	Mindset of stakeholders		X	
		Attitude of users			X
		Rigid to adopt new system		X	

The above matrix shows the impact of different risk factors on Land Records project. The high-impact factors are monitored more attentively for reducing converting the risk into benefit. Like risk analysis, the decision analysis of revenue projects is discussed also.

3.2 Decision Analysis

In Fig. 1, the hierarchical structure of Revenue Administration is described. But, the core structure is described in Fig. 4. Thus, the services under ORTPS are regularly monitored by minister and secretary at state, district administration at district and tahasildars at the tahasil level. The information from state level to district → tahasil → village → case details is available in the software output. The dashboard is available at all levels of Revenue Administration to take decisions at their concerned capacities.

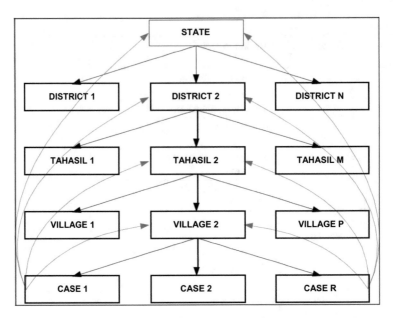

Fig. 4 Minimum path decision tree for Land Records project of Odisha

The state-level information is provided in Table 1. But, with drill-down facility, the administration can see the details at district, tahasil, village and casewise status. Figure 4 shows how the information is available in top to bottom and vice versa [18, 19].

3.3 Cost-Effectiveness Analysis

The cost–benefit analysis is the basis of any e-governance project to find the return of investment. The benefits include both tangible and intangible aspects. For example, the number of cases disposed and the number of RoRs delivered to public in time are tangible benefits. But, the reduction of footfalls in tahasils, maintenance of transparency to reduces unfair activities and other such improvements in the system are having intangible benefits (Fig. 5).

In fact, government spends a lot of money for providing services to the citizens. Different projects, schemes and other socio-economic developmental activities are taken up for service of the society. The above matrix shows that how government to decide which projects to be monitored and how to be emphasized them. If the project has both social and economic benefits, then more importance needs to be given to those projects [20–22].

Generally, Land Record projects are high cost → high social benefit → high economic benefit projects. Besides these, it is the base for many other projects such

Table 1 Mutation case status in ORTPS daily bulletin

S. No.	District name	No. of cases pending till end of October	No. of cases instituted during November	No. of cases for disposal	No. of cases disposed during November	No. of cases pending	No. of cases pending beyond ORTPS time limit	% of cases pending beyond ORTPS time limit
1	Deogarh	461	11	472	8	464	164	35
2	Sambalpur	3,378	26	3,404	26	3,378	1,200	35
3	Malkangiri	847	14	861	4	857	258	30
4	Khurda	34,517	1,314	35,831	3,468	32,363	8,985	27
5	Gajapati	4,006	57	4,063	44	4,019	1,062	26
6	Puri	12,029	865	12,894	861	12,033	2,703	22
7	Sundergarh	3,535	441	3,976	214	3,762	792	21
8	Kalahandi	10,061	721	10,782	667	10,115	2,193	21
9	Jajpur	10,003	931	10,934	388	10,546	2,146	20
10	Bhadrak	11,432	785	12,217	514	11,703	2,256	19
11	Balangir	8,368	970	9,338	264	9,074	1,749	19
12	Kendrapara	8,559	665	9,224	730	8,494	1,675	19
13	Boud	1,815	43	1,858	397	1,461	288	19
14	Nuapada	1,340	153	1,493	252	1,241	228	18
15	Ganjam	44,412	3,014	47,426	2,101	45,325	7,973	17
16	Koraput	4,064	195	4,259	195	4,064	701	17
17	Nayagarh	7,545	943	8,488	868	7,620	1,185	15
18	Bargarh	8,808	348	9,156	485	8,671	1,311	15
19	Angul	4,777	492	5,269	667	4,602	646	14
20	Balasore	17,384	1,530	18,914	1,014	17,900	2,522	14
21	Jharsuguda	1,921	29	1,950	204	1,746	237	13
22	Keonjhar	4,526	523	5,049	498	4,551	537	11
23	Nabarangpur	1,681	138	1,819	180	1,639	191	11
24	Dhenkanal	4,435	198	4,633	602	4,031	437	10
25	Rayagada	3,567	425	3,992	205	3,787	400	10
26	Mayurbhanj	4,253	388	4,641	296	4,345	416	9
27	Sonepur	913	70	983	151	832	77	9
28	Jagatsinghpur	4,956	1,296	6,252	340	5,912	459	7
29	Cuttack	12,758	1,296	13,884	1,193	12,691	996	7
30	Kandhamal	710	110	820	56	764	36	4

Source ORTPS Daily Bulletin of Revenue and Disaster Management Department, Government of Odisha

		Social Benefits for Citizens			Economic Benefits for Citizens		
		High	Medium	Low	High	Medium	Low
Cost for Government Services	Low	Highly Desirable	Desirable	Least Desirable	Highly Desirable	Desirable	Least Desirable
	Moderate	Highly Desirable	Desirable	Least Desirable	Highly Desirable	Desirable	Least Desirable
	High	Desirable	Least Desirable	Not Required	Desirable	Least Desirable	Not Required

Fig. 5 Cost-effectiveness analysis matrix

as industry, agriculture, rural development, urban development and financial institutions. Thus, strengthening of Land Records is strengthening the pillar for many other projects.

These analyses show that e-governance projects may yield better a result with the use of decision science techniques. The timely action of government and monitoring will increase the throughput of the projects.

4   Key Findings of the Paper

Decision science is having high impact on e-governance applications. It makes the decisions accurate and consistent. Mainly, the four basic types of ICT services in government are government to citizen (G2C), government to government (G2G), government to business (G2B) and government to employees (G2E) [23].

- Government to government (G2G): It is the base of government services. If G2G is stronger, then the government functionalities are easier for implementation. Interoperability of government functions becomes easy.
- Government to citizens (G2C): The prime objective of government is citizen services. These services need to be provided in due time, less cost and easy to avail it. The government needs to provide these services in such a manner that they reach the right citizen at right time. On the other hand, the citizen must be educated enough to avail the service without any third party support requirement.
- Government to business (G2B): Ease of doing business is a step of government to welcome industrialization. The growth of business will lead to the growth of society. Thus, G2B is required to develop the economy.
- Government to employee (G2E): The key resource of any government is its manpower. The G2E service must be effective to increase the productivity of each employee.

The following matrix shows the four basic services of government and their importance.

	Government	Citizen	Business	Employee
Government	G2G (Strong Govt. Base)	G2C (Best Services to Citizens)	G2B (Industrial Development)	G2E (Competent Employees)
Citizen	G2C (Best Services to Citizens)			
Business	G2B (Industrial Development)	Government is not directly involved in decision making		
Employee	G2E (Competent Employees)			

Fig. 6 Service matrix and involvement of government

The basic service matrix and the objectives of government are shown in Fig. 6. With the development of ICT, the e-governance became possible. So also the decision science also improved a lot. The methodologies became easy to implement in different sectors.

Here, the analysis of ORTPS is done. Any service relating to e-governance can be analysed for taking effective decisions. Starting from strategic decision to operational-level decisions may be taken. The data analytics will help a lot use decision science to maximum extents [24, 25].

5   Conclusion

The responsibility of any government may be central or state or local, taking right decision at right time. The right decision is possible with vast experience, knowledge, unbiased thoughts, cognition, intuition, etc. For this purpose, committees are formed, meetings are conducted, and conferences and workshops are held to reach a viable justification of taking action. With the development and wide usage of information and communication technology, the availability of information, data analytics and deriving conclusions become easy. Similarly, with the use of data analytics, machine learning and artificial intelligence, the decision science is also easy to implement. The field of e-governance, which is the use of ICT in governance, the role of decision science is very vital. Now, decision science is very much useful for good decision-making in government sectors. The improvement in data analytics and data science provides a better environment of decision-making. It will help to reduce the risk and improvising sustainability of e-governance projects. Here, the case of ORTPS is studied which is one of the most crucial areas in Government of Odisha. This shows that how government can take effective decisions with the use of science and technology. It also indicates that how decision science would be used in other areas of government for the benefit of the public.

Acknowledgements We are thankful to National Informatics Centre which is a premier and successful organization in developing and implementing e-governance application in India. Further, we are thankful to Government of India and Government of Odisha for providing valuable information in the websites.

References

1. http://www.businessmanagementideas.com/decision-making/top-10-techniques-of-decision-making/3377
2. Ministry of Electronics and Information Technology at http://meity.gov.in
3. National Informatics Centre at http://www.nic.in
4. Bhulekh at http://bhulekh.ori.nic.in
5. BhuNaksha at http://bhunakshaodisha.gov.in
6. Revenue and Disaster Management Department, Government of Odisha website at http://revenueodisha.gov.in, Revenue Dash Board at http://bhulekh.ori.nic.in/statedashboard
7. Odisha Right to Public Services Act at http://bhulekh.ori.nic.in/ORTPSA
8. https://decisionsciences.org
9. Bell, D.E., Raiffa, H., Tversky, A.: Decision Making: Descriptive, Normative and Prescriptive Interactions. Cambridge University Press (1988)
10. http://www.ibm.com
11. https://hbr.org
12. https://www.datascience.com
13. Digital India Land Records Modernisation Programme at http://dilrmp.nic.in
14. Weiss, M.S., Indurkhya, N.: Predictive Data Mining: A Practical Guide. Morgan Kaufmann (1997)
15. Government of Odisha at http://odisha.gov.in
16. http://bigdataanalytics.mit.edu
17. Ojha A.: E-Governance in Practice, pp. 33–41. GIFT Publishing
18. Howard, R.A., Abbas, A.E.: Foundations of Decision Analysis. Pearson (2015)
19. https://digitalvidya.com
20. Kahneman, D., Slovic, P., Tversky, A.: Judgement, Under Uncertainty: Heuristics and Biases from Cambridge University Press (1982)
21. Nisbet, R., Elder, J., Miner, G.: Handbook of Statistical Analysis and Data Mining Applications. Elsevier (2009). ISBN: 978-0-12-374765-5
22. Džeroski, S.,Kobler, A., Gjorgijoski, V., Panov, P.: Using decision trees to predict forest stand height and canopy cover from LANDSAT and LIDAR data. In: 20th International Conference on Informatics for Environmental Protection—Managing Environmental Knowledge—ENVIROINFO (2006)
23. e-Governance standardization at http://egovstandards.gov.in/
24. Myatt, G.J.: Making Sense of Data: A Practical Guide to Exploratory Data Analysis and Data Mining. John Wiley (November 2006). ISBN: 0-470-07471-X
25. Cerrito, P.: Introduction to Data Mining Using SAS Enterprise Miner. SAS Press (2006). ISBN: 978-1-59047-829-5

HR Value Proposition Using Predictive Analytics: An Overview

Poonam Likhitkar and Priyanka Verma

Abstract Human resource management undergoes a drastic change due to digitization. In the current competitive environment, talented employee is undoubtedly the most valuable assets of the organization. The concept of data analytics in HRM is increasingly gaining attention and popularity among consultants and practitioners in the field of human resource management. Predictive analytics in the field of human resource management (HRM) leads to achievement of organizational benefits. This study aims to explore the role of predictive analytics in human resource management domain. In doing so, the paper examines relevant paper on HR analytics and how it helps in better decision-making in the organization. This paper also proposes that the use of analytics surely helps in effective decision-making process in the organization without any biasness. The present paper highlights the value of predictive analytics in the field of management, especially in the domain of human resource management.

Keywords Data · Human resource management · Predictive analytics · Retention

1 Introduction

According to report (2017), the adoption of analytics in Indian industry is estimated approximately 2.03 $ billion and is likely to be double by the year 2020 with 23.8% rate of CAGR which is a very healthy rate [1]. According to Bhasker Gupta, CEO and Founder, Analytics India Magazine stated "India is definitely emerging as a growing hub for analytics". Employees are considered as investment for every organization [2], as their performance affects organization effectiveness [3]. Due to high competition, they are more expected to perform better [4]. Human resource management is

P. Likhitkar (✉) · P. Verma
Department of Management Studies, Maulana Azad National Institute of Technology (NIT),
Bhopal, Madhya Pradesh, India
e-mail: poonaml0610@gmail.com

P. Verma
e-mail: drpriyankaverma@rediffmail.com

© Springer Nature Singapore Pte Ltd. 2020
S. Patnaik et al. (eds.), *New Paradigm in Decision Science and Management*,
Advances in Intelligent Systems and Computing 1030,
https://doi.org/10.1007/978-981-13-9330-3_15

defined as acquiring of candidate, their development positioning and maintaining and retaining them in the organization. It is the art of managing workforce of the organization in order to achieve its goals in an efficient manner. Due to urbanization and digitization, organization should be technology-driven to stand in the market which is quite competitive. Employees are undoubtedly the greatest asset of the organization, and human resource management plays a prime role which can influence the proper optimization of resources. A good data management process with companies to align strategies and identify areas of growth and today competitive environment being able to make those decisions before the competition or making the best decisions is crucial for the growth of the business. The paper includes the explanation of Data Analytics, Relevant Literature Review, Findings and Discussion, Conclusion, Originality, and lastly Declaration followed by the References.

2 Data Analytics

Data analytics is the process of examining data sets in which large data sets are analysed to draw some conclusions from the information they contain for the better decision-making processes.

2.1 Descriptive Analytics

It is also called as business intelligence. It will provide you an insight into the past. It basically tells us what has happened. It creates a summary of historical data to give useful information and possibly prepare the data to give useful information and possibly prepare the data for further analysis. It shows the summary through dashboards, charts, diagrams, etc.

2.2 Predictive Analytics

It tells us what could happen? Predictive analytics is basically used to make predictions about future unknown events. Predictive analytics uses several techniques from artificial intelligence, data mining, statistical modelling and machine learning to analyse current data to make predictions about the future. Predictive analytics provides solution to solve problem of talent management faced by the organization.

2.3 Prescriptive Analytics

Which tells us what should we do? Prescriptive analytics attempts to identify the effect of future decisions in order to advise on possible outcomes before the decisions are actually made.

3 Literature Review

This paper contributes and integrates and extends the literature on two different independent fields: analytics and HRM, which was earlier predominantly an information technology (IT) domain [5] and latter being a management domain. HR analytics are treated as a growing trend in the management field [6], as the process of using data analysing tool for measuring the efficacy of HRM systems of the organization. Renowned companies such as Google, Disney, Amazon, Apple and Microsoft are already associated with HR analytics [7]. HR analytics were also considered an opportunity to reduce attrition rate through quantitative methods [8].

3.1 HR Analytics

Predictive analytics can bring a drastic change in the organization success cycle. It can be game changer of the organization, especially in the context of human resource practices such as recruitment and selection, training and development, and talent retention [9]. HR analytics plays a vital role in improving the retention of employees in the organization [10]. Very less descriptive paper is available on analytics in HR stream [11]. Authors highlighted the importance of data analytics in HRM [11, 12]. The employee attrition score of each employee will be assed with predictive model of attrition. This may help organization with preventing the high performers, ensure continuation of them and enhance loyalty towards the organization [13]. Employees' demographic profile, compensation, performance, rewards and recognition, training information and employee survey scores will be used for this attrition analysis [14].

4 Findings and Discussion

To apply predictive analytics, HR managers need to have systematic and organized information of demand and supply which can be adjusted according to business need [15]. Managers must ensure the needs of employees in the upcoming years applying projected requirement and range of loyal employees which will stay in the coming years, thereby enhancing the retention rates in the organization. Predictive analytics helps in bridging the gap between from a number of employees needed and at what level of employees.

4.1 Importance of Analytics in HR

Employee attrition analysis: This is to check the attrition of employees in the organization in a given period of time.

Employee loyalty analysis: It is used to check the loyalty of employees towards the organization.

Employee sentiment analyses: This analysis is used to analyse the opinions, likes and dislikes of employees towards different entities such as organization, people, product and services etc. [16]. It is a widely used analysis, especially in the social media review on any topic which may be negative or positive. Sentiment analysis is of two types: *machine learning approach* which applies the machine learning approach with the use of linguistic approach for analysis, whereas *lexicon-based approach* is based on sentiment lexicons which assembles some known terms used in the statement or reviews. The word in the reviews/statement/feedback has been given ranks such as 5 for like and 0 for dislike [17]. The below tables are drawn from the relevant literature used in the study (Tables 1, 2 and 3).

5 Conclusions

To conclude, this review paper made an attempt to explore the importance of predictive analytics for HR value proposition in the organization. The paper also highlights different techniques which can be used for better decision-making and found to be a potential solution for fair processes and practices of human resource management with the use of different analytical tools. Along with this, both HR practitioners and researchers in the field of human resource management, the field of study suggested necessary directions for further future research to enrich the field. Academicians, researchers and HR managers are encouraged to continue developing better solutions by applying HR analytics in organization for gaining competitive advantage. Hence, employees are undoubtedly greatest asset of the organization and human resource management plays a prime role which can influence the proper optimization of resources.

Table 1 Different predictive analytics marketing models

Models	Application of models	Techniques
Churn modelling (customer attrition model)	Marketing	Classification model includes decision tree, logistic regression and neural network model
Customer lifetime model	Marketing	Classification model
Revenue model	Marketing	Classification model
Market mix model	Marketing	Linear regression, panel data analysis, mixed model
Sales forecasting model	Marketing	Time series, multiple linear regression, panel data analysis, machine learning model
Cross/up-sell model	Marketing	Classification algorithm, linear programming, market basket model
Loyalty model	Marketing	Econometric model, linear regression, hypothesis testing, ANOVA, SEM
Segmentation model	Marketing	Linear regression, decision tree, clustering factor analysis
Survey analytics	Marketing	Mix of classification and regression model
Price elasticity of demand model	Marketing	Linear regression
A/B testing	Marketing	Decision-making model
Campaign analytics	Marketing	Econometric model, principle component analysis, tree models, optimization techniques

Table 2 Applying predictive analytics in the field of HR

HR areas	Techniques
Recruitment and selection analysis	Psychometric tools and algorithm
Training and development	Talent data analysis techniques
Compensation and benefits	Experience fitment grid, Conjoint analysis, Workforce analytics
Performance management system	Work quantity metrics, work quality metrics, work efficiency metrics, organization performance metrics
Retention of workforce	Employee turnover metrics, past resignation record metrics, involuntary turnover of employees metrics, Clustering algorithm

Table 3 Different data mining techniques of HRM

Context	Data Mining techniques in HRM	Author(s)
Employee turnover analysis	Decision trees, logistic regression & neutral network	[9]
Talent Management	Classification, Clustering, Decision tree, Neutral network, Association and Prediction rules	[9, 16]
Employee recruitment and selection	Decision trees, classification & regression tree (CART) algorithm	[18, 19].

6 Originality

Also, the paper has given shed light on the contribution of predictive analytics in the field of human resource management. Lastly, expertises in HR analytics can make a significant difference towards business success.

7 Declaration

The authors received no financial assistance for the present study.

References

1. Online Bureau, http://www.businessworld.in/article/analytics-industry-in-india-growing-at-a-rate-of-23-8-CAGR-study/26-07-2017-122871/
2. Schraeder, M., Jordan, M.: Managing performance: a practical perspective on managing employee performance. J. Qual. Participation **34**(2), 4 (2011)
3. Sundaray, B.K.: Employee engagement: a driver of organizational effectiveness. Eur. J. Bus. Manag. **3**(8), 53–59 (2011)
4. Biswas, S., Varma, A.: Antecedents of employee performance: an empirical investigation in India. Empl. Rela. **34**(2), 177–192 (2011)
5. Pemmaraju, S.: Converting HR data to business intelligence. Empl. Rela. Today **34**(3), 13–16 (2007). https://doi.org/10.1002/ert.20160
6. Rasmussen, T., Ulrich, D.: Learning from practices: how HR analytics avoids being a management fad. Org. Dyn. **44**(3), 236–242 (2015)
7. Morgan, J.: The Employee Experience Advantage: How to Win the War for Talent by Giving Employees the Workspaces They Want, the Tools They Need, and a Culture They Can Celebrate. Wiley, NY (2017)
8. Harris, E.C., Light, D.A.: Talent and analytics: new approaches higher ROI. J. Bus. Strategy **32**(6), 04–13 (2011)
9. Tamizharasi, K., Rani, U.: Employee turnover analysis with application of data mining methods. Int. J. Comput. Sci. Inf. Technol. **5**(1), 562–566 (2014)
10. Malisetty, S., Archana, R.V., Kumari, V.K.: Predictive analytics in HR management. Indian J. Public Health Res. Dev. **8**(3), 115–120 (2017)

11. Verma, S., Mehrotra, R.: Research paper on role of analytics in renovating human resource management. Int. J. Manag. **5**(5), 17–23 (2017)
12. Angrave, D., Charlwood, A., Kirkpatrick, I., Lawrence, M., Stuart, M.: HR and analytics: why HR is set to fail the big data challenge. Hum. Resour. Manag. J. **26**, 01–11 (2016)
13. King, G.K.: Data analytics in human resources: a case study and critical review. Hum. Resour. Dev. Rev. **15**(4), 487–495 (2016)
14. Khatri, B.: Talent analytics: toolkit for managing HR issues. Sai Om J. Commer. Manag. **1**(5) (2014)
15. Waxer, C.: HR Executives: Analytics Role Needs Higher Profile (2013)
16. Manogna, N., Mehta, S.: Talent management in organization using mining techniques. Int. J. Comput. Sci. Inf. Technol. **6**(1) (2015)
17. Elkan, C.: Predictive Analytics and Data Mining, pp. 07–12 (2013)
18. Azar, A., Sebt, M.V., Ahmadi, P., Rajaeian, A.: A model for personal selection with a data mining approach: a case study in a commercial bank. J. Hum. Resour. Manag. **11**(1) (2013)
19. Sivaram, N., Ramar, K.: Applicability of clustering and classification algorithms for recruitment data mining. Int. J. Comput. Appl. **4**(5) (2000)

Determination of Time-Dependent Quadratic Demand by Using Optimal Ordering Policy, Salvage Value Under Partial Backlogging, Time-Proportional Deterioration, Time-Varying Holding Cost

Trailokyanath Singh and Chittaranjan Mallick

Abstract The present paper deals with the deteriorating items under partial back-logging by using Economic Order Quantity (EOQ) of the inventory model. The quadratic function of time is determined by the demand rate as well as cost of holding inventory of each item. Time proportional is decided by deterioration, partially backlogged salvage value of each deteriorated item. The theoretical expressions are obtained for optimum inventory level and total average cost. A mathematical model and an algorithm are developed for the determination of total inventory cost. Various parameters of sensitivity analysis associated with this model are analyzed, by taking a couple of numerical examples.

Keywords Economic ordered quantity · Partially backlogged · Lost sale · Salvage value · Time-proportional deterioration, time-dependent quadratic demand

1 Introduction

In the business scenario, for its practical importance, the deteriorating items in the inventory system have been playing a major role. The classical Harris [9] and Wilson [25] inventory models are based on the constant demand rate. The deteriorating items are defined as the decay or damage of item which are not used in original purposes. Ghare and Schrader [6], the first researcher, who pointed out that the inventory is not only depleted by deterioration, but also by the demand rate. The demand rate

T. Singh
Department of Mathematics, C. V. Raman College of Engineering,
Bhubaneswar 752054, Odisha, India
e-mail: trailokyanaths108@gmail.com

C. Mallick (✉)
Department of Mathematics, Parala Maharaja Engineering College,
Berhampur 761003, Odisha, India
e-mail: cmallick75@gmail.com

of an item is assumed to be a constant in the classical economic ordered quantity model. Therefore, in most of the inventory problems, the solution is obtained by taking deterioration rate as a constant. The linear inventory model for deteriorating items was developed by Dave and Patel [5] with an increasing demand rate, which was further extended by Sachan [20] by allowing the deteriorating items with a varying demand and shortages. EOQ models proposed by Goswami and Chaudhuri [8], Chakrabarti and Chaudhuri [2], and Lin et al. [13] were mainly focusing on deteriorating items with varying demand and shortages. Moreover, EOQ models were also developed by researchers by taking exponential increasing or decreasing demand. Ouyang et al. [17] analyzed the inventory model for deteriorating items with partial backlogging by assuming the exponential declining demand rate. Wee [24] studied the lot size model for deteriorating items with partial back ordering. The optimal ordering model for deteriorating items with time-varying demand and partial backlogging was analyzed by Chang and Dye [3]. The deteriorating inventory model for maximization of profit was studied by Sahoo et al. [21].

The depletion of inventory is due to a constant demand rate, which was invented by Harris and Wilson's in their inventory model. However, in real-life situations, due to deterioration inventory loss arises. Further, new class of models was considered to be time-dependent inventory models. Covert and Philip [4] and Philip [18] developed inventory models for items with time-dependent deterioration rate was a new class of model, which is called Weibull distribution deterioration rate. The optimal policy for deteriorating items with time-proportional deterioration rate was developed by Singh et al. [23] by considering the time-dependent linear demand rate. Some researchers considered inventory system with time-dependent linear and exponential demand and time-dependent holding costs for developing their inventory models. The linear type occurs due to steady increasing or decreasing in demand whereas the exponential type increases highly which are unstable in daily life. The better alternative is choosing the demand which applies to real-life situation as the quadratic demand rate. Khanra and Chaudhuri [12], and Ghosh and Chaudhuri [7] have established their models by considering the demand as quadratic function of time. Generally, shortages are not always completely backlogged. Wee [24], and Hollier and Mak [10] considered the opportunity cost while developing their model in backlogging case. The manufacture for calculating the backlogging rate was the length of waiting time for the next replenishment. An inventory model with quadratic demand, constant deterioration, and partial backlogging is studied by Singh and Pattanayak [22]. The optimal replenishment policies for instantaneous deteriorating items with backlogging and trade credit were developed by Rajan and Uthayakumar [19] under the inflation rate situation.

In real-life situations, deterioration items in the inventory system exhibit to be a complete loss to the inventory system, due to no sale value. Practically, the vendor can offer a reduced unit cost for the deteriorated unit in the inventory to his buyer. Jaggi and Aggarwal [11] formulated the mathematical model for deteriorating items into the deteriorated units by incorporating salvage values. Mishra and Singh [15] studied the inventory model for ramp type demand, time-dependent deteriorating items with salvage value and shortages. An inventory model for deteriorating items with salvage value by assuming two-parameter Weibull distribution deterioration rate parameters was proposed by Mishra and Triparty [14]. Annadurai [1] has incorporated salvage value and developed an ordering policy for decaying items with shortages. Recently, Mishra et al. [16] developed an optimal policy under partial backlogging for deteriorating items with time-dependent demand and time-varying holding cost.

In the present paper, an EOQ model is analyzed by incorporating the salvage value for the deteriorated item and considering the demand rate as the quadratic function of time. Shortages in the system are allowed and partially backlogged. Among the various demands in EOQ models, the most realistic approach is to consider a quadratic demand rate, because it represents both accelerated rise and fall in time demand. The accelerated growth in demand rate is found to occur in the case of spare parts of newly introduced the state-of-the-art aircrafts, computers, etc. whereas the retarded growth in demand rate is found to be occur in the case of spare parts of the absolute aircrafts, computers, etc. The demand rate in these cases is a quadratic function of time, $R(t) = a + bt + ct^2 (a, b \& c > 0)$. At $c = 0$ and $b = c = 0$, the quadratic demand rate changes into a linear and constant demand rate, respectively. Backlogging rate in inventory system is inversely proportional to the next replenishment of the waiting time. In order to minimize the total relevant cost, the shortage point and the length of the cycle is optimized, which is the main objective of the problem. A numerical example is given to illustrate the model. In order to note the changes in the parameters, the sensitivity analysis of the model is examined.

2   Assumptions

In order to develop the proposed mathematical model, the following assumptions are used:

   (i)   Only single items are handled by the inventory system.
   (ii)   Unsatisfied demands are partially backlogged and shortages are permitted to occur.
   (iii)   Lead time is zero.
   (iv)   Replenishment rate is infinite and instantaneous.
   (v)   During the cycle time, the units in inventory deteriorate rate are time-proportional.
   (vi)   During the cycle time, the demand rate for an item is quadratic.
   (vii)   The associated salvage value with deteriorated units is constant, during the cycle time.
   (viii)   There is no repair or replacement of deteriorated units during the cycle time.

3 Modeling

The following notations and mathematical formulation are provided in this paper, to develop and solve the proposed model.

3.1 Notations

The model is developed with the following notations:

$I(t)$: The on hand level of inventory at any time t over $[0, T]$.
$R(t)$: Demand rate at time t is quadratic function of time, i.e., $R(t) = a + bt + ct^2 (a, b \& c > 0)$.
Here parameters a, b and c act as initial, increasing and self increasing demand rates respectively.
$\theta(t)$: Time-proportional, Deteriorating rate at time t is i.e., $\theta(t) = \theta t$, $\quad 0 < \theta << 1$.
$B(t)$: At time t, backlogging rate i.e., $B(t) = \beta$, $\quad 0 \le \beta \le < 1$
I_{max}: Maximum level per cycle.
S: Maximum amount of backlogged demand per cycle.
I_0: Ordered quantity.
$h(t)$: Holding cost/unit of time where $h(t) = h + \alpha t$, $h > 0 \& \alpha > 0$.
c_o: Ordering cost of inventory, \$/order.
c_b: Shortage cost, \$/unit/unit time.
c_d: Deterioration cost, \$/unit.
c_l: Cost of opportunity due to lost sale.
t_b: Time of approaching zero level.
T: Length of ordered cycle.
$Z(t_b, T)$: Total inventory cost.
t_b^*: Optimal cost of t_b.
T^*: Optimal cost of T.
$Z^*(t_b, T)$: Optimal value of $Z(t_b, T)$.

3.2 Mathematical Formulation

The instantaneous states of the inventory level $I(t)$ at any instant during the time interval $[0, t_b]$ is given by

$$\frac{dI_1(t)}{dt} = -[\theta(t)I_1(t) + R(t)], \quad 0 \le t \le t_b \tag{1}$$

where $R(t) = a + bt + ct^2$, $\quad a > 0, b > 0 \& c > 0$ and $\theta(t) = \theta t, 0 < \theta << 1$.

The solution of Eq. (1) with boundary conditions $I_1(0) = I_{max}$ and $I_1(t_b) = 0$ is given by

$$I_1(t) = \left[at_b + \frac{bt_b^2}{2} + \frac{ct_b^3}{3} + \frac{\theta}{2}\left(\frac{at_b^3}{3} + \frac{bt_b^4}{4} + \frac{ct_b^5}{5}\right) \right.$$
$$\left. - \left(at + \frac{bt^2}{2} + \frac{ct^3}{3} + \frac{\theta}{2}\left(\frac{at^3}{3} + \frac{bt^4}{4} + \frac{ct^5}{5}\right)\right)\right] e^{-\frac{\theta t^2}{2}}, \quad 0 \le t \le t_b, \quad (2)$$

by neglecting higher terms of θ, $0 < \theta << 1$.

Using the boundary condition $I_1(0) = I_{max}$ in Eq. (2), we get

$$I_1(0) = I_{max} = at_b + \frac{bt_b^2}{2} + \frac{ct_b^3}{3} + \frac{\theta}{2}\left(\frac{at_b^3}{3} + \frac{bt_b^4}{4} + \frac{ct_b^5}{5}\right). \quad (3)$$

Similarly, during the shortage period, the instantaneous level $I(t)$ at the interval $[t_b, T]$ is given by

$$\frac{dI_2(t)}{dt} = -B(t)R(t), \quad t_b \le t \le T, \quad (4)$$

where $R(t) = a + bt + ct^2$, $a > 0, b > 0 \& c > 0$ and $B(t) = \beta$, $\beta > 0$ is backlogging rate.

By using boundary condition $I_2(t_b) = 0$, the solution of Eq. (4) can be given by

$$I_2(t) = -\beta\left[at + \frac{bt^2}{2} + \frac{ct^3}{3} - \left(at_b + \frac{bt_b^2}{2} + \frac{ct_b^3}{3}\right)\right], \quad t_b \le t \le T. \quad (5)$$

Using the boundary condition $S = -I_2(T)$ in Eq. (2), we get

$$S = -I_2(T) = \beta\left[aT + \frac{bT^2}{2} + \frac{cT^3}{3} - \left(at_b + \frac{bt_b^2}{2} + \frac{ct_b^3}{3}\right)\right]. \quad (6)$$

Thus, the optimum ordered quantity per cycle is

$$I_0 = I_{max} + S$$
$$= at_b + \frac{bt_b^2}{2} + \frac{ct_b^3}{3} + \frac{\theta}{2}\left(\frac{at_b^3}{3} + \frac{bt_b^4}{4} + \frac{ct_b^5}{5}\right)$$
$$+ \beta\left[aT + \frac{bT^2}{2} + \frac{cT^3}{3} - \left(at_b + \frac{bt_b^2}{2} + \frac{ct_b^3}{3}\right)\right]. \quad (7)$$

In view of the above, elements of optimum total inventory cost can be evaluated as:

(i) Holding cost (HC):
   $HC = \int_0^{t_b} h(t)I_1(t)dt$ where $h(t) = h + \alpha t$, $h > 0 \& \alpha > 0$.

Thus

$$HC = h\left[\frac{at_b^2}{2} + \frac{bt_b^3}{3} + \frac{ct_b^4}{4} + \theta\left(\frac{at_b^4}{12} + \frac{bt_b^5}{15} + \frac{ct_b^6}{18}\right)\right]$$
$$+ \alpha\left[\frac{at_b^3}{6} + \frac{bt_b^4}{8} + \frac{ct_b^5}{10} + \theta\left(\frac{at_b^5}{40} + \frac{bt_b^6}{48} + \frac{ct_b^7}{56}\right)\right], \quad (8)$$

by neglecting higher terms of θ, $0 < \theta << 1$.

(ii) Deterioration cost (DC):
$DC = c_p \int_0^{t_b} \theta(t) I_1(t) dt$ where $\theta_0(t) = \theta t$, $0 < \theta << 1$.
Thus

$$DC = c_p \theta\left(\frac{at_b^3}{6} + \frac{bt_b^4}{8} + \frac{ct_b^5}{10}\right), \quad (9)$$

by neglecting higher terms of θ, $0 < \theta << 1$.

(iii) Salvage value (SV):
$SV = \chi \int_0^{t_b} \theta(t) I_1(t) dt$ where $\theta_0(t) = \theta t$, $0 < \theta << 1$.
Thus

$$SV = \chi\theta\left(\frac{at_b^3}{6} + \frac{bt_b^4}{8} + \frac{ct_b^5}{10}\right), \quad (10)$$

by neglecting higher terms of $\theta, 0 < \theta << 1$.

(iv) Shortage cost (SC):

$$SV = \chi \int_0^{t_b} \theta(t) I_1(t) dt$$
$$= \beta c_b \left[\frac{a\left(T^2 - t_b^2\right)}{2} + \frac{b\left(T^3 - t_b^3\right)}{6} + \frac{c\left(T^4 - t_b^4\right)}{12} - \left(at_b + \frac{bt_b^2}{2} + \frac{ct_b^3}{3}\right)(T - t_b)\right]. \quad (11)$$

(v) Lost sales cost (LSC):
$LSC = c_s \int_{t_b}^T R(t) B(t) dt$ where $R(t) = a + bt + ct^2$, $a > 0, b > 0 \& c > 0$
where $a > 0, b > 0 \& c > 0$ and $B(t) = \beta$, $\beta > 0$ where $\beta > 0$ is backlogging rate.
Thus,

$$LSC = (1 - \beta)c_s \left[a(T - t_b) + \frac{b\left(T^2 - t_b^2\right)}{2} + \frac{c\left(T^3 - t_b^3\right)}{3}\right]. \quad (12)$$

(vi) Ordering cost (OC):

$$OC = K_0. \tag{13}$$

The total relevant cost, TVC, during the time interval of the system $[0, T]$ by using (8), (9), (10), (11), (12), and (13) is given by

$$TVC = OC + HC + DC + SC + LSC - SV. \tag{14}$$

Thus, by using (8), (9), (10), (11), (12), (13) and (14), the average total cost $(Z(t_b, T)$ is given by

$$
\begin{aligned}
Z(t_b, T) &= \frac{1}{T}[OC + HC + DC + SC + LSC - SV] \\
&= \frac{K_0}{T} + \frac{h}{T}\left[\frac{at_b^2}{2} + \frac{bt_b^3}{3} + \frac{ct_b^4}{4} + \theta\left(\frac{at_b^4}{12} + \frac{bt_b^5}{15} + \frac{ct_b^6}{18}\right)\right] \\
&\quad + \frac{\alpha}{T}\left[\frac{at_b^3}{6} + \frac{bt_b^4}{8} + \frac{ct_b^5}{10} + \theta\left(\frac{at_b^5}{40} + \frac{bt_b^6}{48} + \frac{ct_b^7}{56}\right)\right] + \frac{(c_p - \chi)\theta}{T}\left(\frac{at_b^3}{6} + \frac{bt_b^4}{8} + \frac{ct_b^5}{10}\right) \\
&\quad + \frac{\beta c_b}{T}\left[\frac{a\left(T^2 - t_b^2\right)}{2} + \frac{b\left(T^3 - t_b^3\right)}{6} + \frac{c\left(T^4 - t_b^4\right)}{12} - \left(at_b + \frac{bt_b^2}{2} + \frac{ct_b^3}{3}\right)(T - t_b)\right] \\
&\quad + \frac{1}{T}\left[(1 - \beta)c_s\left(a(T - t_b) + \frac{b\left(T^2 - t_b^2\right)}{2} + \frac{c\left(T^3 - t_b^3\right)}{3}\right)\right]. \tag{15}
\end{aligned}
$$

The optimum values of t_b and T are obtained from non-linear equations

$$\frac{\partial Z(t_b, T)}{\partial t_b} = 0, \tag{16}$$

and

$$\frac{\partial Z(t_b, T)}{\partial T} = 0 \tag{17}$$

satisfying the sufficient conditions

$$\frac{\partial^2 Z(t_b, T)}{\partial t_b^2} > 0, \tag{18}$$

$$\frac{\partial^2 Z(t_b, T)}{\partial T^2} > 0 \tag{19}$$

and

$$\frac{\partial^2 Z(t_b, T)}{\partial t_b^2} \cdot \frac{\partial^2 Z(t_b, T)}{\partial T^2} - \left(\frac{\partial^2 Z(t_b, T)}{\partial t_b \partial T}\right)^2 > 0. \tag{20}$$

From Eqs. (16) and (17), we have

$$\frac{\partial(Z(t_b, T))}{\partial t_b} = \frac{h}{T}\left[at_b + bt_b^2 + ct_b^3 + \frac{\theta}{3}\left(at_b^3 + bt_b^4 + ct_b^5\right)\right]$$
$$+ \frac{\alpha}{2T}\left[at_b^2 + bt_b^3 + ct_b^4 + \frac{\theta}{4}\left(at_b^4 + bt_b^5 + ct_b^6\right)\right]$$
$$+ \frac{(a + bt_b + ct_b^2)}{T}\left[\frac{(c_p - \chi)\theta_0 t_b^2}{2} - \beta c_b(T - t_b) - (1 - \beta)c_s\right] = 0, \tag{21}$$

and

$$\frac{\partial(Z(t_b, T))}{\partial T} = \frac{\beta c_b}{T}\left[aT + \frac{bT^2}{2} + \frac{CT^3}{3} - \left(at_b + \frac{bt_b^2}{2} + \frac{ct_b^3}{3}\right)\right]$$
$$+ \frac{1}{T}\left[(1 - \beta)c_s\left(a + bT + cT^2\right)(Z(t_b, T))\right] = 0, \tag{22}$$

respectively.

However, if the above simultaneous equations have a solution, then it is easy to verify that this solution satisfies the sufficient conditions for a minimum (which are provided in the Appendix).

4 Numerical Examples

Example 1 Consider $a = 10$, $b = 50$, $c = 20$, $h = 0.5$, $\theta = 0.8$, $\alpha = 20$, $\beta = 0.8$, $c_p = 10$, $c_b = 4$, $c_s = 8$, $\chi = 0.1$, and $K_0 = 2500$ in proper units and from Eqs. (13) and (14), we get the optimum shortage point and the optimum cycle time, respectively, are $t_b^* = 0.704757$ and $T^* = 0.704757$. Substituting the values of t_b^* and T^* in Eqs. (5) and (12), we get the optimum total cost and the optimum order quantity, respectively, are $Z^*(t_b, T) = 1427.05$ and $I_0^* = 267.334$.

The sufficient conditions are $\frac{\partial^2(Z^*(t_b, T))}{\partial t_s^2} = 549.043 > 0$, $\frac{\partial^2(Z^*(t_b, T))}{\partial T^2} = 435.884 > 0$ and $\frac{\partial^2[Z^*(t_b, T)]}{\partial t_b^2} \cdot \frac{\partial^2[Z^*(t_b, T)]}{\partial t_b \partial T} - \left(\frac{\partial^2[Z^*(t_b, T)]}{\partial T^2}\right)^2 = 234,881 > 0$.

Example 2 Consider $a = 500$, $b = 100$, $c = 5$, $h = 0.5$, $\theta = 0.001$, $\alpha = 20$, $\beta = 0.8$, $c_p = 10$, $c_b = 4$, $c_s = 8$, $\chi = 0.1$ and $K_0 = 2500$ in proper units and from Eqs. (13) and (14), we get the optimum shortage point and the optimum cycle time, respectively, are $t_b^* = 0.664064$ and $T^* = 1.64674$. Substituting the values of t_b^*

and T^* in Eqs. (5) and (12), we get the optimum total cost and the optimum ordered quantity, respectively, are $Z^*(t_b, T) = 3043.03$ and $I_0^* = 844.61$.

The sufficient conditions are $\frac{\partial^2 (Z^*(t_b, T))}{\partial t_s^2} = 5866.91 > 0$, $\frac{\partial^2 (Z^*(t_b, T))}{\partial T^2} = 1431.12 > 0$ and $\frac{\partial^2 [Z^*(t_b, T)]}{\partial t_b^2} \times \frac{\partial^2 [Z^*(t_b, T)]}{\partial t_b \partial T} - \left(\frac{\partial^2 [Z^*(t_b, T)]}{\partial T^2} \right)^2 = 7,175,370 > 0$.

5 Sensitivity Analysis

The sensitivity analysis is performed by considering Example 2 and the following results are obtained in Table 1.

The following observations can be made, from the results of the above table

(i) The decreasing order of the parameter a, b, c increases shortage time point and cycle length and decreases average total cost and ordering quantity.
(ii) The decreasing order of the parameter h increases shortage point and ordering quantity and decreases cycle length and average total cost.
(iii) The decreasing order of the parameter θ, α, c_p increases shortage point, cycle and ordered quantity and decreases total cost.
(iv) The decreasing order of the parameter β increases shortage time point and average total cost and decreases cycle length and ordering quantity.
(v) The decreasing order of the parameter c_b, c_s decreases shortage time point and average total cost and increases cycle length and ordering quantity.
(vi) The decreasing order of the parameter χ decreases shortage time point and keeps cycle length, average total cost, and ordering quantity fixed.
(vii) The decreasing order of the parameter K_0 decreases shortage time point, cycle length, average total cost, and ordering quantity.

6 Conclusions

In the present paper, an ordered quantity model for a deteriorating item with time-dependent quadratic demand, time-proportional deterioration, time-varying holding cost, and salvage value is included to determine the optimal ordering cost. When unsatisfied demand is partially backlogged, shortages are permitted to occur in the system. By citing several aspects of managerial importance, the linear demand and exponential demand rates are not suitable for market situation because the former gives the steady increase or decrease whereas the latter gives high exponential in demand rate. On the other hand, the rise of demand rate is seen in case of the state-of-the-art aircrafts, supercomputers, machines, and their spare parts whereas the fall in demand rate is seen in case of obsolete aircrafts, supercomputers, machines, and their spare parts, which is only possible in quadratic demand rate. The salvage value is incorporated in the system to reduce the total cost.

Table 1 Sensitivity analysis

Parameters	Change in parameters	t_b^*	T^*	$Z^*(t_b, T)$	% Change in $Z^*(t_b, T)$	I_0^*	% Change in I_0^*
a	750	0.620502	1.42137	3641.36	+19.6623	1034.50	+22.5622
	525	0.659031	1.62009	3106.41	+2.08279	864.759	+2.45219
	475	0.669281	1.67452	2978.74	-2.11270	822.926	-2.50397
	250	0.726649	1.9913	2352.08	-22.7060	609.159	-27.8300
b	150	0.647543	1.55987	3155.68	+3.701900	846.161	+0.248797
	105	0.662256	1.63715	3054.76	+0.385471	844.236	+0.020733
	95	0.665912	1.65656	3031.18	-0.389414	843.894	-0.019785
	50	0.684679	1.75753	2918.72	-4.085070	842.976	-0.128545
c	7.5	0.662611	1.63903	3049.23	+0.203744	842.699	-0.161363
	5.25	0.663917	1.64594	3043.65	+0.020374	843.923	-0.016350
	4.75	0.664211	1.64752	3042.40	-0.020703	844.198	+0.016231
	2.5	0.665555	1.65466	3036.75	-0.206373	845.462	+0.165983
h	0.75	0.654414	1.64692	3061.01	+0.590858	843.061	-0.118475
	0.525	0.663092	1.64676	3044.85	+0.059809	843.961	-0.011848
	0.475	0.665037	1.64671	3041.20	-0.061374	844.155	+0.011137
	0.25	0.673869	1.64645	3024.49	-0.609261	845.021	+0.113736
θ	0.0015	0.663979	1.64671	3043.12	+0.00295758	844.048	-0.001540
	0.0010	0.664055	1.64674	3043.04	+0.00032862	844.061	0
	0.0009	0.664072	1.64674	3043.02	-0.00032862	844.060	-0.000118
	0.0005	0.664149	1.64676	3042.94	-0.00295758	844.068	+0.008293
α	30	0.560445	1.62096	3169.12	+4.143570	818.423	-3.037460
	21	0.650840	1.64327	3058.88	+0.520862	840.675	-0.401156
	19	0.678193	1.65051	3026.17	-0.554053	847.718	+0.433263
	10	0.875008	1.70949	2800.21	-7.979550	902.795	+6.958500

(continued)

Table 1 (continued)

Parameters	Change in parameters	t_b^*	T^*	$Z^*(t_b, T)$	% Change in $Z^*(t_b, T)$	I_0^*	% Change in I_0^*
β	1.2	0.573359	1.65170	2112.30	−30.5856	1103.05	+30.6837
	0.84	0.655495	1.65166	2952.20	−2.98485	870.558	+3.13923
	0.76	0.672571	1.64053	3133.28	+2.96579	817.551	−3.14077
	0.4	0.749292	1.49471	3909.02	+28.4581	587.942	−30.3437
c_p	15	0.663996	1.64672	3043.11	+0.00262896	844.042	−0.002251
	10.5	0.664057	1.64674	3043.04	+0.00032862	844.06	−0.000118
	9.5	0.664071	1.64674	3043.02	−0.00032862	844.062	+0.000118
	5	0.664132	1.64676	3042.95	−0.00262896	844.079	+0.002132
c_b	6	0.698045	1.45320	3273.49	+7.573370	744.665	−11.7759
	4.2	0.668271	1.62147	3070.73	+0.910277	830.859	−1.56410
	3.8	0.659614	1.67388	3013.99	−0.954312	858.315	+1.68874
	2	0.602706	2.06274	2666.26	−12.38140	1070.99	+26.8854
c_s	12	0.695155	1.56485	3329.58	+9.416600	803.490	−4.806640
	8.4	0.667268	1.63878	3072.65	+0.973372	840.110	−0.468094
	7.6	0.660838	1.65464	3013.20	−0.980273	847.983	+0.464658
	4	0.63079	1.72353	2735.74	−10.09820	882.225	+4.521470
χ	0.15	0.664065	1.64674	3043.03	0	844.061	0
	0.105	0.664064	1.64674	3043.03	0	844.061	0
	0.095	0.664064	1.64674	3043.03	0	844.061	0
	0.05	0.664063	1.64674	3043.03	0	844.061	0
K_0	3750	0.724089	1.97673	3732.23	+22.6485	1035.10	+22.6333
	2625	0.671025	1.68385	3118.09	+2.46662	865.054	+2.48714
	2375	0.65682	1.60844	2966.23	−2.52380	822.524	−2.55159
	1250	0.57229	1.1858	2163.06	−28.9176	593.372	−29.7003

There are several ways to change the proposed model. For example, we may extend the time-proportional deterioration rate to a two-parameter Weibull distribution deterioration rate, a three-parameter Weibull distribution deterioration rate and gamma distribution deterioration rate. In addition, we could consider the demand as a trapezoidal function of time, product quality, stock-dependent demand, stochastic demand pattern, and others. Finally, we could generalize the model to allow quantity discounts and others.

Acknowledgements The authors would like to thank the editor and anonymous reviewers for their valuable comments and suggestions, which improved the presentation of the paper.

References

1. Annadurai, K.: An optimal replenishment policy for decaying items with shortages and salvage value. Int. J. Manag. Sci. Eng. Manag. **8**(1), 38–46 (2013)
2. Chakrabarti, T., Chaudhuri, K.S.: An EOQ model for deteriorating items with a linear trend in demand and shortages in all cycles. Int. J. Prod. Econ. **49**(3), 205–213 (1997)
3. Chang, H.J., Dye, C.Y.: An EOQ model for deteriorating items with time varying demand and partial backlogging. J. Oper. Res. Soc. **50**(11), 1176–1182 (1999)
4. Covert, R.P., Philip, G.C.: An EOQ model for items with Weibull distribution deterioration. AIIE Trans. **5**, 323–326 (1973)
5. Dave, U., Patel, L.K.: (T, Si) policy inventory model for deteriorating items with time proportional demand. J. Oper. Res. Soc., 137–142 (1981)
6. Ghare, P.M., Schrader, G.F.: A model for exponentially decaying inventory. J. Ind. Eng **14**(5), 238–243 (1963)
7. Ghosh, S.K., Chaudhuri, K.S.: An EOQ model with a quadratic demand, time-proportional deterioration and shortages in all cycles. Int. J. Syst. Sci. **37**(10), 663–672 (2006)
8. Goswami, A., Chaudhuri, K.S.: An EOQ model for deteriorating items with shortages and a linear trend in demand. J. Oper. Res. Soc. **42**, 1105–1110 (1991)
9. Harris, F.W.: How many parts to make at once. Factory Mag. Manag. **10**(135–136), 152 (1913)
10. Hollier, R.H., Mak, K.L.: Inventory replenishment policies for deteriorating items in a declining market. Int. J. Prod. Res. **21**(6), 813–826 (1983)
11. Jaggi, C.K., Aggarwal, S.P.: EOQ model for deteriorating items with salvage values. Bull. Pure Appl. Sci. **15**E (1), 67–71 (1996)
12. Khanra, S., Chaudhuri, K.S.: A note on an order level inventory model for a deteriorating item with time dependent quadratic demand. Comput. Oper. Res. **30**, 1901–1916 (2003)
13. Lin, C., Tan, B., Lee, W.C.: An EOQ model for deteriorating items with time-varying demand and shortages. Int. J. Syst. Sci. **31**(3), 391–400 (2000)
14. Mishra, U., Tripathy, C.K.: An inventory model for Weibull deteriorating items with salvage value. Int. J. Logistics Syst. Manag. **22**(1), 67–76 (2015)
15. Mishra, V.K., Singh, L.S.: Inventory model for ramp type demand, time dependent deteriorating items with salvage value and shortages. Int. J. Appl. Math. Stat. **23**(D11), 84–91 (2011)
16. Mishra, V.K., Singh, L.S., Kumar, R.: An inventory model for deteriorating items with time-dependent demand and time-varying holding cost under partial backlogging. J. Ind. Eng. Int. **9**, 1–5 (2013)
17. Ouyang, L.Y., Wu, K.S., Cheng, M.C.: An inventory model for deteriorating items with exponential declining demand and partial backlogging. Yugoslav J. Oper. Res. **15**, 277–288 (2005)
18. Philip, G.C.: A generalized EOQ model for items with Weibull distribution deterioration. AIIE Trans. **6**, 159–162 (1974)

19. Rajan, R.S., Uthayakumar, R.: Optimal pricing and replenishment policies for instantaneous deteriorating items with backlogging and trade credit under inflation. J. Ind. Eng. Int., 1–17 (2017)

20. Sachan, R.S.: On (T, Si) policy inventory model for deteriorating items with time proportional demand. J. Oper. Res. Soc. **35**(11), 1013–1019 (1984)

21. Sahoo, N.C., Singh, T., Sahoo, C.K.: An inventory model of deteriorating item for maximization of profit. J. Inf. Optim. Sci. **37**(1), 111–124 (2016)

22. Singh, T., Pattanayak, H.: An EOQ inventory model for deteriorating items with quadratic demand and partial backlogging. J. Orissa Math. Soc. **33**(2), 109–123 (2014)

23. Singh, T., Mishra, P.J., Pattanayak, H.: An optimal policy for deteriorating items with time-proportional deterioration rate and constant and time-dependent linear demand rate. J. Ind. Eng. Int. **13**(4), 455–463 (2017)

24. Wee, H.M.: Economic production lot size model for deteriorating items with partial backordering. Comput. Ind. Eng. **24**(3), 449–458, (1993)

25. Wilson, R.H.: A scientific routine for stock control. Harvard Bus. Rev. **13**(1), 116–128 (1934)

Inventory Control for Materials Management Functions—A Conceptual Study

Rashmi Ranjan Panigrahi and Duryodhan Jena

Abstract Inventory control is an important aspect of production system as well as business concern. Materials and inventory control is a priority start-up steps for evaluation of current state of your business. Inventory control plays a vital role in increasing productivity through material management function. In the present study, an attempt has been made to conceptualize and understands the significance of inventory control system in relation to material management function. The study reveals that the effectiveness of inventory control measures leads to better material management function.

Keywords Inventory control · Material management · Organizational productivity

1 Introduction

Inventory management or inventory control deals with proper management of ideal resources which fetch future value to the company. Material management function with proper inventory management/control brings strength to the company. Material management principle concerned with flow of materials into an organization, it includes purchasing, production, planning, scheduling, inventory control, etc.

For the survival and growth of an enterprise, it is very important to manage the material through proper inventory management planning.

Every organization needs proper tracking of stock towards smooth running of business activities. It maintains a link between production and distribution system of an organization.

R. R. Panigrahi (✉) · D. Jena
Institute of Business and Computer Studies (IBCS), Siksha 'O' Anusandhan (Deemed to be University), Khandagiri, Bhubaneswar, Odisha, India
e-mail: rashmipanigrahi090@gmail.com

D. Jena
e-mail: duryodhanjena@soa.ac.in

© Springer Nature Singapore Pte Ltd. 2020
S. Patnaik et al. (eds.), *New Paradigm in Decision Science and Management*,
Advances in Intelligent Systems and Computing 1030,
https://doi.org/10.1007/978-981-13-9330-3_17

Path Diagram of Inventory control towards Organizational Productivity

Fig. 1 Path diagram of inventory control towards organizational productivity

The investment in inventories constitutes play significant contribution in current assets (CA) and working capital (WC) in most of the company. The objective of inventory management is to ensure material should be available in sufficient quantity according to the requirement of the business and also need to minimize cost of investment in inventories. Definition of inventory, reasons, purpose and objectives of holding inventory have been studies by many authors such as Anichebe and Agu [1], Water [2], Telsang [3] and Ubani [4], because inventory plays significant role of success material management functioning organization. The requirement of management is to ensure proper management of materials by applying inventory control techniques should bring productivity (Fig. 1).

The above path diagram shows the path how Inventory control leads to organizational productivity through material management function.

By applying inventory management/control techniques of material management, organization brings productive of capital by reducing its cost. Proper techniques help in reducing blockage of capital for a long time period and improve in capital turnover ratios. As per the survey, it reveals that 65% of revenue is accounted due to techniques used in the operations.

2 Review of the Literature

2.1 Inventory Management

According to Chase et al. [5], in a manufacturing organization inventories is classified as raw materials, finished products, component parts, suppliers and work in process.

According to Donald Water view, the term stock consists of all goods and materials stored by an organization for its future use.

Inventory can be classified according to usage and entry point in operation.

Inventory management

- **Raw materials inventories (RMI)**
- **Bought outs components inventories (BOCI)**
- **Work-in-process inventories (WIPI)**
- **Finished goods inventories (FIGI)**
- **Maintenance, repairs, operating suppliers inventories (MROI).**

Raw materials inventories (RMI): RMI is physical/tangible resources used in the production process. The reasons of organization to hold or maintain RMI is providing smooth and uninterrupted flow of materials to the production unit. Proper quantity provides strength to production units to encash market opportunity. Inventory control managers should focus points in relation to RMI are:

1. It helps in placing order in bulk quantities of materials, save the investment.
2. It helps to meet the changes in production.
3. It helps to plan for buffer stock to protect the business from delay of inventories.

Bought outs components inventories (BOCI): it is that part of materials which is available for direct use. It is neither manufactured nor processed by organization, for example, tin packet of coconut oil.

Work-in-process inventories (WIPI): WIPI is the materials that are in the process of converting RMI into finished goods for purpose of creating market value. These are semi-finished inventory, which serves the purpose like,

1. It brings economic lot size (ELS).
2. It differentiates need of varieties of products.
3. It helps to reduce wastages.
4. It maintains uniformity in production irrespective of volume of sales.

Finished goods inventories (FIGI): These are good hold by the company for last time before sale. It serves as buffer stock for the company because of avoiding demand fluctuations. It also helps to reduce the market risk of the products. These types include:

1. It helps to provide adequate quantity to the customer.
2. It helps in promotional programme of the company.
3. This concept brings stability in the level of production.

Maintenance, repairs, operating suppliers inventories (MROI)

This type of inventories involves maintenance, repairs, operating supplies. It includes consumable which is a part of production process, for example, lubricants, pencil and biscuits. It deals with activities like:

1. It deals after sales services to customer.
2. It uses the product to its fullest extent.

Why we hold inventory

Inventories can be hold to meet the requirement of customer on a priorities basis , helps to meet fluctuation in demand, economy in procurement function and helps in reduction of transit cost and time. These are the following needs for holding inventories.

1. **Maintain Stability in production**: Fluctuation arises in the business due to seasonality, production schedule, etc. to address such situation we have to hold the adequate inventories.

2. **To avail price concession**: If the manufacturer wants to purchase the raw material with some price discount then he/she has to take a decision for bulk purchase of inventory and hold it until it completed.
3. **To manage demand in the period of replenishment**: This situation arises when business decision based on led time for procurement of raw materials, which may affect by the factors like nearness, gap of demand and supply, etc.
4. **To avoid loss of orders:** In this competitive business world, no firm wants to compromise in the missing delivery of product. To address this situation, firm has to hold the inventories 100%.

2.2 Material Management

Material management deals with coordinating functions for flow of material from one department to another department. It is based on proper planning and coordination of the concerned department.

The objectives of material management are to provide significant contribution in profitability by minimizing material cost phenomenon. It is only possible through capital investment optimization, support from capacity and personnel, constant provider of customer service.

Due to the gain of importance of material, it is highly required centralization of power towards planning, procuring, preserving, handling, usage and other related aspect of material management. According to Rajeev [6], poor material management practices were characterized by lack of an integrated approach, if there is a gap between physical stock and accounting systems, this study based on 40 SMEs Bangalore, India.

Functions of material management

The major material management are as follows

1. Material requirement planning.
2. Purchasing.
3. Storekeeping.
4. Inventory control.

Material management concept

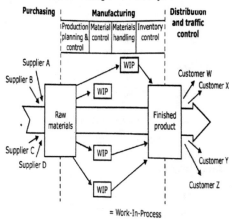

Source Bhat [7, p. 8]

Interface of inventory control with material management

Bhat [7] in the book material management, HPH publisher ISBN: 978-93-5024-806-5 has explained interface of IC with material management. In this article, we took the reference of the above and try to show the relationship of inventory control system with proper material management functions should bring organizations productivity. Inventory control techniques used to optimizing the total cost through proper management of materials. If we control total cost then it automatically includes ordering cost and inventory carrying cost. Inventory control techniques determines what quantity to be order, when order is to be placed with suppliers which is linked to material management, all these above things possible if we should have availability of adequate quantity materials for production and at the same point of time we also ensure for smooth production we have to manage two things, i.e.

a. keeping minimum inventory and
b. Helps to avoid stock-out situations.

Measurement of the effectiveness of inventory control

As per the modern concept, the measurement of performances of inventory management should be needed because it is a key

- To know the status of number of month's holding in store as compared to budget figure.
- To know the status of number of month's finished goods inventory as compared to budgeted sales.

Traditional concept helps us to measures the performances of inventory control (IC) for smooth working of material management.

Performances indicators of inventory control

Different formulas are used to calculate performances of inventory management for effective material management function.

Different ratios are as follows:

1. **Overall inventory turnover ratio:**

$$\frac{\text{Cost of goods sold}}{\text{Average inventory at cost}}$$

2. **Raw materials inventory ratio:**

$$\frac{\text{Annnual consumption of raw materials}}{\text{Average raw materials inventory}}$$

3. **Work-in-process inventory turnover ratio:**

$$\frac{\text{Cost of manufacturer}}{\text{Average WIP inventory at cost}}$$

4. **Finished goods inventory turnover ratio:**

$$\frac{\text{Cost of goods sold}}{\text{Average finished goods inventory at cost}}$$

5. **Week's inventory of raw material on hand:**

$$\frac{\text{Raw material inventory on hand}}{\text{Weekly consumption of raw material}}$$

6. **Week's raw material on order:**

$$\frac{\text{Raw materials on order}}{\text{Weekly consumption of raw materials}}$$

7. **Week's inventory of finished goods on hand:**

$$\frac{\text{Finished goods inventory}}{\text{Weekly sales of finished goods}}$$

8. **Average age of raw materials in inventory:**

$$\frac{\text{Average raw materials inventory at cost}}{\text{Average daily purchase of raw materials}}$$

9. **Average age of finished goods inventory**:

$$\frac{\text{Average finished goods inventory at cost}}{\text{Average cost of goods manufactured per day}}$$

10. **Out of stock index**:

$$\frac{\text{Number of times out of stock}}{\text{Number of times requisitioned}}$$

11. **Spare parts index**:

$$\frac{\text{Value of spare parts inventory}}{\text{Value of capital equipment}}$$

3 Conclusion

The present study we conceptualize the use of inventory control system in effective material management function towards organization productivity. The inventory planning is highly required in material management, so which is clearly found out in the study. On the basis of inventory planning (IP), organization should place purchase requisition and keep the stock in possession before the clearance of stock. Inventory control helps to maintain zero inventory. Thus, it is evident that proper adoption of inventory control tools and techniques brings effective material management for well organization productivity.

References

1. Anichebe, N.A., Agu, O.A.: Effect of inventory management on organisational effectiveness. Inf. Knowl. Manag. **3**(8), 92–100 (2013)
2. Waters, D.: A Practical Introduction to Management Science. Financial Times/Prentice Hall (1998)
3. Telsang, M.: Industrial Engineering and Production Management. S Chand and Company Ltd. (2010). ISBN 81-219-1173-5
4. Ubani, E.C.: Production and Operations Management: Concepts, Strategy and Applications (2012). ISBN 978-978-51063-0-5
5. Chase, B.R., Nicholas, A.J., Jacobs, F.R.: Production and operations managment. Manuf. Serv. **8** (2004)
6. Rajeev, N.: Inventory management in small and medium enterprises: a study of machine tool enterprises in Bangalore. Manag. Res. News. **31**(9), 659–669 (2008)
7. Bhat, S.K.: Material Management, p. 12. Himalaya Publishing House (2011)

ACO-Based Secure Routing Protocols in MANETs

Niranjan Panda and Binod Kumar Pattanayak

Abstract Applications of mobile ad hoc networks (MANETs) have gained increasing popularity among different areas of human life and significantly attracted the attention of the researchers in this field. Routing in a MANET plays the vital role in its real-world application. Further, routing in MANET is associated with a set of issues that need to be addressed before its implementation. The issues include satisfaction of specified quality of service (QoS) requirements and security issues. In this paper, we have addressed security issues as our point of focus. Again, security issues can be addressed using various algorithms and out of all, evolutionary technique-based algorithms appear to be more effective for the purpose. Ant colony optimization (ACO) is one of the most popular evolutionary algorithms for optimization. We have carried out an extensive review of ant colony optimization (ACO)-based secure MANET routing protocols and the contents of this paper can help researchers in the field of MANET to devise secure routing protocols, especially using ACO algorithm for addressing the security issues in a MANET.

Keywords MANET · Multicast routing protocols · Reactive routing · Proactive routing

1 Introduction

In recent years, a substantial evolution in the sphere of wireless communication in MANET has been noticed. Security now a day is one of the major problems that MANET faces today. MANET is based on the wireless links that suffer from contention with impacts of radio communication like noise, fading interference.

N. Panda (✉) · B. K. Pattanayak
Department of Computer Science and Engineering, Siksha 'O' Anusandhan Deemed to be University, Bhubaneswar, India
e-mail: niranjanpanda@soa.ac.in

B. K. Pattanayak
e-mail: binodpattanayak@soa.ac.in

© Springer Nature Singapore Pte Ltd. 2020 195
S. Patnaik et al. (eds.), *New Paradigm in Decision Science and Management*,
Advances in Intelligent Systems and Computing 1030,
https://doi.org/10.1007/978-981-13-9330-3_18

Also, the links possess less bandwidth in comparison with wired networks. The wireless network is a place for legitimate users as well as malicious attackers, with a lack of traffic monitoring and access control mechanism where MANET relies on the implicit trust relationship between the nodes involved to route packets. In MANET, packets are broadcasted leading to a risk of snooping or interfering. Due to the limitation of resource cryptographic schemes that require larger computations may fail to be implemented for the reason that it is easily accessible to the external attackers. Further, a compromised node may be used for launching an internal attack. Threats also increase when MANET packets are transferred over multiple paths instead of a single path and suffer from many attacks like hole attack impersonation or spoofing, denial of service, byzantine, rushing attacks, etc. [1]. Unauthorized and compromised nodes may collapse the network communication entirely or partially by advertising incorrect paths, dropping packets, or draining battery power of nodes. Therefore, sufficient security measures must be taken during the design of routing protocols for MANET keeping in mind that implementation of security mechanisms should be lightweight as it increases the delay and communication cost.

MANETs are designed in a data centric and application-specific manner where the network topology and requirements change regularly. Hence, due to the MANET vulnerabilities like unreliability of wireless link, dynamic topologies, implicit trust relationship between neighbors, lack of secure boundaries, threats from insider compromised nodes, unavailable centralized management facility, restricted power supply, and scalability routing becomes more challenging and requires the development of an adaptive, secure, and efficient routing algorithm to avoid a complete breakdown of the network [2].

The adaptive and dynamic nature of ACO-based algorithms makes it suitable to solve the complex problems during routing. This is a heuristic method as it builds a solution and its representation affects all ants present in the network. ACO provides better solutions than genetic algorithm (GA) and simulated annealing (SA) [3] for choosing a shortest path. The probabilistic movement made by the ants within the system enables them to investigate new paths as well as to re-investigate the earlier traversed paths. The quality of the pheromone accumulation governs the artificial ants to move toward the optimum path, and its dispersion enables the system to overlook the old information and keep away from the quick convergence to the suboptimal solution. This behavior of ACO motivated many of the researchers to design the protocols using ACO metaheuristics.

The rest of the paper is organized as follows. ACO metaheuristics are detailed in Sect. 2. Section 3 presents a discussion of various ACO-based secure routing protocols. A comparison and analysis of the secure routing protocols based on various parameters is done in Sect. 4. Section 5 concludes the paper with future work.

2   Ant Colony Optimization

Swarm intelligence techniques [4] are stimulated by the functioning of the social insects and animals and used for solving various real-life problems. Many techniques are proposed out of which ant colony optimization (ACO) [5] is a standout among the most mainstream and extensively utilized optimization technique in light of the behavior of the biological ants.

ACO concept was introduced in the early nineties, and since then it has become an area of interest for the researchers to solve optimization problems in various research fields. Nowadays, due to the availability of a significant amount of theoretical results, research in this area is becoming more vigorous facilitating further applications of ACO.

In ACO metaheuristics, artificial ants collaborate to find optimal solutions for arduous combinatorial optimization issues. To apply ACO algorithm to resolve a real-life optimization issue, the issue is represented first in the form of construction graph, the limitations of the issue ought to be characterized as the ants can just proceed onward the arcs which interface the nodes, each arc associated with the pheromone trails (a persistent memory about the ant forage process) and a heuristic value (a prior realization about the issue occurrence or run-time data given by different sources), ants pursue a probability decision percept to use the network as a function of neighborhood pheromone trails and the heuristic information. Additionally, it can be cognate to the ant's personal memory and the issue requirements.

The basic ACO algorithm usually can be described as the exchange of three basic procedures constructs ant's solutions, update pheromones, and daemon actions. Many variants of ACO algorithms are proposed in past out of which ant system (AS) [6], MAX–MIN ant system (MMAS) [7] and ant colony system (ACS) [8] are the widely used ones.

3   Security Aware ACO Routing Protocols

ACO-based routing protocols are categorized into various types like basic ACO routing protocols, QoS aware ACO routing protocols, energy aware ACO routing protocols, location aware ACO routing protocols, and security aware ACO routing protocols. But here in our study, we focus only on secure routing protocols. Here, we detail some secure routing protocols for MANET based on ACO routing strategy.

3.1 Secure Antnet Routing Algorithm for Scalable Ad Hoc Networks Using Elliptic Curve Cryptography

V. Vijayalakshmi and T. G. Palanivelu proposed a Secure Antnet Routing protocol based on ACO and Elliptical Curve Cryptography (ECC) [9] described as SAR-ECC [10] for cluster-based networks. In this approach, the trust value of each neighbor node is measured by every node in a cluster. The trust value of a neighbor node is measured as a growing function correlating with the likelihood of packet delivery ratio. The source node during route discovery uses the AntNet mechanism for establishing multiple routes. The source and destination nodes mutually authenticate each other using the concept of ECC. This mechanism estimates the trust value in view of the idea of uncertainty rather than the usual pheromone idea. Also trust value updating procedure and the benefits of combining cluster with ACO is not described.

The authors implemented and simulated the protocol using NS-2. The protocol is analyzed based on the packet forwarding performance, authentication cost ratio, trust estimation, and necessary packet rate versus speed. Results show that this algorithm possesses a high fault tolerance level for up to eighty percentages of adversary nodes per cluster with a sixty percentages of success rate and with a very low failure rate.

3.2 Secure Power Aware Ant Routing Algorithm

S. Mehfuz and M. N. Doja proposed a Secure Power Aware Ant Routing Algorithm named as SPA-ARA [11]. Initially, a source node when requires to start a data transmission toward the destination node and gets no route information available, then the reactive forward ants (FANTs) are sent affixed with a cryptographic message authentication code (MAC). Each intermediate node checks for the validity of MAC and finding it to be correct, checks for the trust value of the sender with the previously fixed threshold value. When both are found to be positive, then a secret key is built up between the two nodes using a two-party key establishment protocol. Otherwise the FANT is discarded or killed. Repeating the procedure at each intermediary node, at the point when the FANT achieves the destination node an analogous backward ant (BANT) is generated. The BANT travels in the reverse of the path travelled by the FANT and at each intermediary node the MAC affixed with the BANT is validated utilizing the beforehand created secret key by FANT. During its travel, also the BANT updates the pheromone table. On reaching the source node, the authentication of nodes is completed by the BANT using the MAC attached to it. Further, the data packets are sent using the route information present on the pheromone table and secured with MAC attached to it. This scheme may declare some genuine nodes as malicious as if a node when measures the trust value of a node, some packet loss due to congestion shows it to drop packets. Also in this paper, nothing has been stated about the overheads occurring due to cryptographic operations.

The authors simulated the SPA-ARA algorithm utilizing Scalable Wireless Ad hoc Network Simulator (SWANS) [12]. Performance is analyzed based on metrics like energy, successful routes found, number of packets dropped and destination location time with respect to pause time, number of malicious nodes, and compared with standard routing protocols like AODV, DSR, and ARA. Analysis of the result shows that SPA-ARA algorithm discovers secure and optimal routes with a proper distribution of energy across multiple paths, extending the longevity of network communication.

3.3 Swarm-Based Detection and Defence Technique

G. Indrani and K. Selvakumar proposed a technique for designing a multipath routing protocol which set up numerous paths connecting the source and destination node utilizing the SI behavior of ACO algorithm and uses the concept of trust values for securing the network by detecting the intrusions in the network. The protocol named as Swarm based intrusion Detection and Defence Technique (SBDT) [13]. Each node possessing a greater trust value, remaining bandwidth and energy, monitors all of its neighbor nodes and is termed as an active node (NA). During transmission of data, if any communicating neighbor node's trust value falls below the threshold, at that point the NA distinguishes it as a malicious node and informs the source node about it. Further, a revocation process may be initiated by the source node to defend against the detected malicious node. Here, the NAs are solely responsible for the novelty of the intrusion detection system (IDS). The authors of this proposal have not given a clear explanation about the updating and estimation of the trust value, whereas it plays the key role to implement these IDS. Also, no explanation is given about the situation when a NA is compromised and behaves maliciously.

SBDT is evaluated by the authors based on the simulation done using NS-2. Results are analyzed, and performance is measured comparing the proposed scheme with another existing algorithm capability-based secure communication mechanism (CAPMAN) [14] according to the simulation metrics like average packet delivery ratio, end-to-end delay and packet drop as for the expanding number of attackers and number of nodes. This routing protocol is found to detect the attacks and handle the packet drops efficiently.

3.4 Detection and Prevention of Blackhole Attack in MANET Using ACO Routing

Sowmya et al. have suggested an idea of detection and prevention of blackhole attack in MANET using ACO routing technique [15] which is named as DPBA-ACO by us in this chapter during further studies. An AntNet routing algorithm which has been discussed before in this chapter is used to discover optimal routes within the network during routing using this protocol. A threshold value is used for the detection of blackhole attack in MANET which can be updated dynamically over a settled brief time interim. Threshold values can be computed as a normal of the difference between sequence numbers present within the routing table and the sequence number present with the BANT. A node suspected to be a blackhole node if it forward a BANT packet with a higher sequence number than the threshold value. Once a node is detected as malicious, it is added into the blacklist and a control packet ALARM containing the blackhole nodes ID as a parameter is sent to the neighbors. By getting the ALARM packets, the network nodes get aware of the blackhole node and put it into their blacklist prohibiting the communication with that node. Hence, the blackhole node gets isolated from the network.

No simulations have been done by the authors to check the effectiveness and examine the working of the routing protocol in correlation with the other routing protocols.

3.5 Autonomous Bio-inspired Public Key Management Protocol

P. Memarmosherefi et al. proposed an approach named as autonomous bio-inspired public key management (ABPKM) [16] for defending attacks those occur in the network layer of the OSI model. This approach mainly combines the features of ACO algorithm to self-organized public key management schemes defending against the misbehavior of nodes and maintaining the correctness of keys. Initially, every node in the network issues a public key to its adjacent nodes along with initializes its trust value. During route discovery, a FANT is transmitted by the source towards the destination which keeps up a chain of public keys encountered on its path. On reaching the destination, the FANT is killed and a BANT is generated to traverse the path of FANT in reverse direction carrying the public key chain to the source. The source node authenticates those public keys and the chain with most noteworthy trust value is selected by the source for further data transmission. Also, the trust value of its neighbors is updated by every node starting from the source node hop-by hop along the chain depending upon the result of authentication at the source node. In [17] the proposal again is re-improved by the addition of an agglomerative hierarchical clustering algorithm for providing more security toward the attacks like Sybil attack. The node features like the total number of groups it belongs to, distance from the destination, its social degree and trust value for chains it participates in are gathered by the source node for clustering purpose. Also, a new parameter aggression value is introduced in this model for evaluating the level of danger within the network. In light of the different experimental and aggression values, the authors conclude that the mobile nodes can detect the attackers more efficiently in comparison with the static nodes.

3.6   ANT-Based Trustworthy Routing

S. Sridhar suggested an ANT-based trustworthy routing in mobile ad hoc networks spotlighting quality of service [18] based on the trust value calculation for every network node and selecting the node with higher trust value in comparison with the threshold value for routing. From this onward, we use the name ANTBTR to identify this routing protocol. ANTBTR protocol provides an optimal and secure communication and also provides an improvement for the QoS parameters like end-to-end delay and packet delivery ratio. Before transmission, this protocol selects the route and nodes participating in the route and checks their trust value against the threshold value. The nodes with trust value not as much as the threshold value is made out of the route, along with alternate nodes are selected to fulfill the purpose. Trust value calculation is made based on a model proposed by the author. The model depends on various parameters of RREQ, RREP, and Data. The parameter based on RREQ is a query request success rate (QRS) that represents the ratio of the neighbor nodes that accepted the RREQ packets sent by the sender with respect to the total number of neighbor nodes, query request failure rate (QRF), is the complement of QRS which means the number of neighbor nodes that did not accept the RREQ with respect to the total number of neighbor nodes. Similarly RREP parameters, query reply success rate (QPS) and query reply failure rate (QPF) describe the rate of success and failure of RREPs received from the nodes by the sender, respectively. Data transmission parameters, data success rate (QDS), and data failure rate (QDF) represent the total amount of data successfully transmitted to the receiver and the data dropped during the transmission with respect to the total amount of data sent by the sender, respectively. Also, the values of the parameters are normalized considering the various constraints that are responsible for the loss of data during transmission. Route discovery takes place between the nodes using the general ACO routing mechanism and the pheromones are updated after completion of one iteration of ants. The detailed mechanism of route discovery and path maintenance is not described by the authors.

The protocol ANTBTR is evaluated using NS-2 simulator for the metrics delay and packet delivery ratio and is found to be better in correlation with the traditional AODV routing protocol. The protocol concentrated only on trust mechanism as a trust-based routing protocol, but no clear explanation is given about how the other factors like overhead and throughput are affected not presented.

3.7   Pair-Wise Key Agreement and Hop-by-Hop Authentication Protocol

K. Shanthi and D. Murugan have suggested a pair-wise key agreement and hop-by-hop authentication protocol (PKAHAP) [19] to provide secure communication in MANET. This protocol uses one round ID-based authenticated group key agreement protocol (IDAGKA) [20] at each node for authenticating the data packets. Basically, the proposed swarm-based authenticated routing scheme consists of two types of ant agents such as FANT and BANT. Initially to start a transmission, the source node selects the routes to the destination based on the ACO technique. Source initiates the FANTs attached with a threshold trust value and sends them to the neighbors. The threshold value is compared with the neighbor node's trust value and the neighbor node is updated in the routing table if it is found to be trustworthy, else finding the trust value of the node to be not as much as the threshold value, it gets eliminated from the routing table. This process repeats until the FANT reaches the destination. On achieving the destination, the FANT changed over into BANT and transmitted in the reverse direction on the path traveled by the FANT toward the source. During its travel BANT updates the pheromone values across the nodes and when it achieves the source, in light of the information contained by it, source decides the most trusted path to communicate. During data transmission, pair-wise keys are set up between every set of two nodes to provide security at each instance. To make the communication more secure, session keys are generated implicitly and used for encryption of the message transmitted during the communications between two nodes which avoids the situation of compromised or impersonated keys.

Authors simulated their proposed routing protocol using NS-2 for various performance metrics like delay, packet delivery ratio, packet drop, control overhead and throughput. The results are compared with the lightweight hop-by-hop authentication protocol (LHAP) [21] and are found to be extremely better, in sense of delay and routing overhead with variable amount of nodes. The route maintenance and link failure mechanisms are not clearly explained in this scheme.

4 Comparison and Analysis of Security Aware ACO Routing Protocols

The ACO-based routing protocols that we have discussed before are compared in this section based on various parameters and represented in a tabular form in Table 1.

5 Conclusion and Future Work

In this paper, security aware ACO routing protocols for MANET have been discussed in detail and analyzed based on their working principles, prospective, and consequences. In our discussion, we have concentrated only on the security feature as a major concern during routing protocol design, whereas other factors like mobility, reliability and QoS may also be considered. Comparison and analysis of the reviewed routing protocols have been presented, which shows that ACO-based routing protocols perform better in comparison with the traditional routing protocols. Finally, it can be concluded that the present MANET scenario requires a hybrid,

Table 1 Comparison of various ACO routing protocols

Year	Authors	Protocol	Routing method	Ant agents used	Trans method	Techniques used	Simulation tools used	Protocols compared	Performance metrics used
2007	Vijayalakshmi and Palanivelu	SAR-ECC	Reactive	FANT, BANT	Unicast	Trust value in a clustered approach	NS-2	Not compared	Authentication cost, stability, packet rate
2008	Mehfuz and Doja	SPA-ARA	Reactive	FANT, BANT	Both	HMAC and two-party key establishment protocol for ensuring security and MRB to perceive energy efficiency	SWANS	AODV, DSR, ARA	Dest loc time, packet dropped, energy standard deviation, successful routes
2012	Indirani and Selvakumar	SBDT	Reactive	FANT, BANT	Not specified	Trust values used for securing the network by detecting the intrusions	NS-2	CAPMAN	False detection accuracy, Packet drop, delay, PDR
2012	Sowmya et al.	DPBA-ACO	Reactive	FANT, BANT	Not specified	BANT sequence number and a dynamically updated threshold value	Not simulated	Not compared	No matrices

(continued)

Table 1 (continued)

Year	Authors	Protocol	Routing method	Ant agents used	Trans method	Techniques used	Simulation tools used	Protocols compared	Performance metrics used
2013	Memarmoshrefi et al.	ABPKM	Reactive	FANT, BANT	Unicast	Self-organized public key management schemes to provide security against attacks	QualNet	Not compared	Packet drop, delay
2015	S. Sridhar	ANTBTR	Reactive	RREQ, RREP	Broadcast	Trust-based mechanism	NS-2	AODV	Delay, PDR
2017	K. Shanthi and D. Murugan	PKAHAP	Reactive	FANT, BANT	Broadcast	One round ID-based authenticated group key agreement protocol (IDAGKA)	NS-2	Not compared	Throughput, delay, packet drop, control overhead, PDR

robust, efficient, dynamic, and secure multipath routing algorithm to overcome the drawbacks of existing routing protocols. The contents of this paper provide the basic idea about the ACO-based secure routing techniques and may become useful to the researchers for devising novel secure routing protocols.

References

1. Panda, N., Pattanayak, B.K.: Defense against co-operative black-hole attack and gray-hole attack in MANET. Int. J. Eng. Technol. 7(3.4), 84–89 (2018)
2. Panda, P., Gadnayak, K.K., Panda, N.: MANET attacks and their countermeasures: a survey. Int. J. Comput. Sci. Mobile Comput. (IJCSMC) 2(11), 319–330 (2013)
3. Alhamdy, S.A.S., Noudehi, A.N., Majdara, M.: Solving traveling salesman problem (TSP) using ants colony (ACO) algorithm and comparing with tabu search, simulated annealing and genetic algorithm. J. Appl. Sci. Res. (2012)
4. Dorigo, M., et al. (eds.): Ant Colony Optimization and Swarm Intelligence: 6th International Conference, ANTS 2008, Brussels, Belgium, September 22–24, 2008, Proceedings, vol. 5217. Springer (2008)
5. Dorigo, M., Blum, C.: Ant colony optimization theory: a survey. Theoret. Comput. Sci. 344(2–3), 243–278 (2005)
6. Dorigo, M., Maniezzo, V., Colorni, A.: Ant system: optimization by a colony of cooperating agents. IEEE Trans. Syst. Man Cybern. B (Cybern.) 26(1), 29–41 (1996)
7. Stützle, T., Hoos, H.H.: MAX–MIN ant system. Future Gener. Comput. Syst. 16(8), 889–914 (2000)
8. Dorigo, M., Gambardella, L.M.: Ant colony system: a cooperative learning approach to the traveling salesman problem. IEEE Trans. Evol. Comput. 1(1), 53–66 (1997)
9. Kapoor, V., Abraham, V.S., Singh, R.: Elliptic curve cryptography. Ubiquity 2008(5), 7:1–7:8 (2008)
10. Vijayalakshmi, V., Palanivelu, T.: Secure antnet routing algorithm for scalable adhoc networks using elliptic curve cryptography. J. Comput. Sci. 3(12), 939–943 (2007)
11. Mehfuz, S., Doja, M.N.: Swarm intelligent power-aware detection of unauthorized and compromised nodes in MANETs. J. Artif. Evol. Appl. 2008 (2008)
12. Barr, R.: Swans-Scalable Wireless Ad Hoc Network Simulator (2004)
13. Indirani, G., Selvakumar, K.: Swarm based detection and defense technique for malicious attacks in mobile ad hoc networks. Int. J. Comput. Appl. 50(19), 1–6 (2012)
14. Jia, Q., Sun, K., Stavrou, A.: Capman: capability-based defense against multi-path denial of service (dos) attacks in manet. In: 2011 Proceedings of 20th International Conference on Computer Communications and Networks (ICCCN). IEEE (2011)
15. Sowmya, K., Rakesh, T., Hudedagaddi, D.P.: Detection and prevention of blackhole attack in MANET using ACO. Int. J. Comput. Sci. Netw. Secur. 12(5), 21 (2012)
16. Memarmoshrefi, P., Zhang, H., Hogrefe, D.: Investigation of a bioinspired security mechanism in mobile ad hoc networks. In: Proceedings of the WiMob, pp. 709–716 (2013)
17. Memarmoshrefi, P., Zhang, H., Hogrefe, D.: Social insect-based sybil attack detection in mobile ad-hoc networks. In: Proceedings of the 8th International Conference on Bioinspired Information and Communications Technologies, pp. 141–148 (2014)
18. Sridhar, S., Baskaran, R.: Ant based trustworthy routing in mobile ad hoc networks spotlighting quality of service. Am. J. Comput. Sci. Inf. Technol. (AJCSIT), 3(1), 64–73 (2015)

19. Shanthi, K., Murugan, D.: Pair-wise key agreement and hop-by-hop authentication protocol for MANET. Wireless Netw. **23**(4), 1025–1033 (2017)
20. Reddy, K.C., Nalla, D.: Identity based authenticated group key agreement protocol. In: International Conference on Cryptology in India. Springer, Berlin (2002)
21. Zhu, S., et al.: LHAP: a lightweight hop-by-hop authentication protocol for ad-hoc networks. In: Proceedings of the 23rd International Conference on Distributed Computing Systems Workshops, 2003. IEEE (2003)

An Approach Toward Integration of Big Data into Decision Making Process

Bindu Rani and Shri Kant

Abstract In today's era of big data, large volume of data is generated from different sources having different qualities. It has set a path of rapid change and new discoveries in business decision making. Business organizations either big or small are yearning valuable and accurate information in decision making process. To achieve this, organizations are analyzing internal data in association with external data and producing big data value chain. Due to rapid growth of such data, efficient platform and appropriate data mining algorithms need to be developed extracting value from these datasets. This paper is a brief survey of approaches used for big data analytics in decision making process with current challenges and frameworks developed. Based on that, we are proposing an approach to increase the efficiency of decision making process in big data analytics. Some future research directions are also proposed to translate knowledge into improved decision making and performance.

Keywords Big data · Data analytics · Data mining · Decision making

1 Introduction

Over the years, data is increasing at a rapid pace, According to Industrial Development Corporation (IDC) and Egan, Marino Corporation (EMC) Corporation, in 2020 rate of generation of data will be 44 times more in comparison of year 2009 and will grow at rate of 50–60% yearly [1]. This large volume extensively varied with high-velocity data is called as big data. Recently, industries are keen to discover new values, gain deeper insights by exploiting hidden values of big data and to devise

B. Rani
Department of Computer Science and Engineering, School of Engineering and Technology,
Sharda University, Greater Noida, India
e-mail: bindu.var17@gmail.com

S. Kant (✉)
Research and Technology Development Centre, Sharda University, Greater Noida, India
e-mail: shrikant.ojha@gmail.com

© Springer Nature Singapore Pte Ltd. 2020
S. Patnaik et al. (eds.), *New Paradigm in Decision Science and Management*,
Advances in Intelligent Systems and Computing 1030,
https://doi.org/10.1007/978-981-13-9330-3_19

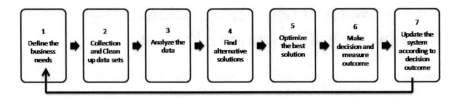

Fig. 1 Steps for decision making process in business organization

intelligent decisions. According to data scientist, decisions should be data-driven decision on the basis of evidence rather than intuitions. But they need to use data in a structured format to improve their decision making process. To achieve data in structure format, first data is collected and stored, some methods are applied to perform analysis on data and transform the results into intelligent decisions, and then decisions are converted into proactive outcome. There should be appropriate environment incorporating a data warehouse to store data, a processor for analysis and making intelligence and a user-friendly interface. Figure 1 shows steps for decision making process in any business organization.

However, data has become a new factor for production, but it presents two challenges for industries. First, challenge is to adopt revolution in collection and measurement of data then implementation of new technologies by business leaders. Second, the industries as a whole must also comply with new tools for big data analytics. Hence, it is necessary to understand the incorporation of big data tools and technologies with decision making process to produce value for industries.

In a survey [2], authors state that in big data technology, USA is much ahead than Europe and Europe is not exploiting the real value of big data across its industrial sectors. USA has paid and continuously paying great efforts to big data. Obama Administration announced a USD 200 million investment to launch the "Big Data Research and Development Plan," in March 2012, an initiative of second major scientific and technological development after the "Information Highway" initiative in 1993. Hence, integration of big data analytics into decision making process can generate profitable business values, lead the decision itself and can generate alternatives or predict the results adequately [3].

The aim of this paper is to provide an overview on usability of big data in decision making process. Section 2 covers background of topic, Sect. 3 discusses the challenges faced during big data analytics in the context of decision making. In Sect. 4, an integrated framework is discussed and an approach to increase the efficiency of model is proposed. Conclusion and future directions are drawn in Sect. 5.

2 Background

Big data is a ubiquitous term that has transformed science, engineering, health care, business and ultimately society itself. Besides volume, big data has too complex structure to be handled by traditional processing techniques.

McKinsey defines big data as "the datasets whose size is beyond the ability of typical database software tools to capture, store, manage, and analyze" [4, 5]. This definition makes McKinsey as one of the most important players usher big data in the next frontier for innovation, competition and productivity [6]. Gartner as another global consulting company defines big data as the "high-volume, high-velocity and high variety information assets that demand cost-effective, innovative forms of information processing for enhanced insight and decision making" [7].

Elgendy et al. [8, 9] mentioned big data is the data that cannot be managed and processed by traditional methods not only due to size of data to be processed by standalone machine but also most traditional processing tools and methods cannot be directly applied to big data processing because big data processing is distributed processing.

In 2001, initially Laney defined three characteristics of big data as 3Ds (data volume, data variety and data velocity) [10]. Later these 3Ds have been changed into 3Vs: volume (large size), variety (data in different formats) and velocity (speed at which data is generated). Then 3Vs extended into 4Vs by adding the characteristic, veracity (data coming from different sources and in different forms is correct or not). 4Vs are extended to 5Vs by adding the characteristic, value (every data can generate value). Then finally adding 5 more Vs to big data as validity (validity of data for intended purpose), variability (inconsistency in data), venue (distribution of heterogeneous data from multiple sources), vocabulary (schema, data models and semantics) and vagueness (ambiguity in the meaning of big data). These 10Vs are 10 big challenges of big data.

One essential feature is value, somewhere in data or extracting value from data. Big data value chain is a process to find value or hidden information from big data. In literature, big data value chain has been defined in terms of different phases, but there are four main phases for big data value chain: data generation, data acquisition, data storage and data analysis [11].

In consideration with decision making support system in big data analytics, data is converted into information, information into knowledge and then knowledge into wisdom to make intelligent decision. Decision support system is a media to process, analyze, share and visualize the important information to improve organizational knowledge. Hence, data visualization/interpretation and decision making are necessary but distinct steps from analysis in big data value chain [9].

Decision science supports decisions at every phase of big data processing as shown in Fig. 2.

In the literature, It is found that there are four essential means to strengthen the decision making process with big data analytics

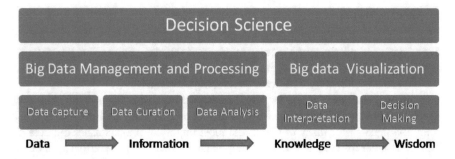

Fig. 2 Big data value chain

1. Information should be transformed into transparent and usable form with high frequency.
2. As more data is in digital form, organization should collect more accurate and detailed information to boost performance and for better management decisions.
3. Big data allows segmentation of customers and therefore much more precisely tailored products or services.
4. Decision making depends on data analysis. Appropriate analytics can improve decision making process.

3 Challenges During Big Data Analytics for Decision Making

Most organizations have different goals for adopting big data analytics. But the primary goal is to enhance customer satisfaction and other goals are reduction in cost, better future marketing, etc. The usage of big data matters over sector to sector due to the variation in opportunities and challenges. According to big data executive survey—2017, telecom, financial/banking and media sectors have invested more in data initiatives [12].

Data scientists are dealing with various challenges to extract knowledge from information. These challenges occur at every step of big data value chain such as data acquisition, recording, extraction, cleaning, management, analysis and visualization [13, 14].

Analyst and deployment experts also face several challenges in implementation of big data technologies called as technological challenges. Table 1 shows challenges faced during big data analysis for decision making.

Apart from all the above challenges, one big challenge is the lack of talented people in organizations. This is big challenge to have shortage of people who have profound knowledge about analytical skills and use of analytics in effective decision making.

Table 1 Challenges and issues

Challenges	Issues
Big data management	Collection, storing, integration and provenance of data with different standard and from different datasets. Managing the data properly to be usable, trustworthy
Big data cleaning	Minimization of complexity of big data by cleaning noise, errors and incomplete data
Big data aggregation	Synchronization of external data sources and distributed big data platforms with the internal infrastructures of an organization
Big data imbalancing	Classification of imbalanced datasets
Big data redundancy, reduction and compression	Selection of appropriate data redundancy, reduction and compression algorithms to decrease the cost of system
Big data analytics	Need of advanced analysis algorithms to extract valuable knowledge and to find hidden patterns
Big data analytics bottleneck on data analysis algorithms	Synchronization between different analysis algorithms executing parallel
Big data reporting	Displaying statistical data in the form of values in easily understandable manner
Big data communication cost	Since big data analytics is based on distributed and parallel processing, reduction in communication cost is major issue
Big data scalability	Variability of data streams can bring unpredictable changes called drift. This affects the accuracy in classification of data trained from past history. Therefore, data mining methods should be scalable to handle drifts
Big data privacy	Leak of personal privacy data during storage, transmission and usage, even if acquired with the permission of users
Big data safety	Authentication, validation and access control to data

Hence to deal with all big data challenges, there is need to exploit efficient, adaptable and flexible technologies in advanced big data analytics to efficiently manage vast amount of data despite the type of data format as textual or multimedia content.

4 Integrated Framework of Big Data Analytics with Decision Making Process

In the view of decision making, the process of big data analytics is interpreted in Fig. 3. In spite of several challenges, different tools and technologies are being used to support decision making for each phase of big data processing. Techniques are batch processing, real-time processing and hybrid processing techniques [15]. Some tools are mathematical tools, optimization tools, visualization tools, cloud computing, etc. The challenge is to map these tools and technologies with big data analysis. By integration of big data paradigms and decision making tools and techniques, intelligent decisions can be achieved. Researchers are constantly finding how big data analytical tools must be integrated with decision making process to find more valuable insight.

On the basis of decision science methodology, a Big-Data Analytics and Decision framework (B-DAD) was designed to map big data tools, architectures and analytics to different decision making process [8] as shown in Fig. 4. The first phase of decision making process is intelligent phase to identify sources of big data and to collect data from various sources. Tools can range from traditional database management systems to distributed file systems as Hadoop Distributed File System (HDFS) such as Cassandra and NoSQL (non-relational database). After acquiring and storing, data is organized, transformed and processed. This is achieved using several techniques as middleware technique, extract, transform, load (ETL) technique and data warehousing technique such as Hadoop and MapReduce. The second phase is design phase used to model planning, data analytics and analyzing. Traditional data mining such as classification, clustering, regression techniques, association rules and advanced analytical techniques such as artificial intelligence and machine learning can be used in design phase. Several analytical tools can be used in this step such as

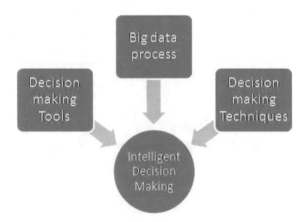

Fig. 3 Integration of big data process with decision making tools and techniques

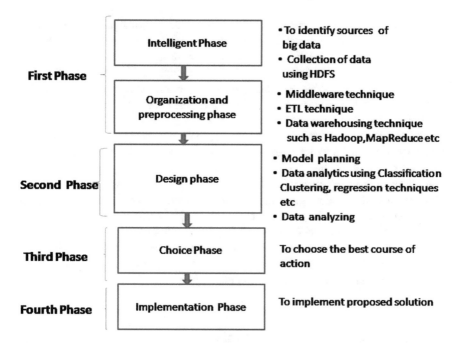

Fig. 4 Big data analytics and decisions framework

(high-performance analytic appliance) HANA, Greenplum and Aster. In analyzing step, output of previous step and result of analytics is analyzed.

Next phase is choice phase to choose the best or most appropriate course of action. Last phase is implementation phase to implement the proposed solution. Hence, big data tools and technologies are used to monitor the results of decision in addition to provide real-time feedback on the outcomes of implementation.

4.1 Proposed Approach

In any organization, data can be used either to support managers or to automate decision making process. According to the survey, 58% on average use big data for decision support and 29% for decision automation. Hence, the job of increasing the efficiency to automate decision making process is far from over [16]. In support of this, we are proposing an approach to increase the performance of decision making process.

Storage of large volume of data is not a problem but how to analyze data is a challenging task. Several techniques have been proposed to data analysis, but clustering is one of the well-known data mining techniques used in design phase of decision making process [17]. It is used to analyze data and to understand the "new" input unlabeled data using unsupervised process. Clustering analysis is broadly being used in the field of market research, health care, image processing, social media, etc. The objective is to classify the data into homogeneous groups such that same groups consist of similar types of data while different groups consist of dissimilar types of data as much as possible. Similarity measure parameters are used to group data points. This similarity measure is maximized for the data points in same group and minimized for data points in different groups.

To analyze real-life data, no unique algorithm for clustering can be applied. Scaling, correct parameterization, parallelization and cluster validity are some problems in using clustering techniques. In view of all aforementioned problems, different algorithms have been proposed and implemented. Clustering algorithms have been classified into six categories, partitioning, hierarchical, density-based, model-based, grid-based and evolutionary algorithms, but these algorithms still face technological challenges that multiply with the complexity of algorithms.

Our future research work will be in the direction of designing efficient clustering algorithm to enhance the performance of big data analysis and to get better insight into big data value for effective decision making process.

5　Conclusion and Future Aspects

Big data analytics is revolutionizing the approaches of organizations for using sophisticated information technologies to gain insight knowledge to take proactive decisions. This data-driven approach is remarkable, as the data is being generated exponentially via various living (normal users) and non-living media (sensors, web media, etc.).

A necessary step in this field is to find new sources of data to improve decision making capabilities. With the advent of big data, transformations are not only in the field of management and technology, but cultural and learning environment are also being transformed into organizations.

As in this mobility world, most of the data is user-generated data outcome of user decisions, actions and interactions, there is a need to design a framework-decision learning to integrate learning with strategic decision making. In literature, existing learning approaches did not focus on user's interaction and decisions. It can be useful in analyzing user's behaviors in a wide range of domains, from social networking sites like Facebook or Twitter to crowdsourcing sites like Amazon Mechanical Turk or Topcoder to online question-and-answer (Q&A) sites like Quora or Stack Overflow.

In the future, every organization will use big data analytics efficiently and effectively by providing solutions for various challenges occurred in analysis of big data. To compete in this world, organizations have to prepare themselves to deal with high technology standards in big data-driven decision making era.

References

1. Intel: Big Data Analaytics. http://www.intel.com/content/dam/www/public/us/en/documents/reports/data-insightspeer-research-report.pdf (2012)
2. Cavanillas, J.M., Curry, E., Wahlster, W.: New Horizons for a Data-Driven Economy a Roadmap for Usage and Exploitation of Big Data in Europe. Springer Open
3. Chen, H., Chiang, R.H.L., Storey, V.C.: Business intelligence and analytics: from big data to big impact. MIS Q. **36**(4), 1165–1188 (2012)
4. Chen, C.P., Zhang, C.Y.: Data-intensive applications, challenges, techniques and technologies: a survey on big data. Inf. Sci. **275**, 314–347 (2014)
5. Iqbal, M., Kazmi., S.H.A., Manzoor, A., Soomrani, A.R.: A study of big data for business growth in SMEs: opportunities & challenges. In: International Conference on Computing, Mathematics and Engineering Technologies—iCoMET, IEEE (2018). 978-1-5386-1370-2/18/$31.00
6. McKinsey: Big data: the next frontier for innovation, competition, and productivity. [Online] (2011). Available http://www.mckinsey.com/business-functions/business-technology/our-insights/big-data-the-next-frontier-for-innovation
7. Furht, B., Villanustre, F.: Introduction to big data. In: Big Data technologies and Application, pp. 3–11. Springer International Publishing, Cham (2016)
8. Elgendy, N., Elragal, A.: Big data analytics in support of decision making proces. Procedia Comput. Sci. **100**, 1071–1084 (2016)
9. Renu, R.S., Mocko, G., Koneru, A.: Use of big data and knowledge discovery to create data backbones for decision support systems. Procedia Comput. Sci. **20**, 446–453 (2013)
10. Khan, N., Yaqoob, I., Abaker, I., Hashem, T.: Big data: survey, technologies, opportunities, and challenges. Sci. World J. (2014). (Hindawi Publishing Corporation)
11. Chen, M., Mao, S., Liu, Y.: Big data: a survey. Mob. Netw. Appl. **19**(2), 171–209 (2014)
12. Oussous, A., Benjelloun, F.-Z., Lahcen, A.A., Belfkih, S.: Big data technologies: a survey. J. King Saud Univ. Comput. Inf. Sci. (2017). http://dx.doi.org/10.1016/j.jksuci.2017.06.001
13. Gartner: Big data [Online]. Available: http://www.gartner.com/it-glossary/big-data/ (2016)
14. Laney, D.: 3-D Data Management: Controlling Data Volume, Velocity and Variety. Application Delivery Strategies by META Group Inc. Retrieved from (2001). http://blogs.gartner.com/doug-laney/files/2012/01/ad949-3D-Data-Management-Controlling-Data-Volume-Velocity-and-Variety.pdf
15. Hey, A.J., Tansley, S., Tolle, K.M.: The Fourth Paradigm: Data-Intensive Scientific Discovery, vol. 13. Microsoft Research, Redmond, WA (2009)
16. Kościelniak, H., Puto, A.: Big data in decision making processes of enterprises. In: International conference on communication, management and information technology (ICCMIT) (2015)
17. Fahad, A., Alshatri, N., Tari, Z., Alamri, A., Khalil, I., Zomaya, A.Y., Bouras, A.: A survey of clustering algorithms for big data: taxonomy and empirical analysis. IEEE Trans. Emerg. Top. Comput. **2**(3), 267–279 (2014)

Big Data Analysis and Classification of Biomedical Signal Using Random Forest Algorithm

Saumendra Kumar Mohapatra and Mihir Narayan Mohanty

Abstract The healthcare industries generate a huge amount of data due to computer-aided diagnosis system. In this paper authors have analyzed these huge amounts of data for the classification of the ECG signals using random forest algorithm. A four-step classification model is designed in this work. First, the raw ECG signals are filtered using the median filter and a 12 order low-pass filter. Then the features from the clean ECG signals are extracted using wavelet transform. These wavelet features are then taken for the classification using random forest algorithm. The proposed random forest classification model is giving around 87% classification accuracy which is quite good. This method can be applied in the medical sector for early detection and better diagnosis of any disease.

Keywords Big data · Data mining · Machine learning · ECG · Wavelet · Filtering · Low-pass filter · Ensemble learning · Decision tree · Random forest

1 Introduction

Either structured or unstructured data of large amount refers to big data, and its analysis is required for specific processing. Its key features are specified as 3Vs: (i) volume, (ii) velocity, and (iii) variety. Here, volume refers to the amount of data and velocity means that the data is created rapidly whereas variety refers that various types of data are created. These high volumes of data can be generated from different

S. K. Mohapatra
Department of Computer Science and Engineering, ITER, Siksha 'O' Anusandhan (Deemed to be University), Bhubaneswar, India
e-mail: saumendra.mohapatra.27@gmail.com

M. N. Mohanty (✉)
Department of Electronics and Communication Engineering, ITER, Siksha 'O' Anusandhan (Deemed to be University), Bhubaneswar, India
e-mail: mihirmohanty@soa.ac.in

© Springer Nature Singapore Pte Ltd. 2020   217
S. Patnaik et al. (eds.), *New Paradigm in Decision Science and Management*,
Advances in Intelligent Systems and Computing 1030,
https://doi.org/10.1007/978-981-13-9330-3_20

sources such as educational institutions, online marketing agencies, banking sector, medicals, and social networking sites [1, 2].

The healthcare industry generates a large volume of data due to the computer-aided diagnosis systems. It is one of the complicated task to manage these high volumes and complex electronic health datasets. Discovering the useful information from these large datasets is a challenge. Big data analytics system gives a complete knowledge discovering platform for the large volume of data generated from the medical industry [3, 4]. It enables the analysis of big amount of datasets collected from different patients, identifies the correlations and cluster among the datasets, and it also creates a analytical model by using different data mining techniques. In healthcare sector, big data analysis integrates multiple scientific areas such as medical imaging, biomedical signal processing, health informatics, bio informatics, and sensor informatics [5, 6].

In this work, we have done the analysis and classification task for the ECG signals. The ECG arrhythmia signals are classified using random forest algorithm with wavelet features. Total five arrhythmia classes are classified using this ensemble learning technique.

Different techniques have been applied by the researcher for getting a better accuracy result, and some of them are cited. In [7], authors have introduced a data resampling approach for arrhythmia detection. They have taken random forest classifier for their classification purpose. This random forest classifier was also used by other researchers for the classification of different disease. The distance of this classifier was also modified using non-parametric method [8, 9]. Congestive heart failure, diabetes, and other cardiac approach were also detected using this ensemble classifier. Some authors were also compared this method with other method like KNN, SVM, decision tree, and neural network [10]. Random forest algorithm was viewed as a better classifier because of its incredible execution. To discover extra methodology for rising the variety of trees is an exciting task. This technique was also presented in big data approach. Out of bag error was introduced for the analysis of big data problem [11, 12].

Whatever is left of the paper is sorted out as following way: Proposed strategy is portrayed in Sect. 2. The result obtained by our work is presented in Sect. 3, and Sect. 4 concludes the work.

2 Proposed Methodology

A four-step classification model is developed and is shown in Fig. 1. The data may be collected from the patients directly, or similar data can be collected from different databases.

The ECG data is taken from the MIT-BIH Arrhythmia Database. It is an open access database in which different ECG signals can be freely accessed [13]. It contains

Fig. 1 Proposed work

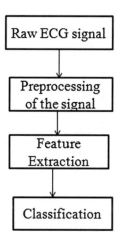

48 half-hour two-channel ECG data with 360 Hz sampling frequency. The signals have been recorded at Boston's Berth Israel Hospital (BIH). Each record in the database has a clarification document in which each ECG beat has been distinguished by cardiologist specialist.

3   Preprocessing of the Signal

The ECG signals are generally corrupted by different kinds of noise. This noise consequently leads to a wrong diagnosis or classification system. The noise in ECG signals generally included power line interface, muscle expansion interference, baseline interface, electrode contact noise, and some noise generated by the electronic ECG monitoring device [14]. Here, we have denoised the raw ECG signal by applying median filter with order of 12. As the signal is low frequency, the low-pass filter (LPF) is considered to remove the noise at the time of acquisition.

After applying median filter, we have applied a 12-order low-pass FIR filter with cutoff frequency 35 Hz. The filtered signal after applying these two filters is presented in Figs. 2 and 3.

4   Feature Extraction

After obtaining the raw ECG signals in digitalized form, signals are preprocessed. Now after this, we need to extract the useful features from this filtered signal. DWT is a process for the disintegration of input signals into sets of function, called wavelets [15]. It utilizes filter banks for the development of multitier solution analysis, which enhances calculation effectiveness. It decomposes the signal into detailed and approx-

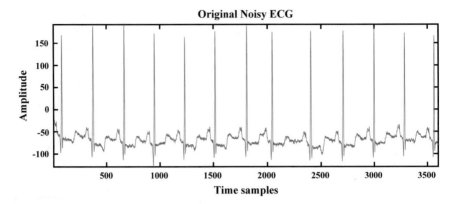

Fig. 2 Original noisy ECG

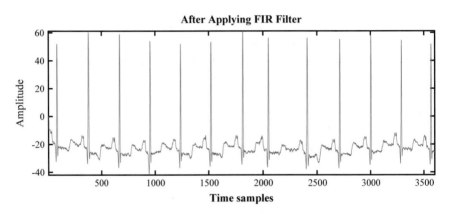

Fig. 3 Clean ECG after applying median and FIR filter

imation coefficient by correspondingly high and low pass filtering the signal [16–18]. Consider $y(t)$ is the signal. It is decomposed using wavelet transform with 'dB2' mother wavelet type. Mathematically,

$$w_h(a, b) = \sum_{t=-\infty}^{\infty} y(t)h(2n - k) \tag{1}$$

$$w_l(a, b) = \sum_{t=-\infty}^{\infty} y(t)g(2n - k) \tag{2}$$

where $h(\cdot)$ and $g(\cdot)$ are the shifted version of the impulse response of the mother wavelet. The parameters a and b are the dilations and positions of function. In Fig. 4, the extracted QRS features have been presented.

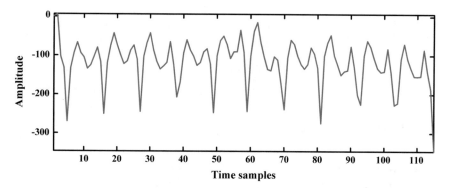

Fig. 4 QRS complex

5 Random Forest

Random forest is the combination of decision tree algorithm. Decision tree is a treelike structure to make any decision. It is a combination of number of decision tree classifier. It basically works on ensemble learning method where a number of learners are trained to solve a particular problem. This method tries to construct a set of assumption and combine them to use [19, 20]. Here, θ_m is random vector which free from earlier random vectors. The tree is developed by the training data, and it generates a classifier $h(y, \theta_m)$; here, y is the input vector. The vote for most accepted class happens when a big amount of tree is generated. We call these procedures random forests. This classifier consists of a group of treelike classifiers $\{h(y, \theta_m), m = 1, \ldots\}$ where $\{\theta_m\}$ are autonomous identically circulated arbitrary vectors, and every tree radiates a vote for the most accepted class on input y.

A group of classifiers $h1(y)$, $h2(y)\ldots hM(y)$ is given and the training data drawn at random from the circulation of the random vector X, Y. The margin function can be defined as:

$$\text{mrg}(Y, X) = \text{avg}_m I(h_m(Y) = X) - \max_{k \neq X} \text{avg}_m I(h_m(Y) = k) \qquad (3)$$

Here, $I(.)$ is the indicator function. The margin evaluates the degree to which the standard amount of votes at Y, X for the proper class exceed the average vote for other class. Generalization error can be evaluated as:

$$PE^* = P_{Y,X}(\text{mrg}(Y, X) < 0) \qquad (4)$$

Here, the subscripts Y, X point out that the possibility is more than the Y, X space.

Table 1 Performance of random forest classifier

Performance	Training set (%)	Testing set (%)
Sensitivity	100	100
Specificity	100	89
Accuracy	100	87

6 Classification Result

For estimating the execution of any machine learning method, distinctive methodologies are utilized. The first approach is to separate the entire dataset into two training set and testing set. Both these two sets should be selected separately from each other. The classifier performance can be measured by sensitivity and specificity. Sensitivity is applied to specify the classifier performance for recognizing the positive samples, and it can be defined by:

$$\text{Sensit.} = \frac{\text{TrP}}{\text{TrP} + \text{FlN}} * 100 \tag{3}$$

TrP is amount of true positive samples, and FlN is the amount of false negative samples. Sensitivity characterizes amount of patients having cardiac problem. Specificity is used to calculate the classifier performance for recognizing patients without having cardiac disease, and it is calculated by:

$$\text{Spec.} = \frac{\text{TrN}}{\text{TrN} + \text{FlP}} * 100 \tag{4}$$

Here, TrN is number of true negative samples, and FlP is the number of false positive samples. The overall accurateness of the model is calculated by:

$$\text{Acc.} = \frac{\text{TrP} + \text{TrN}}{N} * 100 \tag{5}$$

A random forest classifier is designed to classify five types of arrhythmia disease from the ECG signals collected from different people. The original data is divided into training and testing set with a ratio of 70 and 30% respectably. Then the random forest model is built with the training set with 500 tree size. After successfully building the model, it is trained with the training set data. The training and testing performance of the classifier model is presented in Table 1. Figure 5 shows the performance of the random forest classifier.

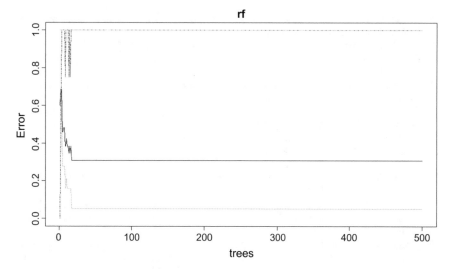

Fig. 5 Performance of the classifier

7 Conclusion

Big data analysis is one of the useful concepts in medical sector for better diagnosis of a disease. Arrhythmia is one of major disease, and it is one of the vital task for accurate detection of the disease. In this work, a random forest classifier model is designed to detect and classify arrhythmia from the ECG signals. We have archived around 87% classification accuracy which is quite good. In future, different methods can be applied for better classification and detection.

References

1. Trnka, A.: Big data analysis. Eur. J. Sci. Theol. **10**(1), 143–148 (2014)
2. Voorsluys, W., Broberg, J., Buyya, R.: Introduction to cloud computing. In: Cloud Computing: Principles and Paradigms, pp. 1–41. (2011)
3. Ristevski, B., Chen, M.: Big data analytics in medicine and healthcare. J. Integr. Bioinf. (2018)
4. Wang, L., Alexander, C.A.: Big data in medical applications and health care. Am. Med. J. **6**(1), 1 (2015)
5. Viceconti, M., Hunter, P.J., Hose, R.D.: Big data, big knowledge: big data for personalized healthcare. IEEE J. Biomed. Health Inform. **19**(4), 1209–1215 (2015)
6. Wang, Y., Kung, L., Wang, W.Y.C., Cegielski, C.G.: An integrated big data analytics-enabled transformation model: application to health care. Inf. Manag. **55**(1), 64–79 (2018)
7. Özçift, A.: Random forests ensemble classifier trained with data resampling strategy to improve cardiac arrhythmia diagnosis. Comput. Biol. Med. **41**(5), 265–271 (2011)
8. Hu, C., Chen, Y., Hu, L., Peng, X.: A novel random forests based class incremental learning method for activity recognition. Pattern Recogn. **78**, 277–290 (2018)

9. Tsagkrasoulis, D., Montana, G.: Random forest regression for manifold-valued responses. Pattern Recogn. Lett. **101**, 6–13 (2018)
10. Ozcift, A., Gulten, A.: Classifier ensemble construction with rotation forest to improve medical diagnosis performance of machine learning algorithms. Comput. Methods Programs Biomed. **104**(3), 443–451 (2011)
11. Abellán, J., Mantas, C.J., Castellano, J.G., Moral-García, S.: Increasing diversity in random forest learning algorithm via imprecise probabilities. Expert Syst. Appl. **97**, 228–243 (2018)
12. Mohapatra, S.K., & Mohanty, M.N.: Analysis of resampling method for arrhythmia classification using random forest classifier with selected features. In: 2018 2nd International Conference on Data Science and Business Analytics (ICDSBA), pp. 495–499, IEEE (2018)
13. Genuer, R., Poggi, J.M., Tuleau-Malot, C., Villa-Vialaneix, N.: Random forests for big data. Big Data Res. **9**, 28–46 (2017)
14. Moody, G.B., Mark, R.G.: The impact of the MIT-BIH arrhythmia database. IEEE Eng. Med. Biol. Mag. **20**(3), 45–50 (2001)
15. Cuomo, S., De Pietro, G., Farina, R., Galletti, A., Sannino, G.: A revised scheme for real time ecg signal denoising based on recursive filtering. Biomed. Signal Process. Control **27**, 134–144 (2016)
16. Alickovic, E., Kevric, J., Subasi, A.: Performance evaluation of empirical mode decomposition, discrete wavelet transform, and wavelet packed decomposition for automated epileptic seizure detection and prediction. Biomed. Signal Process. Control **39**, 94–102 (2018)
17. Mohapatra, S.K., Palo, H.K., Mohanty, M.N.: Detection of arrhythmia using neural network. Ann. Comput. Sci. Inf. Syst. **14**, 97–100 (2017)
18. Mohanty, M.N., & Mohapatra, S.K.: ECG biometrics in forensic application for crime detection. Med. Leg. Update **19**(1), 749–756 (2019)
19. Breiman, L.: Random forests. Mach. Learn. **45**(1), 5–32 (2001)
20. Jones, Z., Linder, F.: Exploratory data analysis using random forests. In: Prepared for the 73rd Annual MPSA Conference. (2015)

An Intelligent Framework for Analysing Terrorism Actions Using Cloud

Namrata Mishra, Shrabanee Swagatika and Debabrata Singh

Abstract Terrorism, having spread its roots throughout the world, has become a grieve matter of concern globally. In simple words, terrorism is the use of violence to create fear and alarm. Terrorism being a major challenge, the paper deals with how terrorism can be combated using the fastest growing and most effective technology of computer science, i.e. cloud computing. An understanding of the history, nature and mechanism will contribute towards the eradication of terrorism. Here, we will use a directed graph that will provide us with an interconnected network of terrorists' activities based on the data collected from the Global Terrorism Database (GTD) for a certain period. We will develop an analytical model which would connect the distributed data on terrorists' activities to aid decision-making by counter-terrorist security agents around the world.

Keywords Cloud computing · Global terrorism · Intelligent data · Web mining · Surveillance

1 Introduction

Terrorism, having spread its roots throughout the world, has become a matter of grieve concern. It has created a global network to attack government, human beings, organization, transport system, etc. with an evil intention. Terrorism has found a complex definition nowadays, but in simple words, it can be defined as the use of violence to create fear and alarm. Right to live is a fundamental right and terrorism violates the fundamental rights of people. After 11 September 2001, there has been a

N. Mishra (✉) · S. Swagatika · D. Singh
Department of CSE, ITER, S 'O' A, Bhubaneswar, India
e-mail: namratamishra1702@gmail.com

S. Swagatika
e-mail: shrabaneeswagatika@soa.ac.in

D. Singh
e-mail: debabratasingh@soa.ac.in

© Springer Nature Singapore Pte Ltd. 2020
S. Patnaik et al. (eds.), *New Paradigm in Decision Science and Management*,
Advances in Intelligent Systems and Computing 1030,
https://doi.org/10.1007/978-981-13-9330-3_21

significant change in the world. Fear, anxiety and hatred are dominating most of our daily lives. Thus, talks about terrorism have become part and parcel of our lives and the media play a vital role in delivering the latest news on this issue, thereby keeps people informed. A wonderful increment in fear monger occasions can be found in Pakistan. Four noteworthy territories, i.e. Balochistan, Khyber Pakhtunkhwa, Punjab and Sindh have turned out to be a place of refuge for the psychological militants. Similarly, Iraq, Syria and Iran have also witnessed a rise in the number of terrorist attacks. Every day, the heinous acts of the terrorists create a headline in every newspaper. This has created a feeling of fear amongst the common population. Many innocents have lost lives. Many places have suffered destruction. People are forced to live in the realm of fear. Violence has badly affected society. Stepping out from home includes a shivering heart with a fear of whether one will be able to return back to home safely. People are no more afraid of the thunder sound since bomb blasts drums every time in their mind.

The United Nations has shown major concern to fight with the increasing terrorism in the world. Various government schemes have also been implemented to check terrorism. No positive result from the above followed by failure from the use of brute military forces leads the world to rely upon the technological advancement of the twenty-first century to fight terrorism. Technology nowadays plays an important role in our daily lives. Technology has constantly improved its applications and our way of communication. It has become almost impossible to imagine life without the use of technology. Basically, technology symbolizes the use of machines, techniques and power sources to achieve the effective and productive result. Development goes side by side with the advancement of technology. It has shown its victory in almost every field, may it be the medical sector or educational field or household. It has changed the social, economic, ethical, political as well as cultural picture of the world. Our prosperity has led us to this technical world. Various researches have been undertaken to find solutions to fight terrorism using the most effective technology of computer science, i.e. cloud computing.

There lies a huge after impact of a terrorist attack. The impacts can be summarized in the following diagram.

Cloud computing represents the delivery of hardware and software resources on demand over the Internet. On the other hand, the IoT concept envisions a new generation of devices (sensors, both virtual and physical) that are connected to the Internet and provides different services for value-added applications (Fig. 1).

2   Motivation

Terrorism needs to be eradicated from the world as soon as possible. An understanding of the history, nature and mechanism will contribute towards the eradication of terrorism. Checking terrorism at every corner of the world can only contribute towards checking of terrorism worldwide. Cloud computing, being the smartest technology of recent times can be used effectively to fight terrorism globally.

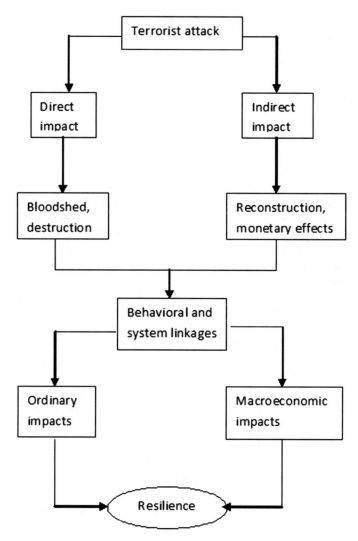

Fig. 1 Flowchart of after impacts of terrorist

3 Objective

The world has become a victim of increasing terrorism. Every day many people lose their lives, buildings get destroyed and most remarkably the progress rate of a country faces a barrier due to the unchecked rise of terrorism. The unethical act of terrorism against humanity has slowed down the progress. Huge amount of money is spent worldwide for recover purposes post a terrorist attack. This money can be used in developmental activities. Therefore, for the well-being of a country, terrorism needs to be eliminated from roots.

4 Literature Review

Gundabathula and Vaidhehi [1] analyse terrorism challenges faced in India. From the research, it is identified that there are different methods and approaches to combat terrorism. Various data mining techniques such as clustering, asocial rule mining, social network analysis and classification have been used to model frameworks and implement various models. The author here uses clustering and associate rule mining to find patterns and similarities in terrorism data. Associate rule mining is a procedure used to find frequent patterns, associations and correlations from the data sets.

Using the GTD data, the unethical actions of terrorists are modelled using machine learning algorithms like J48, Naïve Bayes, IBK, ensemble approach using VOTE. The data has been pre-processed by using various pre-processing techniques, and the class imbalance problem in the data has been solved using the over-sampling and under-sampling techniques. This pre-processed data is used to create classification models. Out of all the models, the J48 and VOTE ensemble algorithm gave the highest possible accuracy. By taking the highest classified model, prediction can be made to find the terrorist groups responsible for various attacks if the perpetrator is unknown. The study also reveals the terrorist risk faced by the states and shows the perpetrator groups with the highest frequent attacks in the particular state. From the results, the conclusion is drawn that India faces more of domestic terrorism than the outside threats.

Steps followed to develop the prediction system:

1. Collection of data: Data for the research has been taken from the Global Terrorism Database (GTD). The GTD data is used for the prediction purpose. It is an open-source database having over 170,000 terrorist incidents across the globe. The author has dealt with data of terrorist attacks that have taken place in India.
2. Pre-Processing: The raw data needs to be processed before it is used as input in the prediction model creation. After pre-processing, the data becomes clean and fit to use. There are four major steps involved in pre-processing of data. Those are data cleaning, data integration, data transformation and data reduction.

   - Data cleaning—The process that detects and removes inaccurate records.
   - Data integration—It is a procedure that consolidates information from different sources and gives a brought together perspective of the information.
   - Data transformation—It is a procedure in which information is changed into shapes which are fitting for mining.
   - Data reduction—It is a technique which is applied to get a reduced form of data set.

Fig. 2 Flow of model creation

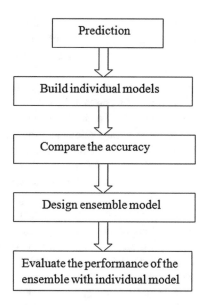

Steps in Data pre-processing:

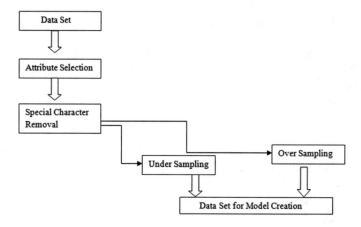

The author focuses on studying the nature of the terrorist groups based on the past data, and if the group involved in a particular incident is not known then it is not useful anymore. Hence, these records are discarded.

Model Creation: This is the most important step of the prediction system creation. After the required pre-processing and analysis of the data, a prototype of the model is created based on the data. Here, we have discarded all the unnecessary information. Model creation again involves several steps to be followed. Several individual methods are built and their accuracy is compared so as to obtain a model that gives the optimal result. Then, the ensemble model is designed (Fig. 2).

The result showed that the domestic terrorism has much impact on the country. Most of the terrorist incidents involve domestic terrorist groups. Therefore, counter-terrorism agencies should concentrate more on fighting the domestic terrorism in the country.

Goteng and Tao [2] utilized guided diagram to concoct an interconnected system of terrorists' exercises in view of information got from the Global Terrorism Database (GTD) from 2005 to 2015. The digraph used here helped in obtaining the locations of the terrorist organizations and the relationship between them. This work visualized the effectiveness of mathematical models and cloud technologies applied together to combat global terrorism.

To start the undertaking of understanding the system of psychological militants, the creator utilized the idea of digraph to make graphical situations of performing artists (fear-based oppressors, security staff), activities (correspondence, transportation, development, assault and exchange), targets (structures, transport frameworks, individuals and associations) and assets (arms, telephones, online apparatuses, cash). Coordinated diagram or digraph is where the edges have course, i.e. the chart has requested match of vertices.

After deriving a connection between the terrorist groups, the method of attack and the actual target, the author developed a cloud terror algorithm which is a measurable model to foresee if fear-based oppressors are arranging assault in view of development, correspondence and exchange acquired from the data accessible in GTD. For example, phone calls and emails were traced to obtain communication details. Movement was tracked through transport system and transaction involved arms and money transfer.

The CTA algorithm developed involves the following steps:

The first step involves calculating the average of frequency of movement, transaction and communication for a given sample of terrorist.

It is calculated as:

$$avg = (\textstyle\sum_{(i=0)}^{n} a\_i)/n$$

This implies three mean values will be acquired for every fear terrorist.

Each average value obtained is contrasted with the frequency so as to discard them listing no possible threats. This formulation is given as:

b_(i) = { (a_i if a_i≥avg@   0   else)} where bi is the new value after the comparison.

The third step is to find the probability of each value for the three criteria for each terrorist by normalizing the values obtained from step 2 using the equation:

⟦tran⟧_(i =) b_i/(∑_(i=0)^n b_i) where ⟦tran⟧_i stands for transaction.

The formulation for the normalized probabilities multiply by the weights is given as:

p_i= ⟦tran⟧_i * ⟦weight⟧_trans + ⟦tran⟧_i * ⟦weight⟧_phone + ⟦tran⟧_i * ⟦weight⟧_mov

The frequency of communication, transactions and movement among the community of terrorists is seemed upon by way of the CTA to locate their possibility of making plans an assault based totally on a distinct threshold. The paper [2] used AWS EC2, S3, RDS and virtual private cloud for the implementation of the cloud gadget.

The exercises learnt in this examination are in four categories. The first includes trouble in acquiring and examining complex information on terrorists. It is extremely unlikely we could have acquired close solid information other than GTD. Governments and specialists should make data about terrorists and their exercises open to the world. The second lesson learnt is the expansion being used of innovation to manage worldwide psychological warfare. Utilizing innovation for insight gathering, fingerprints catch, scientific confirmations, cash clothing discovery, guest checks and strengthening of security faculty is critical to control the present development of worldwide terrorism. The third exercise learned is creating numerical models that can be useful in anticipating and ceasing the brutal demonstrations of fear-based oppressors. The issue is not gathering information about fear-based oppressors, yet making the information assembled accessible and to security faculty at the correct time. That is the reason the advancement of savvy frameworks utilizing calculations ought to be a piece of the innovations and ought to be utilized to battle terrorism. The last exercise learnt is the preposterousness that both counter-terrorism offices and the terrorists are utilizing the most recent advances for their demonstrations, each endeavouring to outmanoeuvre the other. This is the place it ends up fascinating and additionally entangled relating which innovations best suit the war against terrorism. A blend of information driven insightful scientific and diagnostic models actualized with the cloud-based innovations to interface counter-fear-based oppression specialists is one of the surest approaches to battle terrorism globally.

The prototype of the model when tested proved to be very effective, thereby showing that the system can be of utmost significance for the usage of technologies in the fight against worldwide terrorism.

Disha Talreja, Jeevan Nagaraj, Varsha NJ and Kavi Mahesh [3] worked on learning towards the prediction of the perpetrator involved in a terrorist attack with the help of various machine learning algorithm including k-means clustering, Boruta analysis, C4.5, etc.

The author believes prediction of the terrorist group involved in an attack is the foremost process to check terrorism. When we find the association behind an attack, necessary steps can be taken for reaching to the culprit. Most common procedures for finding the terrorist group behind an attack are tracking of email, extracting information from telephone signal, social network analysis, etc. The paper concentrated on utilizing GTD information alongside suitable classifier calculations to foresee the terrorists assemble that is in charge of an assault.

Steps involved in predicting the terrorist group that is behind an attack:

Data cleaning was the first step in the process. It helped extracting structured data relevant to the terrorist incidents in India including longitude and latitude of the location of the attack.

Table 1 Discretizations of terrorism target for different terrorism events

Event of terrorism	Terrorism target discretization
Target killing	Political, sectarian, civilian, foreigner, military, ethnic, criminal
Terror attack	Militant, sectarian, foreigner, civilian, political, criminal
Military operation	Militants, civilians, ethnic

Then, the data was partitioned using k-modes clustering since it involved categorical data set. Attributes such as 'attacktype1', 'targettype1' and 'targetsubtype1' were found out in this step.

Finally, factor analysis of mixed data (FAMD) was performed on the data set for extracting the features that are most contributing.

The three machine learning models, i.e. decision tree algorithm, random forest algorithm and super vector machine were used. It was concluded in the paper that SVM gave the utmost accuracy of nearly 80% for the GTD data set before 2014. It was observed that in case of most of the major terrorist attacks, their perpetrators were identified correctly.

Research by Dr. Tariq Mahmood and Khadija Rohail [4] presents some results of the analysis of event–target combinations. It shows a rise in the frequency of these combinations after 2006, particularly for the Khyber Pakhtunkhwa province.

Dr. Tariq Mahmood and Khadija Rohail [5] analyse terrorist activities in Pakistan and present methods to deal with growing terrorism using data mining techniques. The optimal algorithm, CLOPE is used for clustering purposes. K-means, agglomerative hierarchical clustering has become mostly used clustering techniques; therefore, the author has used CLOPE clustering algorithms.

CLOPE is used in case of categorical data. It is helpful in increasing the intra-cluster overlapping of categorical values.

The author then created four different data sets, where one data set contains instances for one of the four provinces among Balochistan, Khyber Pakhtunkhwa, Punjab and Sindh.

The analysis from clusters obtained annually is shown in Table 1.

For each event (i.e. target killings, terror attacks and military operations), a target list was identified. It concluded that the political persons are the target in most cases. Then, for each unique event–target combination, the clusters containing instances related to the combination only, e.g. clusters containing only target killing—political instances were obtained. It highlighted the significance of a particular combination for a given province, and it was influenced by the increased combination frequency.

A brief analysis about the terrorist events and the target gives the following conclusion:

- Aerial attacks and murder in military operations are most remarkable.
- Aerial attacks are a significant way used in terror attacks but are less in comparison with military operations.
- Suicide attacks, shoot outs and bomb blasts are also commonly used.
- Bomb blasts and shelling are also significant in the terror attacks.

It is concluded with the result that when we have a clear understanding and brief information about the attacks, the target and the methods of the target can be easily known and hence the future attacks can be kept in check. The analysis of terrorism is possible with a sound judgement along with well-integrated technical capabilities.

Arie Perliger and Amit Pedahzur worked to analyse the impacts of social network on terrorism and political violence [6]. The paper concludes with the fact that social network analysis plays a major role in studying terrorist activities and can be helpful to check it.

Alexander Gutfraind researched how the pattern of terrorist activities can be analysed using mathematical models [7]. The paper shows mathematical models are very helpful in such scenarios. There are no such biological models that can contribute to terrorism research.

Martha Crenshaw observed current scenario of terrorism [8] and claimed that terrorism might serve as a useful test case for the general theory of violence.

Valdis E. Crebs researched to map the network of terrorist cell [9]. The research claimed that to obtain a network of terrorist, task and trust ties between the conspirators must be identified accurately.

Stanley Wasserman and Katherine Faust [10] analysed social network and discussed various methods for the analysis which can be useful to analyse social network activities to keep an eye on terrorist activities.

Jeff Victroff made research on theories to study the psychology of terrorism [11]. The author says modifiable psychological as well as social factors are mostly responsible for the growth of terrorism.

Wingyan Chung and Wen Tang's research on online surveillance of U.S. domestic terrorism [12] showed the trend of native work of terrorism grows apace, utilizing web collections to support online observation that can help intelligence and security personnel to keep an eye on the evil activities online. During this analysis, the authors have developed a gathering of US act of terrorism websites and have conducted a fundamental analysis of the sites' substance and usage. A unique approach was developed to extract matter from websites. The in progress works embody finding hidden patterns from a group of domestic terrorism websites of the USA and revealing attention-grabbing usage and content patterns. This work ought to contribute to the realm of online security police investigation victimization website information.

The analysis goal is to modify various uses of this assortment in the discovery of pattern from newly found suspicious sites. Here, twenty terrorist related websites were collected with 44.5 gigabytes of web content. These sites were later confirmed by the Southern poverty law centre to be related to extremist organizations.

5   Comparison of Models

Sl. No.	Author	Techniques/methods used
1.	Varun Teja and Gundabathula	Clustering and associate Rule Mining
2.	Gokop L. Goteng and Xueyu Tao	Digraph, CTA algorithm
3.	Disha Talrejn, Jeevan Nagraj, Varsa NT and Kavi Mahesh	K-means algorithm, Boruta analysis, Decision tree algorithm, Random Forest algorithm, Super Vector Machine
4.	Dr. Tariq Mahmood and Khadija Rohail	CLOPE
5.	Arie Perliger and Amit Pehdahzur	Social network analysis
6.	Stanley Wasserman and Katherine Fauge	Social network analysis

6   Conclusion

Terrorism has become a severe challenge for the world, and it can only be curbed with intelligence, power and technology applied all together. Keeping in mind, the advancement in technology and need to fight terrorism using technology will be beneficial in order to check global terrorism. Higher usage of technology will serve as the ultimate drug for the till now incurable disease, called Terrorism. Technology does not only give us smart tools to fight terrorism, but also provide us with smarter ways to deal with problems and to find smart and effective solutions. Since cloud computing has always given positive and effective result in every research done in the field, using this technology to check terrorism is an appreciable idea. Though some research has been done towards using cloud computing to combat terrorism, the area needs to be explored more. Terrorism is violating the fundamental rights. It is highly condemnable act against humanity, and it should be completely erased from the world. Terrorism is a major hindrance in the progress path of a society and in the country as a whole; therefore, now it is time to wake up from sleep and run to the battlefield with all the weapons; our technology has ever given us and eliminates the fear causing word 'terrorism'.

References

1. Gundabathula, V.T., Vaidhehi, V.: An efficient modelling of terrorist groups in india using machine learning algorithms. Indian J. Sci. Technol. **11**(15) (2018)
2. Goteng, G.L., Tao, X.: Cloud computing intelligent data-driven model: connecting the dots to combat global terrorism. In: 2016 IEEE International Congress on Big Data (BigData Congress), San Francisco, CA, pp. 446–453 (2016)

3. Talreja, D., Nagaraj, J., Varsha, N.J., Mahesh, K.: Terrorism analytics: learning to predict the perpetrator. In: 2017 International Conference on Advances in Computing, Communications and Informatics (ICACCI), Udupi, pp. 1723–1726 (2017). https://doi.org/10.1109/icacci.2017.8126092

4. Mahmood, T., Rohail, K.: Analyzing terrorist events in Pakistan to support counter-terrorism-events, methods and targets. In: 2012 International Conference on Robotics and Artificial Intelligence (ICRAI), pp. 157–164 (2012)

5. Corner, E., Gill, P., Mason, O.: Mental health disorders and the terrorist: a research note probing selection effects and disorder prevalence. Stud. Conflict Terrorism 39(6), 560–568 (2016)

6. Perliger, A., Pedahzur, A.: Social network analysis in the study of terrorism and political violence. PS: Polit. Sci. Polit. 44(1), 45–50 (2011)

7. Gutfraind, A.: Understanding terrorist organizations with a dynamic model. In: Mathematical Methods in Counterterrorism 2009, pp. 107–125. Springer, Vienna

8. Crenshaw, M.: Current research on terrorism: the academic perspective. Stud. Conflict Terrorism 15(1), 1 (1992)

9. Krebs, V.E.: Mapping networks of terrorist cells. Connections 24(3), 43–52 (2002)

10. Wasserman, S., Faust, K.: Social Network Analysis: Methods and Applications. Cambridge University Press (1994)

11. Horgan, J.: From profiles to pathways and roots to routes: perspectives from psychology on radicalization into terrorism. Ann. Am. Acad. Polit. Soc. Sci. 618(1), 80–94 (2008)

12. Chung, W., Tang, W.: Building a web collection for online surveillance of US domestic terrorism. In: 2012 IEEE International Conference on InIntelligence and Security Informatics (ISI), IEEE, pp. 195–195 (2012)

Research on Optimization of J Company Warehouse Storage Process

Jintia Ge and Mengfan Liu

Abstract Based on the analysis of the current situation of the warehousing business process of J company, this paper optimizes it with the idea of intelligent warehousing and constructs the Petri net of the optimized warehouse storage process to verify the feasibility of optimizing the warehouse storage process and improve the intelligence level of the warehousing business of J company.

Keywords Warehouse storage process · Process optimization · Petri net

1 Introduction

As one of the two pillars of logistics, warehousing has always been the focus of logistics industry [1]. To improve the intelligence level of the company's warehousing system and build intelligent warehousing, it is imperative for regional logistics centers, which can effectively connect upstream and downstream enterprises. In order to better adapt to the characteristics of various types and forms of goods meet the requirements of customers for storage, enhance the mechanization and automation level of storage, optimize the whole business process of storage, and realize the intelligence of storage is the only way.

This paper will take J company, a comprehensive modern logistics park, as an example to optimize its warehouse storage process with the idea of intelligent warehousing.

J. Ge (✉) · M. Liu
School of Business, University of Jinan, 250002 Jinan, P.R. China
e-mail: gejintian@163.com

M. Liu
e-mail: liu_mengfan@163.com

© Springer Nature Singapore Pte Ltd. 2020
S. Patnaik et al. (eds.), *New Paradigm in Decision Science and Management*,
Advances in Intelligent Systems and Computing 1030,
https://doi.org/10.1007/978-981-13-9330-3_22

2 Analysis on the Current Situation of J Company

2.1 Warehouse Storage Process

On the basis of arrival notice, J company conducts acceptance inspection on the goods arriving and classifies the goods qualified for acceptance according to different categories and different inventory units.

Then, the staff will be printed in advance tray label paste on the pallet, with the help of handheld PDA collecting number in the form of finished goods and code library work, place cargo in the receiving area at the same time, and according to the actual quantity of goods received feedback to management system through the system, update the order and inventory.

Finally, the forklift truck is used to transport the goods from the receiving area to the corresponding storage area to complete the warehouse storage process.

2.2 Problems with Warehouse Storage Process

2.2.1 Long Waiting Time for Warehousing

The current warehousing process of J company requires that the printed pallet label be attached to the pallet. Although the accuracy of information is guaranteed during the warehousing, warehousing and delivery of goods, the printing and pasting of pallet label increase the waiting time for goods to enter the warehouse.

2.2.2 Unreasonable Allocation of Cargo Space

The allocation of the cargo space is determined by the warehouse keeper, who is highly dependent on the experience of the warehouse keeper. It will take some time to determine the cargo location, and it is easy to find the wrong goods placement.

2.2.3 The Operation of Forklift Is Easy to Make Mistakes

The operation route of the forklift is decided by the staff when entering the warehouse, and there will be problems such as roundabout operation, wrong route selection, and crossing of operation line, which increase the unnecessary time consumption in the process of warehousing.

3   Optimized Warehouse Storage Process

The optimized warehousing business process is shown in Fig. 1.

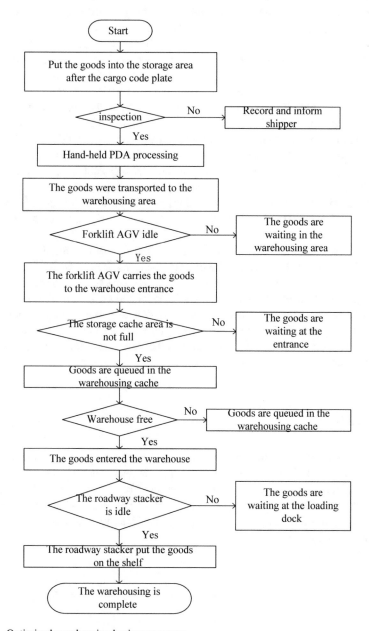

Fig. 1 Optimized warehousing business process

4 Petri Net Model

According to the optimized warehouse storage process, Petri net [2] diagram of the optimized warehouse storage process is established, as shown in Fig. 2.

Table 1 can be obtained by sorting out the meanings of various places and transitions in the Petri net model [3] of warehouse storage process [4].

For the convenience of analysis, this paper divides identification into two parts, one representing logistics layer and the other representing information layer [5]. Where, Mn = (P1, P2, P3, ..., P16, P15, P17) is used for logistics layer, M0i = (Pi1, Pi2, Pr0, Pr1, Pr2, Pc0, Pc1, Pa0, Pa1, Pw0, Pw, Px0, Px1) is used for, information layer; Logistics and information layer is respectively, and the initial state, M0 = (1, 0, 0, 0, 0, 0, 0, 0, 0, 0, 0, 0, 0, 1, 0, 0, 0, 0), M00 = (1, 0, 1, 0, 0, 1, 0, 1, 0, 1, 0, 1, 0), analyze the Petri net model, to observe whether there is any transitions, make Ms = (0, 0, 0, 0, 0, 0, 0, 0, 0, 0, 0, 0, 1, 0, 0, 0, 1) and M0s = (0, 0, 1, 0, 0, 1, 0, 1, 0, 0, 1, 1, 0) can reach. Through analysis, the Petri model accessible tree of warehousing business process is obtained, as shown in Fig. 3.

Through analysis can be found that the layer of logistics initial state M0 = (1, 0, 0, 0, 0, 0, 0, 0,0,0,0,0,1,0,0,0,0) after the transitions = t1, t2, t3, t4, t5, t6, t7, t8, t9, t10, t11, t12, t13, t14, t15, t16 to Ms = (0, 0, 0, 0,0, 0, 0, 0, 0, 0, 0, 0, 1, 0, 0, 0, 1) can reach; Layer at the same time, the information of the initial state M00 = (1,0,1,0,0,1,0,1,0,1,0,1,0) can also be through the transitions = t1, tr1, tc1, tr2, ta1, tw, tx1 and makes identity M0s= (0, 0, 1, 0, 0, 1, 0, 1, 0, 0, 1, 1, 0) can reach. The above analysis indicates that both the logistics layer and the information layer of the warehouse storage process can realize the ideal state of goods and information, and there is no bad state in between. Therefore, the Petri net model of the warehouse storage process is accessible and smooth, and the warehouse storage process can be carried out smoothly.

5 Conclusion

This paper first analyzes the current status of the warehouse storage process of J company and finds that the warehouse storage process is not efficient. In order to build intelligent storage, optimize the warehouse storage process. In order to ensure the feasibility of the optimized process, the Petri net is drawn and validated by the form of accessible tree for the construction of intelligent storage has a certain reference significance.

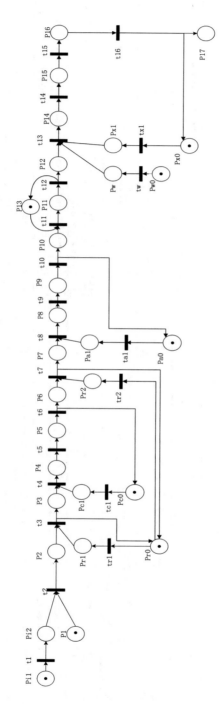

Fig. 2 Petri net

Table 1 Meanings of places and transitions

Place	Meanings	Transitions	Meanings
Pi1	Arrival notice	tr1	Employees go to designated locations
Pi2	Warehouse received notice	tr2	Employees go to designated locations
Pr0	Employees free	tc1	Forklift to designated position
Pr1	The employee arrives at the designated position and is ready to unload	ta1	Forklift AGV to the specified location
Pr2	The employee arrives at the designated position and is ready to unload	tw	Distribution of goods
Pc0	Forklift spare	tx1	The roadway stacker goes to the designated warehousing platform
Pc1	Forklift to the designated position, ready to fork tray	t1	Delivery information processing
Pa0	Forklift AGV idle	t2	Check incoming orders
Pa1	Forklift AGV arrives at the specified position, ready for forklift	t3	Employees of discharge
Px0	The roadway stacker is idle	t4	Forklift tray
Px1	The roadway stacker arrives at the depot	t5	Forklift truck transportation
Pw0	Goods free	t6	Forklift truck unloaded
Pw	Empty space	t7	Staff handles incoming goods and updates cargo information
P1	The goods arrived	t8	Forklift AGV forklift tray to the warehouse entrance
P2	Warehouse entry order check complete, ready to unload	t9	Forklift AGV transports goods to the warehouse entrance
P3	Finish discharging and wait for shipment	t10	Forklift AGV begins unloading the tray at the entrance
P4	The forklift fork is removed and ready for transportation	t11	The storage cache area is not full and the tray enters the storage cache area
P5	Forklift to storage area	t12	The loading dock is free, and the goods enter the loading dock

(continued)

Table 1 (continued)

Place	Meanings	Transitions	Meanings
P6	The unloading of the forklift is finished, and the goods are waiting for warehousing	t13	The roadway stacker forked the tray
P7	The goods have finished warehousing.	t14	The pallets are transported to the designated place
P8	Forklift AGV forklift cargo is finished, waiting for transportation	t15	The roadway stacker began to unload
P9	The forklift AGV arrives at the warehouse entrance, and the goods are waiting to be unloaded	t16	The roadway stacker is idle and waiting for instructions
P10	The goods are at the entrance to the warehouse		
P11	The goods are located in the warehousing cache		
P12	The goods are located at the loading dock, waiting for the roadway stacker to pick them up		
P13	Release a store in the warehousing cache		
P14	Roadway stacking machine fork end, waiting for transportation		
P15	The pallet arrives at the specified position and is ready to unload		
P16	The stacker was unloaded and the roadway stacker was idle		
P17	Warehousing business completed		

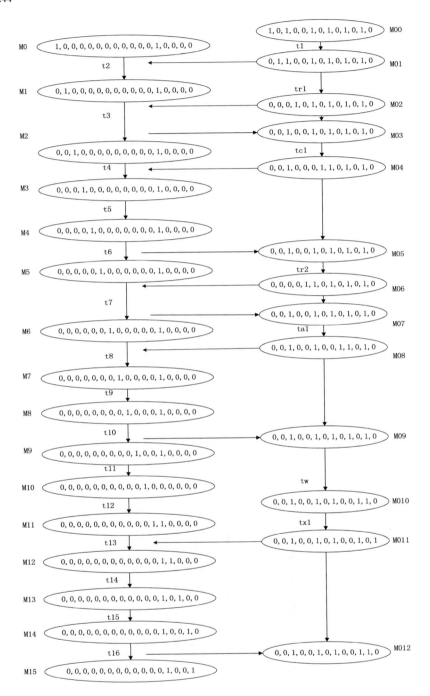

Fig. 3 Accessible tree

References

1. Zhitai, W.: New Edition of Modern Logistics. Capital University of Economics and Business Press (2018). (in Chinese)
2. Giua, A., Silva, M.: Modeling, analysis and control of discrete event systems: a Petri net perspective. IFAC PapersOnLine **50**(1), 1772–1783 (2017)
3. Bi, H., Fei, D., Qi, M.: Business process modeling based on Petri net. Electron. Technol. Softw. Eng. **6**, 22 (2018). (in Chinese)
4. Zhang, Z., Li, F.: Study on optimization of warehousing and warehousing process of enterprise A based on Petri net. Logistics Technol. **40**(9), 141–145 (2017). (in Chinese)
5. Huang, D.: Modeling and application of logistics storage system based on Petri net. Wuhan University of Science and Technology (2012). (in Chinese)

The Influence of Psychological Contract on the Willingness to Share Tacit Knowledge

Jian-Hua Wu and Hao Yan

Abstract The management of knowledge sharing has become an important part of business management. From the perspective of psychological contract, this paper takes knowledge workers as the research object and explores the influence of psychological contract on its willingness to share knowledge. Using the empirical method, through the questionnaire survey, the impact of transactional psychological contract, relational psychological contract, and developmental psychological contract on the tacit knowledge sharing willingness of knowledge workers is collected. After several analyses and verification hypotheses, the following conclusions are drawn: Transactional psychological contract has a negative correlation with the tacit knowledge sharing of knowledge workers; relational psychological contract has a positive correlation with the tacit knowledge sharing willingness of knowledge workers; developmental psychological contract has positive influence on tacit knowledge sharing willingness among knowledge workers.

Keywords Knowledge workers · Transactional psychological contract · Relational psychological contract · Developmental psychological contract · Knowledge sharing

1 Introduction

With the rapid development of technology and society, knowledge is recognized as a key strategic and competitive resource for enterprises [1]. Knowledge sharing is increasingly important to the development of the enterprise. The results of Frappaolo's research show that 42% of the total knowledge owned by companies comes from the tacit knowledge of employees in the organization [2]. HISLOP has found that

J.-H. Wu (✉) · H. Yan
School of Management, Wu Han Textile University, Wuhan 430073, China
e-mail: Rosewujianhua@sohu.com

H. Yan
e-mail: yanhao1208@foxmail.com

© Springer Nature Singapore Pte Ltd. 2020
S. Patnaik et al. (eds.), *New Paradigm in Decision Science and Management*,
Advances in Intelligent Systems and Computing 1030,
https://doi.org/10.1007/978-981-13-9330-3_23

when employees' psychological contracts change, their attitude toward knowledge sharing will also change. Knowledge workers have a strong desire for self-realization and pay more attention to the platform and space for development [3].

Theoretical significance, most of the existing psychological contract research focuses on the impact of psychological contract on employee satisfaction, turnover rate and organizational performance, and the research objects are all employees. The research on knowledge-based employees, which is an important object in the organization, is obviously insufficient. The existing knowledge sharing research mainly involves the related role of knowledge sharing and incentive mechanism [4]; then, it is to introduce the experience of organizational knowledge sharing success. There are few related studies on the willingness of employees to share knowledge in enterprises. Therefore, this paper takes knowledge workers as research objects and explores the impact of psychological contracts on employees' willingness to share knowledge. It has a certain significance for the development of psychological contract theory and knowledge sharing theory.

Realistic meaning, the process of knowledge sharing involves complex psychological factors and cultural characteristics such as the environment. It is difficult to achieve the desired effect by blindly introducing foreign incentive strategies [5]. Today, when material life is gradually enriched, simply relying on material incentives can no longer meet the needs of employees. Through psychological contracts, employees are encouraged to engage in active knowledge sharing and transform personal knowledge into organizational knowledge, which enhances organizational competition.

Based on the previous studies, this paper takes Li Yuan's point of view that psychological contracts include transactional dimensions, relational dimensions, and developmental dimensions [6]. It is difficult to spread and copy tacit knowledge, which is the key to knowledge sharing. The knowledge sharing content selected in this paper is also tacit knowledge sharing. Knowledge-based employees are the main object of knowledge sharing. Combining the analysis of psychological contract and knowledge sharing, this paper selects knowledge-based employees as the research object and explores the influence of the three dimensions of psychological contract on the willingness to share tacit knowledge.

2 Psychological Contract

Early researchers believe that psychological contract is the expectation of both the organization and the employee. Rousseau's research suggests that a psychological contract refers to, in the context of the employment relationship, a subjective understanding and belief in the individual's obligations to each other [7]. Li Yuan, a Chinese scholar, pointed out that the essence of a psychological contract is the psychological experience that employees get reward or payment when they provide labor for the organization. Some scholar propose a psychological contract consisting of two dimensions of trading and relationship. Rousseau verifies that the psychological contract does have a transaction dimension and a relationship dimension. The former emphasizes the visible exchange of wealth and material incentives. The latter is an intangible, non-measurable content, such as good development and promotion opportunities. Due to different countries, cultural backgrounds, employment relationships, etc., there are also differences in the division of psychological contract dimensions. Combining China's time background and social characteristics, Chinese scholars found that the transaction responsibility dimension, the development responsibility dimension, and the interpersonal responsibility dimension together constitute the responsibility of the organization to employees [8]. A typical characteristic of knowledge workers is that they have a strong desire for self-realization. Li Yuan named it as a development dimension [9].

2.1 Knowledge Sharing

Knowledge can be divided into two categories: explicit knowledge and tacit knowledge [10]. Tacit knowledge is that the knowledge owners' subjective judgment on objective things takes a long time to accumulate, and it is difficult to describe in a specific and visual language, so there will be greater difficulty in spreading and sharing. Knowledge sharing is one of the important links in knowledge management. It refers to the process of knowledge spreading between individuals and groups. Knowledge management refers to the organization's desire to transform tacit knowledge into explicit knowledge to further enhance the competitiveness of the organization. Therefore, an important goal of knowledge management is to promote knowledge transfer and sharing within the organization.

2.2 Hypothesis

2.2.1 Transactional Dimensions and Knowledge Sharing

O'Neill et al., combine knowledge sharing with psychological contract and believe that the level of employee turnover and the willingness to share knowledge will have an impact on knowledge sharing. Dong Yue confirmed through research that the transactional psychological contract has a negative impact on the will of knowledge sharing; Hu Bingkun, with emotional trust as the intermediary, selects technical employees as the research object and finds that transactional psychological contract can play a negative prediction role for tacit knowledge sharing. The personal knowledge of employees is a valuable asset of the company. Based on the characteristics of the transactional psychological contract, employees will refuse knowledge sharing for their own value in the enterprise. The stronger the transactional psychological contract, the weaker the willingness to share knowledge. In summary, the Hypothesis 1 is proposed: the transaction-type psychological contract has a negative correlation with the willingness to share tacit knowledge.

2.2.2   Relational Dimensions and Knowledge Sharing

According to the social exchange theory, American scholar Blau pointed out that when people think that sharing can bring them good colleagues and high reputational evaluation, a strong sense of respect and achievement will make them participate in the behavior of knowledge sharing. Highly recognized, thereby, enhancing the willingness to share knowledge. He Mingrui et al., found that the relationship-type psychological contract has a positive correlation effect on knowledge-based knowledge sharing willingness; Dong Yue used organizational identity as a mediator variable to verify that the relationship psychological contract does have a positive impact on employees' knowledge sharing behavior; Hu Bingkun took the enterprise technicians as the research object and found that there is no significant influence between the relational psychological contract and the tacit knowledge sharing. In summary, we propose Hypothesis 2: The relationship-type psychological contract has a positive correlation with the willingness to share tacit knowledge.

2.2.3   Developmental Dimensions and Knowledge Sharing

The career development of employees is influenced by organizational performance and individual performance. Organizations and individual employees need to make contributions to each other's learning and development. Hu Bingkun found through empirical research that the developmental psychological contract has a positive predictive effect on the willingness of employees to share tacit knowledge. When employees believe that their contribution can make the development of the organization better and then enhance their professional development and social recognition, employees will actively recombine and innovate knowledge through knowledge sharing, enhance their competitiveness while achieving their own success of the business, their willingness to share knowledge will also increase with the development of the contract. Therefore, Hypothesis 3 is proposed: The developmental psychological contract has a positive correlation with the will of knowledge sharing.

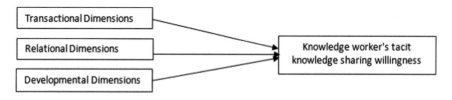

3   Method

The research object of this paper is knowledge-based employees, so the survey targets include knowledge workers such as enterprise technicians and managers. The survey was conducted in the form of online questionnaires. A total of 258 questionnaires were collected, including 12 invalid questionnaires and 246 valid questionnaires. The questionnaire qualification rate was 95.3%. The specific survey data is as follows: According to the gender, men and women accounted for 58.54 and 41.46%, respectively. From the age point of view, 98.78% of the respondents were over 20 years old, and those between 21 and 30 years old accounted for 82.93%. The group is younger. From the academic point of view, 78.05% of the populations are undergraduate or above, which is in line with the high-education characteristics of knowledge-based employees. From the perspective of the nature of enterprises, private enterprises are the mainstay, accounting for 71.95% of the total number.

3.1   Measurement Tools

For the scale of psychological contract, the main reference is the "Psychological Contract Scale" compiled by Rousseau. The scale has been recognized by many scholars and has strong authority, including transactional psychological contracts and relationships. Li Yuan believes that because of the strong self-realization desire of knowledge workers, the psychological contract of knowledge workers should also include the developmental psychological contract [9], and the development of localized developmental psychological contract scale is combined with the Chinese cultural background. Therefore, this article adopts Li Yuan's scale on developmental psychological contracts. The scale includes three types of transactions: the transaction type, the relationship type, and the development type of the psychological contract, each with three questions, six questions, seven questions, and a total of 16 questions. For the tacit knowledge sharing questionnaire, Lin's research scale of tacit knowledge sharing is mainly used, which has strong authority. At the same time, combined with the research of Bock et al., five questions about the willingness to share tacit knowledge are obtained.

4 Result

Using SPSS24.0 to analysis four variables. Reliability analysis, this paper uses Cronbach' α coefficient to analyze the reliability of the questionnaire, the results are as follows: transaction type, relational type, developmental type, tacit knowledge sharing willingness α coefficient are 0.754, 0.820, 0.924, 0.840. According to the relevant definition of reliability analysis, when the alpha coefficient is between 0.7 and 0.8, the questionnaire has higher reliability; four dimensions, the α coefficient of psychological contract and tacit knowledge sharing will be higher than 0.7, indicating that the questionnaires of all four variables have high reliability.

The relevant scales selected in this paper are authoritative scales widely recognized by scholars at home and abroad. Li Yuan and other scholars have also carried out relevant verification and modification in the social and cultural background of China, so they have good content validity.

For structural validity, the test is performed by factor analysis. Through SPSS software data analysis, it is concluded that the factoric load factor of the developmental dimension, transactional dimension, relational dimension, and tacit knowledge sharing will be greater than 0.75, 0.63, 0.7, 0.73 and the commonality is greater than 0.6. It is known that the three dimensions of psychological contract and the willingness to share knowledge are consistent with the corresponding relationship of 16 items. From the Bart sphere test, the KMO coefficient is 0.86, which is greater than 0.6, and the significance of the test P is 0.00, less than 0.05. The comprehensive questionnaire shows that the questionnaire has good validity.

5 Hypothesis Testing

There is a significant correlation between the transactional psychological contract dimension and the tacit knowledge sharing willingness. The significant P value is less than 0.05, and the correlation coefficient is $r = -0.420$, and the absolute value is greater than 0.4. Conclusions that have a negative correlation with the willingness of knowledge workers to share tacit knowledge. So, Hypothesis 1 is true. The significant P value between the relational psychological contract and the tacit knowledge sharing will be less than 0.05, and the correlation coefficient $r = 0.405 > 0.4$, that is, the influence of the relational psychological contract on the knowledge sharing willingness is significantly correlated and positively correlated. From this, we can see that Hypothesis 2 is true. The significant P value between the developmental psychological contract and the tacit knowledge sharing will be less than 0.05, and the correlation coefficient $r = 0.485 > 0.4$, that is, the developmental psychological contract is significantly correlated and positively correlated with the tacit knowledge sharing willingness. So Hypothesis 3 is true.

6 Discussion

In terms of theoretical contribution, this paper proves through empirical research that the various dimensions of psychological contract have an impact on the employee's willingness to share knowledge. Based on the current theoretical research, a supplementary explanation is given. The research confirms that the developmental psychological contract has a significant correlation effect on employees' tacit knowledge sharing willingness, enriching the research on psychological contract and knowledge sharing.

Practical implications: In the case of knowledge management, especially tacit knowledge sharing, in the case of a strong transactional psychological contract, the organization can fully stimulate employees' willingness to share knowledge through material incentives, and at the same time quickly redeem the contract rewards. For the strengthening of the relationship-type psychological contract, the organization needs to enhance the sense of belonging of employees, for example, through the construction of organizational culture and the creation of sharing the atmosphere, to promote the promotion of employees' willingness to share knowledge. If we want to enhance the willingness of employees to share knowledge through developmental psychological contracts, organizations should focus on the incentives of the spiritual level, guide employees to form a mindset of knowledge sharing, and provide career planning for employees in a targeted manner, giving employees a broader perspective growth space and platform.

7 Limitations and Future Research

This study has the following shortcomings: (1) For the selection of samples, mainly concentrated in the population of 20–30 years old, private enterprises account for the majority of the proportion; the conclusion of this paper is applicable to other age groups and employees of different natures need further argumentation; (2) The hypothesis of the model does not apply the mediator variable. In different contexts and contexts, the psychological contract may indirectly affect the employee's knowledge sharing behavior through some intermediate variables.

The outlook for the future is as follows: (1) The psychological contract will change with the changes of individuals and the related factors such as organization and external conditions. It is a dynamic development. At present, the current research is not strong enough, so in the future, we can pay attention to how psychological contract develops with organizational changes. (2) As a subjective personal feeling, psychological contract will be influenced by social background and cultural factors. The cultural traditions and thinking concepts of China and Western countries are quite different. The influence of China's unique culture, such as Confucian culture, on psychological contracts is also an important research direction.

Acknowledgements This research was supported by Ministry of Education Humanities and Social Sciences Project (14YJA630068).

Statement

This article was written by the author independently, without plagiarism, plagiarism, fraud, etc., in violation of academic ethics, and other infringements. Individuals and groups that have made important contributions to the research of this thesis have been clearly identified in the text. The legal results arising from this paper are entirely borne by the author of this article.

References

1. Ipe, M.: Knowledge sharing in organizations: a conceptual framework. Hum. Resour. Dev. Rev. **2**(4), 337–359 (2003)
2. Huang, R., Zheng, L.: Tacit Knowledge Theory. Hunan Normal University Press, Changsha (2007)
3. Hislop, D.: Linking human resource management and knowledge management via commitment: a review and research agenda. Empl. Relat. **25**(2), 182–202 (2003)
4. Haoren, Yan, Shenghua, Jia: On knowledge characteristics and enterprise knowledge sharing mechanism. Res. Dev. Manag. **14**(3), 16–20 (2002)
5. Li, T., Wang, B.: Research on knowledge sharing in knowledge worker organizations in China. Nankai Manag. Rev. **5**(2), 15–18 (2003)
6. Rousseau, D.: Psychological and implied contracts in organizations. Empl. Responsib. Rights J. **2**, 121–139 (1989)
7. Millward, L.J., Hopkins, L.: Psychological contracts, organizational and job commitment. J. Appl. Soc. Psychol. **28**(16), 1530–1556 (1998)
8. Li, Y., Sun, J.: Psychological contract in employment relationship: cognitive differences of "organization responsibility" in contracts from the perspective of organization and employees. Manag. World **11**, 101–110 (2006)
9. O'Neill, B.S., Adya, M.: Knowledge sharing and the psychological contract: managing knowledge workers across different stages of employment. J. Manag. Psychol. **22**(4), 411–436 (2007)
10. Dong, Y.: The influence of psychological contract on knowledge sharing: the mediating effect of organizational identity. Knowl. Econ. **2**, 21–24 (2016)

Research on the Role of Government Management in Chinese Public Indemnificatory Housing System

Wen Long

Abstract Public housing for any national governments is an important problem that is related to people's livelihood. Government plays a vital role in public housing. On the basis of reality, governments try to solve public housing with system, policy, management and etc. in different public housing management stages to give full play to the positive leading on finance, taxation, financial policy, also actively mobilize various, multi-level and multi-type of social economic subject participate in the process of public housing construction, establish and perfect the construction of public housing policy, capital, management system, and eventually improve the overall living conditions of all the residents.

Keywords Government management function · Management roles · Public housing management system

In the process of current social development, every national government has attached great importance to people's public housing issue, which concerns the national economy and people's livelihood for any country and government and is also a political and economic issue that attracts much attention. At the present stage of national development, whether the government, especially its management function, can properly solve people's housing problem and meet the housing needs of the people in different classes is related to the sustained and healthy development of national economy, the stable and harmonious economic life of the people at all levels of society and the overall level of their living and working in peace and contentment. Taking a comprehensive view of the current public housing policies all over the world, all the governments of developed and developing countries have taken the effective, proper and coordinated solution for housing problem of the residents at different income levels as an extremely important goal of government responsibility management, among which the top priority is the public indemnificatory housing issue. In the process of management, the governments around the world improve the housing conditions and living standards of residents at different levels through various ways

W. Long (✉)
Hunan Vocational College of Modern Logistics, Changsha, China
e-mail: 806135262@qq.com

© Springer Nature Singapore Pte Ltd. 2020   255
S. Patnaik et al. (eds.), *New Paradigm in Decision Science and Management*,
Advances in Intelligent Systems and Computing 1030,
https://doi.org/10.1007/978-981-13-9330-3_25

such as systems, policies and management methods and ultimately achieve the goal of improving the overall living conditions of all residents.

Since the implementation of the economic policy on reform and opening up in the late 1970s, Chinese economy has maintained a high-speed growth, and people's living standards have been constantly improved. However, this period of rapid transformation is also a stage when the economic conflict and the contradiction among the different strata in China are prominent. Chinese government has gradually established a new residential housing management system that can adapt to the development of the current market economic system during the long-term process of deepening and promoting the reform and management of housing system. In this system, the influence of market-oriented management on the allocation of housing resources is becoming more and more obvious. Under this background, the overall living conditions of urban residents have indeed been greatly improved than that before the reform and opening up. However, the truth is that the successful transformation of the state economic management system and the proper protection of the legitimate interests of all social strata will directly affect the perfection of the market economic system and the harmonious and orderly development of the whole society. The "public" attribute of housing which is a quasi-public product is not reflected though highly marketized and commercialized housing distribution. The government plays an indispensable managerial role of undertaking the most important management responsibility in solving the problems on this public product. In view of the contradictions and difficulties highlighted in the process of national development, especially to solve the housing problems for most middle- and low-income families, a multi-level, multi-dimensional and all-round indemnificatory housing management system has been gradually established by Chinese government. However, due to uncertainties in the positioning and concept of public housing security, imperfect design of management system mode and unstable development situation in the government management, it is still difficult to adequately solve the prominent housing difficulties in China.

Throughout the management mode of the public indemnificatory housing in other countries and combining Chinese actual situation, there should be a scientific and effective mode of government management responsibility in public housing security rather than the general and vague generalization. The public housing security system in western countries is different from that in China: Western governments focus on forming more perfect operational rules, taking the social risk decentralized policy system as a management mode of public housing security, strictly dividing the specific responsibilities of national government, enterprise units and individuals at different levels, thus establishing the mature, fixed and effective management system. However, the current government management in China still shows some randomness in the establishment of public housing security system. There is no scientific and reasonable strategy in the system formulation, which is difficult to quickly deal with the existing social risks and various problems in the transition period of market economy. Whether according to the research theory of welfare state and government management function, or based on the development course of government housing system management and the actual situation of current economic development in China, it is of great practical significance to rebuild a new and more scientific, feasible and effective responsibility framework for the government public housing security management. Based on the content system of the basic responsibility for public housing security management of governments in various countries, the responsibility framework for public housing security management should be composed by planning responsibility, system building, financial responsibility and supervision responsibility of public housing management and other necessary responsibility sections. However, in different management system modes, the mode, focus and system of management responsibility of each government are distinct. During the process of public housing security management, whether the government can show the initiative and enthusiasm of management responsibility is an important indicator to measure the development level of society, economy and social security in a country. The high efficiency of government management is conducive to improving the living environment of social members, coordinating social relationships, alleviating the contradictions of social distribution, improving the overall level of social security, promoting the harmonious development of all social strata and achieving social equity.

The governmental management in the designing process of public housing security model must not just stay in the level of system formulation. In the process of public housing security management, how can the government effectively and truly undertake its deserved management responsibility is the key factor of making it to play an important role in promoting the sound development of society and economy and residents' living and working in peace and contentment. Therefore, no matter before, during and after the management operation in the all-round dynamic process of government management, the government should assume the responsibility of planning, resource supply, organization and implementation, distribution and supervision. In this process, it is certainly insufficient for the performance of government management responsibility to rely on the government's self-restraint. It is necessary to establish a set of perfect supervision and management mechanism to promote the government to fulfill its management responsibility more fairly, efficiently and effectively. This supervision and management mechanism calls for the establishment of a comprehensive and perfect system of legislative supervision, judicial supervision and public opinion supervision.

Under the current developing situation, the government must establish a perfect management system for the public housing security system, such as how to undertake the management responsibility, what responsibilities should be assumed and how to achieve an ideal division and cooperation of responsibilities with other responsible subjects, which has important realistic significance. At present, the management types for public housing policies of the government in various countries are divided into complete government regulation, complete market spontaneous equilibrium, government regulation aftermarket equilibrium, direct government regulation and indirect government regulation. The following conclusions can be drawn by various research and analysis: It is highly efficient to solve public housing issues by complete market spontaneous equilibrium, but it also highlights the phenomenon of unfair housing use; complete government regulation reflects neither fairness nor efficiency in the process of solving public housing issues; the effect of other management types is almost the same as that of complete government regulation. The relative experiments show that in each management type, if people from all levels can truly participate in the housing market, and the government can implement indirect control policies in the management process, the balance of housing market can be achieved fairly and effectively. According to the current multi-level, all-around and three-dimensional security system in China, the government should undertake different management responsibilities for the different components of this system. Firstly, the government must take the initiative to assume the comprehensive, integrated and direct management responsibilities; Secondly, the government should also bear the responsibilities of external supervision and comprehensive coordination. However, according to the current actual situation, there are still housing difficulties in most people with poor housing conditions under various government management modes. In addition, the living conditions of low-income residents have not been properly improved, and the problems of public indemnificatory housing have not been properly solved. The root cause of these problems is that the specific responsibilities of the government in public housing security management have not been set and the different division of management has not been implemented. At present, the urgent tasks are to formulate relevant laws and policies, establish a scientific and perfect system for government management responsibility and clarify "what the government must do and how to do it." Only by clarifying the specific contents of government management responsibility, defining the rights and obligations of government management and implementing the corresponding legal consequences, can the government be quickly and effectively promoted to play an important role in public housing security management.

The public housing security system established by government really involves the vital interests of every citizen. Citizens have the right to know, supervise, criticize and advise. Only when the people at all levels actively participate in it and the work of public housing security can be subject to the supervision of the whole society, can the expected management objectives be effectively achieved and the principles of fairness, openness, justice and efficiency be embodied. People's supervision over the government's public housing security management should include: the formulation of public housing security system, the source of land for public housing construction, the planning of public indemnificatory housing construction, the fund guarantee for public indemnificatory housing construction, public housing security supply and distribution, follow-up management, property management and other links and to ensure that each link is fair, open, just and is under the "sunshine." First of all, an efficient integrity system must be established to effectively supervise the actual situation and effect on the performance of government's management responsibility. At the same time, the relevant clean government system must be constructed. Moreover, it is necessary to improve the open system of government affairs, standardize the process of government management, rationalize it scientifically and make it efficient. Meanwhile, it should ensure that the channels for complaint and supervision are smooth and effective and enable the people to give full play to their active role in supervision and the supervision by public opinion.

Based on the comprehensive research and analysis, when the Chinese government plays a management role in public housing security, it must take full account of the actual situation of China's development, clarify government's responsibilities in the management process, combine the security level with the security capability and strengthen the government's responsibility in formulating and improving the legal system from the perspective of social fairness and justice. Besides, it is necessary to speed up the corresponding legislative process, put public housing security work in an important position of economic and social development and fix the housing security system scientifically in the form of laws. At the same time, the government should give full play to the positive guiding role of fiscal, tax, financial and other policy means, actively mobilize various, multi-level, multi-type social and economic subjects to participate in the process of public indemnificatory housing construction, establish and perfect the construction, fund and policy management system of public housing security, vigorously promote the improvement and development of the public housing management system to ensure the smooth progress of public housing security construction in China.

References

1. Ren, W.: Transformation and Reconstruction of Public Authority in the Period of Social Transition in China, vol. 3. The World and China Institute (2006)
2. Li, T.: Research on Government Responsibility in the China's Social Security System, vol. 8 (2007)
3. Wen, L.: Research on Governmental Responsibility in Public Housing Security, vol. 6 (2012)
4. Ma, J.: Research on Construction of Chinese Indemnificatory Housing System, vol. 5 (2011)

A Proposed Model Design Under the Concept of Supply Chain Management for Construction Industry in Mainland China

Luo-er Liao

Abstract This design is a model with supply chain management principle to enhance the operational effectiveness for construction industry in Mainland China which aims at providing an innovative management method based on integrating engineering techniques, individuals performance and the project information throughout supply chain management concept to contributing to construction enterprises for the purpose of enhancing the operational effectiveness as well as cost reduction.

Keywords Model · Supply chain management · Construction industry

1 Introduction

The study of proposed design model was initiated by the Vin Tan wharf construction project in Vietnam which is operated by CCCC Overseas Company, and this model aims at combining internal systems with external systems to optimize the operational effectiveness under the concept of supply chain management theory which is covering with suppliers from upstream and downstream of industrial chain to develop relationship management. To be an expected result, the proposed model would come out with a practical operation to integrate the aspect of engineering and commercial considerations, which could increase the operational effectiveness in this industry.

Currently, for manufacturing industrial field, there is part of project management models have been designed for a long period, Amy J.C. Trappey, Tzu-An Chiang*, and Sam Ke (2010) mentioned a study which aimed at developing an intelligent and smart workflow management system (IWMS) to manage complicated operating processes of in onefold project, even though, controllable operational flows and available resources are considered in project execution, there are boundaries existing in organizational management with uncertain dynamically balances with cooperative

L. Liao (✉)
Hunan Vocational College of Modern Logistics, Changsha, China
e-mail: liaoluoer@126.com

© Springer Nature Singapore Pte Ltd. 2020
S. Patnaik et al. (eds.), *New Paradigm in Decision Science and Management*,
Advances in Intelligent Systems and Computing 1030,
https://doi.org/10.1007/978-981-13-9330-3_27

partners. The IWMS makes a contribution in enhancing the effectiveness of project management without consideration of other parties in the industrial chain.

Sako [10] from University of Oxford introduced a very interesting research on business models for strategies innovation and technologies creation: The business management models are more appropriate to be existed without innovation, which is opposite to this proposed designed model. Mari Sako thinks of that even without innovative technologies and methods, the new business models still can play their own right to ensure business success.

Jaselskis and Ruwanpura [8] provided a study with utilizing applications of information techniques based on an independent system to realize RT project management, which consists of monitoring and analyzing functions in construction project management through applying for imagery gathered by digital cameras on pivotal location to record and transmit video through monitoring system, which aims at reducing manpower and realizing intelligent production; however, the complexity and unpredicted accidents in construction area would reject this method with a firm hand as well as lack of early warning system for this industry.

Chatzimouratidis et al. [3] referred to individual training items of construction site training with programmed self-instruction and online training courses in journal, but for this proposed designed research model, the remote classes would only be suitable for the employees who are working in particular district, which is difficult to adapt to diversity of construction project and lack of promptness.

The Elimam and Dodin [5] researched on integrating supply chain operations by scheduling processes optimization, and the most common definition of the term SC is "A system of suppliers, manufacturers, distributors, retailers and customers where material flows downstream from suppliers to customers and information flows in both directions." And each node enterprise could get a relatively equitable interest across the whole value chain. In fact, the majority of profits are taken by sub-contractors and designers, which means that operational methods of construction companies like EPC pattern is only able to acquire extreme low margin, which is discrepant to the theory. To make a change, designing a new management model is urgent, which is relying on improving project and service quality as well cost reduction with high efficiency.

Above all, it seems that construction industry must take a breakthrough with supply chain aspect. The management model should drive construction enterprises devote more efforts to not only inner operation but also outside business relationship management. To realize intensive management and increase core competitiveness for construction enterprises, the designed model with SCM concept would make it possible.

2 Design of Proposed CPWM-Model

The prior issue that affects the model feasibility is to establish a way to organize supply chain management theory with this industry. It needs to collect the finished project information and data in order to analyze data for the success of model adaptation with cost decreasing. Then, another issue for model establishment is the information collection, which is difficult for financial departments to offer accurate data as well as to gather information quickly and continuously. Namely, information cannot be the role to locate potential problem. As a consequence, it is obvious that information management is a big issue. There are two main forces pushing the development and innovation in construction industry to expand overseas market, which are customers and suppliers. The CCCC is confronting with various challenges from competitions not only from domestic but also overseas. In reality, construction industry companies take high risks in (legal, cost, quality, brand) and huge responsibilities for unreasonable low payback, which push this industry comes up with a new designed model under this current situation. It seems that applies an innovative [4] project management model for reality is an urgent affair. For example, the main purposes of CCCC are expanding overseas market, adapting to new competition, increasing profit and improving customer satisfactory.

First of all, ETI (Engineering Technology Innovation) system plays an innovative role of CPWM-Model, which is established for the purpose of utilizing new technologies and management methods to achieve a goal of cost advantage. The ETI system includes two key functions: 1. The R&D (Research and Design) aims at evaluating and searching possibilities of engineering technical development for appropriate construction project according to the technical aspect; 2. PA Group (Program Action Group) is designed for organizing construction plans and evaluating construction plan cost on economic layer (Fig. 1).

Then, HRM System is another component of CPWM-Model, which plays a value-added role. This system is mainly in charge of employee training and recruiting items. In addition, employee performance rating and payroll methods are also included [2]. With establishment of staff database, the employee' personal working experience would be recorded for reference to reanalyze their potentials for solving the issue of labor panic [1]. Also the payrolls and promotion channels for each employee under a fair condition would also be set in HRM for the purpose of encouraging people to relocate quicker and providing more opportunities with staff to find out suitable posts. To cooperate with the CPWM-Model, HRM will be divided into four sections: such as training and recruitment, employee performance, payroll evaluation, and occupation promotion (Fig. 2).

The role of SCM system is responsible for [6] evaluating and selecting suppliers; moreover, supply chain integration is the key function of this system. To some extent, to strengthen the relationship among suppliers and customers highly depends on SC integration to reorganize relationship network quickly. Information sharing is better for all parties to build trust and establish long-term strategic partnership. A common goal would be set for information and inventory sharing, which would effectively

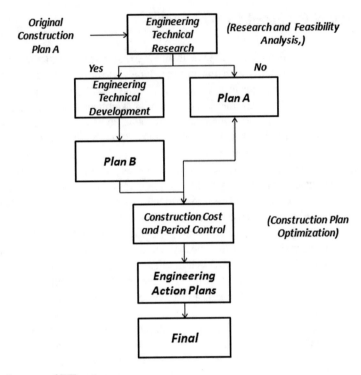

Fig. 1 Structure of ETI system

reduce the cost and wastage. Forecasting is another feature for this industry. Before, the suppliers and sub-contractors are drawing up logistics and construction plans according to the instructions given by EPC, which causes unpredictable delay for construction period, information could be transited thoroughly across the whole chain to earn the leading time for preparation through this system. All node parties make it visible upon business capacities and market understanding. All parties in this system commonly realize collaborative planning forecasting replenishment together to develop their effectiveness and decrease systematic risks such as period reduction, materials logistics, cost reduction, accurate prediction, services improvement and information, and inventory sharing (Fig. 3).

Finally, PI System is a role of connection of CPWM-Model. For the [7] purpose of transmitting information across the whole supply chain, the PI system establishes an intelligent information database to analyze and process, which is obvious to achieve the goals of getting feedback as soon as possible: The instruction from headquarter as well as responses is quite time-consuming. Moreover, the information efficiency is dramatically increasing. All nodes not only from inside but also outside are integrated by PI system. Moreover, there are some advantages for using this system such as comprehensive [9] development, operational effectiveness, and economic optimization. PI has the ability of monitoring and managing employee training matters

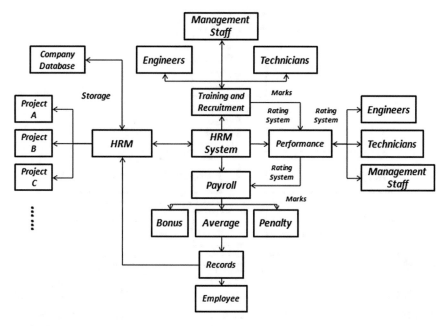

Fig. 2 Structure of HRM system

to develop specialized technical level in order to make engineering innovation work consistently. And PI is the bridge across the CPWM-Model (Fig. 4).

3 Case Simulation

For the purpose of verifying the feasibility of CPWM-Model in real construction project, a case simulation would be conducted by CCCC in Vinh Tan, Tuy Phong District; Binh Thuan Province, Vietnam. Except for that, it could be analyzed through a series of data to prove the impact of CPWM-Model adaption, and sorts of comparisons would be made for testifying model valid or not. First of all, the project consists of five parts, for instance, the auxiliary port basin dredging, an east breakwater with length of approximately 1,300 m, a 30,000DWT coal unloading terminal, a length of 1,343 m west breakwater and a 3,000DWT multipurpose terminal. Before CPWM-Model implementation, the estimated cost is more than 41 million US dollars. In detail, the coal terminal requires 162nos of Φ1000 steel pipe piles with 29–31 m long, 205nos of Φ800 steel pipe piles, and 21nos of Φ1000 bored piles. Moreover, it is predicted that the final cost would be even much higher than estimated cost according to the experience before. Through actuarial and assessment of this project, the general contract is worth around 43 million US dollars, the profit would be nearly around 4% low. With the simulation of CPWM-Model, the cost of steel pipe pile in

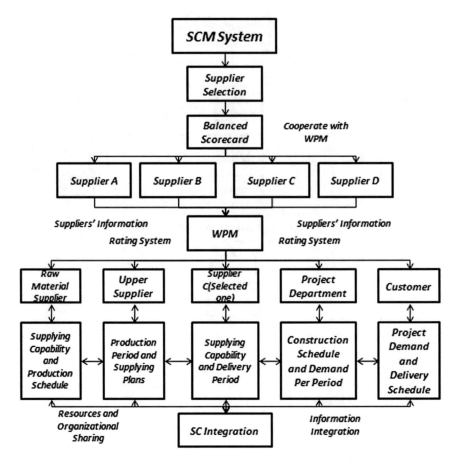

Fig. 3 Structure of SCM system

unit would be decreasing at 15–20% through ETI because of innovative technical creativities and the labor and machinery expenses are nearly decreasing 50%. On the other hand, the breakwater construction item requires about 1.14 million m^3 of ripraps, which is anticipated that there would be a loss with more than 38,000 Chinese-podes during the period of breakwater core stone settlement. However, VT Project integrates its supply chain through CPWM-Model thoroughly in this case, which is dramatically decreasing the loss from 10 to 3% and the overall cost of this construction project would be around 38,000,000 US dollars (Tables 1 and 2).

It can be seen from the tables above, the construction period is almost one month less than before, which also develops the brand reputation of CCCC, and more important is that the revenue of VT project is increasing from 4 to 12.4% after CPWM-Model simulation, which apparently justifies the function and efficiency of CPWM-Model in this project.

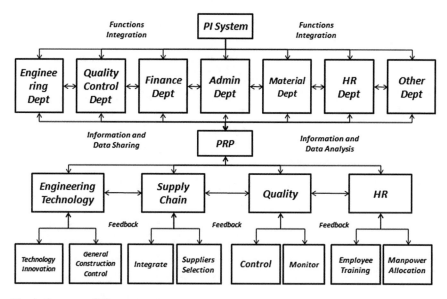

Fig. 4 Structure of PI system

4 Conclusion

Comparing to traditional management method, the new conceptual CPWM-Model testifies its own effectiveness in VT project, which is worth to be further studying on model adaptability and stability, which is promising to take construction industry to a new management stage. To utilize the SCM into construction industry is a bold attempt, which completely changed traditional construction project management with business operational methods and strategies in terms of concept of supply chain management ideology. With CPWM-Model, profit maximization and management optimization are realizing at last [4], which are driven by functional departments integration to operate better in financial performance. Speaking of information management, the model efficiently eliminates the barriers between departments to ensure implementation, which activates both sides from inside and outside across the chain. It also may connect all systems in model itself. In addition, the qualified suppliers would be selected correctly in a short period to guarantee project quality, which is in a stable situation all the time. As it is known that, the quality control is not only inspection matters, CPWM-Model would rather teach others than inspect. On engineering aspect, technicians could utilize this to update and trace data from project to reschedule their plans conveniently, which is beneficial for employees. On this circumstance, a good working environment is helpful for keeping in a high efficiency, which makes human resource management better. Daily works are arranged reasonably and objectively. In the meantime, this model is also highly supportive for

Table 1 Vinh Tan Construction period and cost in CPWM-Model

Construction period and cost (Days)		
Items	Simulation after implementation	Simulation before implementation
I Dredging section		
1. Basin area	60	82
2. Front berth	32	43
3. Approach channel	22	38
Construction period note: simultaneous	75	92
Overall cost (USD)	**4,150,123**	**5,216,312**
II Marine structure		
1. East breakwater		
(1) Dredging	22	45
(2) Concrete	131	154
(3) Anti scour rock	12	23
(4) Under layer rock	23	27
(5) Armour unit	14	18
(6) Other	30	31
2. West breakwater		
(1) Dredging	131	157
(2) Concrete	139	166
(3) Anti scour rock	84	108
(4) Under layer rock	125	130
(5) Armour unit	16	22
(6) Other	70	81
3. 3,000DWT multi		
(1) Dredging	79	95
(2) Pile	126	155
(3) Concrete	163	187
(4) Other	7	10
4. Bridge approach		
(1) Pile	34	48
(2) Concrete	22	28
(3) Super structure	92	128
(4) Other	5	9
Total days	374	426
Total cost (USD)	**28,223,814**	**30,427,977**

(continued)

Table 1 (continued)

Construction period and cost (Days)		
Items	Simulation after implementation	Simulation before implementation
III Equipment		
1. Fender accessories	62	86
2. Bollard accessories	85	107
3. Crane rail	22	38
4. Other	12	17
Total days	110	138
Total cost (USD)	**4,972,422**	**5,160,453**
1. Fire control	20	24
2. Water system	58	70
3. Power system	51	62
4. Environment system	10	11
Total days	90	118
Total cost (USD)	**192,343**	**202,852**
V Auxiliary systems		
1. Fire control	10	13
2. Water system	35	41
3. Power system	39	50
4. Environment system	3	7
5. Oil fence	3	7
Total days	62	83
Total cost (USD)	239,714	278,647
VI Corrosion protection		
Total days	57	66
Total cost (USD)	**24,529**	**25,258**
VII Navigation aids		
Total days	14	21
Total cost (USD)	**7,234**	**7,479**
VIII Environment		
Total days	7	14
Total cost (USD)	**42,530**	**42,897**
IX Tug boats		
Total days	3	7
Total cost (USD)	**223,500**	**360,700**

(continued)

Table 1 (continued)

Construction period and cost (Days)

Items	Simulation after implementation	Simulation before implementation
X Other items		
Total days	137	180
Total cost (USD)	**101,224**	**116,377**
Summary of days	398	436
Summary of cost	38,255,412	41,807,213

Table 2 Financial data analysis under CPWM

Assets	Model simulation	Pre-simulation
Current assets		
Cash flow	20,301,633	18,247,624
Accounts receivable		
Inventory		
Prepaid cost		
Short-term investment		
Total current assets	20,301,633	18,247,624
Fixed (Long-term) assets		
Long-term investment	5,220,000	
Property, plant, and equipment	7,627,000	7,350,000
(Less accumulated depreciation)	(2,079,400)	
Intangible assets		
Total fixed assets	13,174,600	7,350,000
Other assets		
Deferred income tax		
Other		
Total other assets	–	–
Total assets	33,476,233	25,597,624
Liabilities and owner's equity		
Current liabilities		
Accounts payable		
Short-term loans	2,455,000	2,455,000
Income taxes payable		
Accrued salaries and wages		
Unearned revenue		
Current portion of long-term debt		
Total current liabilities	2,455,000	2,455,000

(continued)

Table 2 (continued)

Assets	Model simulation	Pre-simulation
Long-term liabilities		
Long-term debt	2,455,000	2,455,000
Deferred income tax		
Other		
Total long-term liabilities	2,455,000	2,455,000
Owner's equity		
Owner's investment	22,110,170	20,800,622
Retained earnings	4,027,532	
Other		
Total owner's equity	26,137,702	20,800,622
Total liabilities and owner's equity	7,338,531	4,797,002

developing their professional skills to activate their motivations. The CPWM-Model produces a win-win situation not only for enterprise but also for individuals.

Even though, CPWM-Model proves its effectiveness, the potential risks objectively exist. Because of traditional management methods have been familiar to each one for a long period, it has to be considered that people are used to familiar managerial methods instead of CPWM-Model, which causes that model adaptability issue. The concept of CPWM-Model designation aims at managing construction project with open and transparent condition in common situations. Although new business patterns such as BOT (Build Operate and Transfer) and BT (Build and Transfer) projects are not proven by this model, which are uncertain to be testified yet, and it remains to be a question that whether they are the potential threats for CPWM-Model. Theoretically, the CPWM-Model is designed for commonly project management in construction industry, which remains to be discussed in the future to make a complement. For CPWM-Model stability, potential risks and threats are not obviously to be located in a short period. As is known, CPWM-Model is a systematic intelligent model with supply chain management principle, which is combined with systems and networks that are complicated for individuals to be aware of. As a consequence, early warning mechanism becomes necessary and urgent during its process. It could be anticipated that maintenance cost would be relative high and time-consuming from the start. Moreover, lack of confidence is another risk for this model, any mistake or inaccurate information would seriously damage the confidence and courage for people keeping trust to it any longer by this model, which indicates that a pre-warning system should be cooperating with model synchronously.

Acknowledgements This work is supported by the Hunan Provincial Natural Science Basic Research Program—General Project (17C1109): Research on construction project delicacy management and operational reengineering based on supply chain management.

References

1. Carden, L.L., Egan, T.M.: Human resource development and project management. Hum. Resour. Dev. Rev. **7**(3), 309–338 (2008)
2. Charles, M., Ye, K.: Human resource management in the construction industry. Int. J. Proj. Manag. **13**(2), 235–293 (2008). [Peer Reviewed Journal]
3. Chatzimouratidis, A., Theotokas, I., Lagoudis, I.N.: Decision support systems for human resource training and development. Int. J. Hum. Resour. Manag. **23**(4), 662–693 (2012)
4. Childerhouse, P., Towill, D.R.: Arcs of supply chain integration. Int. J. Prod. Res. **49**(24), 7441–7468 (2011)
5. Elimam, A.A., Dodin, B.: Project scheduling in optimizing integrated supply chain operations. Int. J. Proj. Manag. Surv., 447–456 (2013)
6. Fabi, B., Pettersen, N.: Human resource management practices in project management. Int. J. Proj. Manag. **10**(2), 81–88 (1992). [Peer Reviewed Journal]
7. Huemann, M., Keegan, A., Turner, J.R.: Human resource management in the project-oriented company. Int. J. Proj. Manag. **25**(3), 315–323 (2007)
8. Jaselskis, E., Ruwanpura, J.: Innovation in construction engineering education using two applications of internet-based information technology to provide real-time project observations. Int. J. Constr. Dev. Res. **335**, 147–187 (2016)
9. Lee, S.-K., Yu, J.-H.: Success model of project management information system in construction. Autom. Constr. **25**, 82–93 (2012). [Peer Reviewed Journal]
10. Sako, M.: Operation effective performance in project management [J]. J. Organ. People Perform. **342**, 155–171 (2012).

Empirical Analysis of Technological Sophistication of Exports in China's Manufacturing Industries from the New Perspective of TFP

Taosheng Wang, Huimin Bi and Zhenjun Cai

Abstract In order to evaluate the difference of export technology sophistication and its influence factors among manufacturing industries, this paper proposes a new method based on TFP for calculating the technological sophistication of exports in China's 30 manufacturing industries and then constructs a panel data model to analyze empirically the influencing factors of the technological sophistication of exports in China's manufacturing industry. The results show that the export technology content in technology-intensive industries is higher and rising fastest, the export technology content in capital intensive industries is lower but rising more obviously, and the export technology content in labor-intensive industries is lower and rising slowly. Human capital, the scale, and the development level of the industries and fixed capital stock have a significant effect on increasing the technological sophistication of exports in China's manufacturing industry. R&D expense and FDI have certain positive effects on increasing the technological sophistication of exports in the technology-intensive industries and the capital intensive industries in an inconsistent and insignificant way, and on the contrary, import trade, scale economy and profitability capacity of the enterprises may have negative effects on increasing the technological sophistication of exports in manufacturing industries.

Keywords Export product · Technological sophistication · Influencing factors · Comparison of different manufacturing industries · Empirical analysis

1 Introduction

In recent years, China has been ranked as one of major trading countries rather than being a trading power in the world. The export volume of China's manufacturing industry ranks the first in the world, which accounts for 90% of China's total export volume. The export growth rate in China's manufacturing industry is

---

T. Wang (✉) · H. Bi · Z. Cai
School of Business, Hunan International Economics University, Changsha, China
e-mail: wtsjs@126.com

© Springer Nature Singapore Pte Ltd. 2020
S. Patnaik et al. (eds.), *New Paradigm in Decision Science and Management*,
Advances in Intelligent Systems and Computing 1030,
https://doi.org/10.1007/978-981-13-9330-3_28

over twice the average rate in global manufacturing industry. However, assessing the international competitiveness of a country not only concerns its export capacity but also its export products [10], especially the products' quality and technological sophistication. How about the technological sophistication in China's manufacturing industry? How to calculate it as precise as possible? What are the main influencing factors of the technological sophistication in China's manufacturing industry? How to raise the technology level of China's manufacturing industry? All these questions have gained great attention from researchers across the world and are still under heated discussion.

Studies at home and aboard have been carried out into the calculation model of exports' technological sophistication. Rodrik [15] and Hausmann et al. [10] constructed the measuring method based on comparative advantages and income per capita. Given the assumption that exports' technological sophistication of one country is positively correlated to its income per capita, this method assumes that exports' technological sophistication of the exporting country equals to the weighted average value of the country's income per capita, with the weight being the revealed comparative advantage of the country's exports. This method is suitable for estimating the technological sophistication of a comprehensive package of exports by fully taking into consideration the influence of export's comparative advantage. However, this method may miscalculate exports' technological sophistication by ignoring the decisive influences of different factor endowment, technical progress, and technical efficiency in different countries and industries; then this method may overestimate the export's technological sophistication by paying too much attention to the export's comparative advantage of each country; besides, this method cannot be used to estimate the export's technological sophistication of intra-industrial industries of a country since the same income per capita may result in underestimation of the export's technological sophistication between and among industries.

Tentative studies have been done in calculating the technological sophistication of China's exports, which leads to two different results. On the one hand, some researchers hold that the technological sophistication of China's exports has increased rapidly. It is believed that the technological sophistication of China's exports is more than twice as that of other countries (in which countries the income per capita is relatively the same as that in China) and is almost the same as that of such world trading powers as the USA, Japan, and Germany, etc. [16]. On the other hand, other researchers hold that the technological sophistication of China's exports has increased slowly [2] and even fails to catch up with that of China before its reform and opening-up. Furthermore, when the influence of processing trade is excluded, there is no obvious increase [3, 11] or even a decrease in the technological sophistication of China's exports. Obviously, when calculating the technological sophistication of China's exports, researchers in the world mentioned above emphasize on the overall evaluation at the national level rather than the calculation and analysis of the exports' technological sophistication in each manufacturing industry, which deserves further exploration in this paper.

Recently, some researchers have preliminarily explored the influencing factors of technological sophistication of China's exports. Relevant results show that there is a clear positive relationship among international transfer of technology, innovation capabilities and export complexity [9, 12, 14], FDI, R&D, and intellectual property protection help to improve export technological sophistication [3], offshore input servitization can also drive the increase of export complexity [5] and giving full play to native comparative advantages also has a similar effect [1]. In addition, the export complexity is also related to the products density [4]; however, the impact of exchange rate changes may be uncertain [17]. Obviously, the previous literature studied the influence factors of export technology sophistication from the national level, but did not go deep into the industry level to explore the influence factors of export technology sophistication, which left a certain space for the exploration of this paper.

To answer the research questions and solve the remaining problems mentioned above, this paper will take the following approaches: Firstly, using the Malmquist index based on DEA, we select input–output vectors of 30 industries in China's manufacturing industry in 2000–2013, construct the best production practice frontier in each period and industry, estimate Total Factor Productivity (TFP) in China's manufacturing industries (Solow residue); then, we revise the model of calculating export technology sophistication proposed by Hausmann et al. [10] in two aspects. On the one hand, we substitute TFP in manufacturing industries for national income per capita, which helps fully reflect technical progress and changes in technical efficiency among manufacturing industries. On the other hand, we substitute the proportion of overall labor productivity in each manufacturing industry for the proportion of overall labor productivity in the manufacturing industry, with the aim of paying more attention to the weight of the comparative advantages of labor productivity in each manufacturing industry and less attention to the weight of the comparative advantages of export scale in each manufacturing industry. Therefore, based on TFP in manufacturing industries and the revealed comparative advantage, we construct the model for calculating the technological sophistication of China's exports, which is used to calculate the technological sophistication of exports in China's 30 manufacturing industries; finally, we construct the panel data model and testify the influences of input of research and development, input of fixed assets, labor capital, and FDI, etc. on the technological sophistication of exports in China's manufacturing industries through empirical research.

Compared with the previous studies, the features and innovations are as follows: Firstly, from the brand new perspective of exploring TFP in industries, we propose a new approach to calculate the domestic technological sophistication of exports in a country's industries; secondly, to lay the statistical foundation of our new approach, we calculate the TFP in China's 30 manufacturing industries and the decomposed indexes of technical efficiency and technical progress, using the Malmquist index based on DEA.

2 Calculation of the Technological Sophistication of Exports in China's Manufacturing Industries

2.1 Calculation Methods of TFP

Although the concept of TFP was firstly introduced in 1950, there is no consensus on the connotation of TFP among researchers at home and abroad. This paper defines TFP as the increasing output ascribed to technical progress and technical efficiency rather than factors such as capital accumulation or increased labor. Namely, it is a "productivity residual" in which factor inputs are excluded from the calculation. Such a residual was discovered by Solow in 1957 and thus is also called "Solow residue."

This paper adopts the Malmquist index based on DEA in calculation, which was firstly introduced by Sten Malmquist [13] and further extended and improved by Farrell [7], Forsund and Hjalmarsson [8], and Fare et al. [6] and was widely applied to TFP calculation. Advantages of this method are as follows: On the one hand, it contributes to a more precise dynamic analysis by decomposing TFP into technical progress and technical efficiency change; on the other hand, it is more appropriate to multi-sample panel data analysis since it can calculate TFP with marginal production function and distance function based on linear optimization.

In this paper, we regard each industry in manufacturing industry as a production Decision-Making Unit (DMU) and construct the best production practice frontier in China's manufacturing industry for each time period t.

Following Fare et al. [6], for each time period $t = 1,\ldots, T$, the production technology St models the transformation of inputs Xt into outputs yt.

$$St = \{xt, yt : xt \text{ can produce yt}\} \tag{1}$$

Following Fare et al. [6], we use technology in period t is the reference technology and defines the Malmquist productivity index as

$$M_p^t\left(x^t, y^t\right) = D_0^t\left(x^{t+1}, y^{t+1}\right)/D_0^t\left(x^t, y^t\right) \tag{2}$$

Alternatively, one could define a period-$(t + 1)$-based Malmquist index as

$$M_p^{t+1}\left(x^{t+1}, y^{t+1}\right) = D_0^{t+1}\left(x^{t+1}, y^{t+1}\right)/D_0^{t+1}\left(x^t, y^t\right) \tag{3}$$

In order to avoid choosing an arbitrary benchmark, referring to the typical Fisher ideal indexes, we specify the Malmquist productivity change index from period t to period $t + 1$ as the geometric mean of the two Malmquist productivity indexes in Formulations (2) and (3):

$$M_P\left(x^{t+1}, y^{t+1}, x^t, y^t\right) = \left[\frac{D_0^t\left(x^{t+1}, y^{t+1}\right)}{D_0^t\left(x^t, y^t\right)} \times \frac{D_0^{t+1}\left(x^{t+1}, y^{t+1}\right)}{D_0^{t+1}\left(x^t, y^t\right)}\right]^{1/2}$$

$$= \frac{D_0^{t+1}\left(x^{t+1}, y^{t+1}\right)}{D_0^t\left(x^t, y^t\right)} \times \left[\frac{D_0^t\left(x^{t+1}, y^{t+1}\right)}{D_0^{t+1}\left(x^{t+1}, y^{t+1}\right)} \times \frac{D_0^t\left(x^t, y^t\right)}{D_0^{t+1}\left(x^t, y^t\right)}\right]^{1/2} \quad (4)$$

In Formulation (4), $\left(x^{t+1}, y^{t+1}\right)$ represents the input and output vectors in period $t + 1$ while $\left(x^t, y^t\right)$ represents the input and output vectors in period t. If $M_P\left(x^{t+1}, y^{t+1}, x^t, y^t\right) > 1$, TFP grows; if $M_P\left(x^{t+1}, y^{t+1}, x^t, y^t\right) = 1$, TFP remains unchanged; and if $M_P\left(x^{t+1}, y^{t+1}, x^t, y^t\right) < 1$, TFP decreases.

2.2 Modifications to Methods of Calculating the Technological Sophistication of Exports

To calculate the technological sophistication of exports in a country's manufacturing industries, some modifications are required to the calculation method proposed by Rodrik [15] and Hausmann et al. [10].

First of all, according to Rodrik [15] and Hausmann et al. [10], the technological sophistication of exports is defined as

$$\text{EXPY}_k = l\left(\frac{x_{jl}}{X_j}\right)\text{PRODY}_l \text{ subject to } \text{PRODY}_k = \sum_j \frac{x_{jk}/X_j}{\sum_j \left(x_{jk}/X_j\right)} Y_j \quad (5)$$

In Formulation (5), j refers to the country or area, k refers to the product, X_j refers to the total export value of the country or area $\left(X_j = \sum_k x_{jk}\right)$, Y_j refers to the real GDP per capita in the country or area j, and l refers to a comprehensive package of exports from the country or area j. Formulation (5) illustrates the comparative advantages of one country's exports in the global market, and the internal relations between the comparative advantages and the country's productivity per capita, which is used to calculate the technological sophistication of exports between and among countries.

It is obvious that Y_j will lose its original meaning by directly using Formulation (5) to calculate the technological sophistication of exports from a country, since GDP per capita remains unchanged as for the export products from a country within a given period. In that context, we calculate the proportion of the export volume in the industry to the total export volume in the country and then multiply the proportion of the export volume in the industry to the total export volume in the country by the proportion of the world export volume in the industry to the world export volume. By multiplying the latter we can calculate the export technology sophistication in a country's industries. Thus we believe it is debatable or even incorrect to calculate the technological sophistication of exports from a country by using directly Formulation

(5), since the technological sophistication of exports does not equal to the revealed comparative advantage.

We make some tentative modifications to Formulation (5) in order to calculate the technological sophistication of exports in a country's industries. Firstly, we introduce and substitute TFP based on "Solow residue" for GDP per capita (Y). Thus the technological sophistication of exports from a country can be defined as

$$\text{PRODY}_{ik} = \sum_{jm} \frac{x_{jik}/X_{jm}}{\sum_{jm}\left(x_{jik}/X_{jm}\right)} \text{TFP}_{ji} \qquad (6)$$

As seen in Formulation (6), x_{jik} refers to the export volume of product k in industry i in the country or area j, X_{jm} refers to the export volume in the manufacturing industry of the country or area j, $X_{jm} = \sum_k x_{jik}$, and TFP_{ji} refers to TFP in industry i in the country or area j. Formulation (6) illustrates the internal relations between the comparative advantages of a country's manufacturing exports in the world market and the country's TFP. To certain extent, compared with GDP per capita, TFP based on "Solow residue" can better reflect the technical progress and technical efficiency of exports from a country and is therefore more suitable for calculating the technological sophistication of exports from a country. Meanwhile, TFP changes in different industries, so TFP can better reflect the different technological sophistication of exports in different industries.

To calculate thoroughly the technological sophistication of a comprehensive package of exports in the manufacturing industry (EXPY_{ki}), we weight PRODY_{ik} by factor productivity (FP) in industries, which can be defined as

$$\text{EXPY}_{ji} = l_i\left(\frac{fp_{ji}}{\text{FP}_{jm}}\right)\text{PRODY}_{ik} \qquad (7)$$

In Formulation (7), fp_{ji} refers to productivity of factors (labor and capital) in a manufacturing industry i in the country or area j, FP_{jm} refers to productivity of factors (labor and capital) in the manufacturing industry in the country or area j. TFP and FP reflect the productivity of a country's products in a complementary way, with the former from the perspective of the "soft" factors such as technical progress and managerial effectiveness, (Solow residue) and the latter from the perspective of the "hard" factors such as labor and capital. Thus TFP and FP can reflect the productivity and the technological sophistication of products in a country, an area, an industry or in an industry in an all-around way.

In sum, we believe that Formulations (6) and (7) can be used to calculate the different technological sophistication of exports in different industries of a country, the different technological sophistication of exports in different industries of different countries, and the different technological sophistication of exports from different countries (when we define TFP as the TFP in a country).

2.3   Data Source and Data Processing

In this paper, we make the exploring study on the exports of China's manufacturing industries in 2000–2013. However, the classification standard of China's manufacturing industries fails to go with the international classification standard, according to which the trade data are classified based on Standard International Trade Classification (SITC) and The Harmonization Code System (HS-Code). Thus data required in this paper cannot be collected from UNcomtrade database or WTO statistical database. To solve the problem of collecting data, we propose the following solutions:

Firstly, we give the correspondence of the tribit encoding product code (SITC Rev3) with China's manufacturing industries, to unify the classification standard of China's manufacturing industries and that of SITC.

Secondly, according to Formulations (6) and (7), we select the relevant data of the export volumes in the manufacturing industries both in China and in other countries in 2000–2013 from UN comtrade database.

Thirdly, based on Formulations of (4) as well as the input–output data of China's manufacturing industries in 2000–2013, we calculate TFP in China's manufacturing industry using the software of DEAP2.1.

2.4   Calculation of the Technological Sophistication of Exports in China's Manufacturing Industries

According to Formulations (6) and (7), based on the relevant data of China's manufacturing industries in 2000–2013, we can calculate the technological sophistication of exports in China's manufacturing industries in the sample period (see Table 1).

As seen in Table 1, the average technological sophistication index of exports in China's manufacturing industries in the sample period is 0.9148, but there is a significant difference among the average indexes in different industries. In detail, the average technological sophistication indexes of four industries are higher than 1.5, namely the Industry of Garments, Shoes, and Hats Production, the Industry of Instruments, Meters, Cultural and Office Machinery, the Industry of Communication, Computers and Other Electronic Equipment and the Industry of Tobacco Products; the average technological sophistication indexes of ten industries are between 1.0 and 1.5, namely the Industry of Electric Equipment and Machinery, the Industry of Metal Products, the Industry of Ferrous Metals Processing, the Industry of Textile Manufacturing, the Industry of Ordinary Equipment, the Industry of Raw Chemical Materials and Chemical Products, the Industry of Transportation Equipment Manufacturing, the Industry of Nonferrous Metals Processing, the Industry of Furniture Manufacturing and the Industry of Processing of Waste Resources and Materials; the average technological sophistication indexes of seven industries are lower than 0.5, namely the Industry of Beverage Production, the Industry of Leather, Furs and

Table 1 The technological sophistication index of exports in china's manufacturing industries in 2000–2013

Industries	2000	2004	2008	2012	Mean
Farm products and by-food processing	0.730	0.733	0.689	0.701	0.714
Food production	0.596	0.542	0.486	0.454	0.503
Beverage production	0.348	0.323	0.248	0.274	0.290
Tobacco products	1.519	1.707	1.694	1.693	1.613
Textile manufacturing	1.370	1.370	1.399	1.446	1.404
Garments, shoes, hats production	1.502	1.699	1.726	1.685	1.692
Leather, furs and down products	0.715	0.557	0.351	0.242	0.460
Timber, bamboo, cane, palm, straw products	0.627	0.529	0.746	0.883	0.704
Furniture manufacturing	1.085	1.200	1.238	1.233	1.173
Papermaking and paper products	0.017	0.019	0.027	0.023	0.021
Printing and record medium reproduction	0.213	0.313	0.374	0.349	0.323
Cultural educational and sports goods	0.418	0.492	0.534	0.622	0.507
Petroleum, coking and related processing	0.821	0.753	0.599	0.715	0.714
Raw chemical materials and chemical products	0.955	1.069	1.335	1.420	1.204
Medical, pharmaceutical products	0.390	0.296	0.289	0.320	0.308
Chemical fiber manufacturing	0.588	0.699	0.791	0.853	0.725
Rubber products	0.584	0.632	0.684	0.743	0.662
Plastic products	0.227	0.257	0.276	0.319	0.270
Nonmetal mineral products	0.725	0.769	0.986	1.042	0.891
Ferrous metals processing	1.266	1.366	1.592	1.756	1.489
Nonferrous metals processing	0.767	1.091	1.166	1.116	1.058
Metal products	1.389	1.556	1.458	1.435	1.446
Ordinary equipment	1.102	1.208	1.443	1.343	1.232
Special equipment	0.745	0.770	0.993	1.297	0.986
Transportation equipment manufacturing	0.803	1.045	1.268	1.241	1.089
Electric equipment and machinery	1.203	1.327	1.545	1.435	1.415

(continued)

Table 1 (continued)

Industries	2000	2004	2008	2012	Mean
Communication, computers, electronic equipment	1.484	1.499	1.661	1.856	1.617
Instruments, meters, office machinery	1.320	1.691	1.733	1.766	1.662
Handicraft and other production	0.003	0.009	0.010	0.016	0.008
Processing of waste resources and materials	1.138	1.210	1.328	1.269	1.265
Mean	0.822	0.891	0.956	0.985	0.915

Down Products, the Industry of Papermaking and Paper Products, the Industry of Medical and Pharmaceutical Products, the Industry of Plastic Products, the Industry of Printing and Record Medium Reproduction and the Industry of Handicraft and Other Production; the average technological sophistication indexes of the rest nine industries are between 0.5 and 1.0 (See Fig. 1).

In the sample period, the average technological sophistication indexes of exports in Industries shows an overall tendency of increasing, which increases by 0.013 on average per year. However, different industries show different increase rates. The average technological sophistication indexes of exports in six industries enjoy the highest increase rates (increases by 0.039): the Industry of Special Equipment, the Industry of Communication, Computers and Other Electronic Equipment, the Industry of Instruments, Meters, Cultural and Office Machinery, the Industry of Raw Chemical Materials and Chemical Products, the Industry of Nonmetal Mineral Products and the Industry of Ferrous Metals Processing. The average technological sophistication indexes of exports in 18 industries are between 0 and 0.03, which increases by 0.012 on average per year. The average technological sophistication indexes of exports in the rest six industries decreases by 0.013 on average per year.

From the perspective of the three classifications of industries based on different factor intensities (see Fig. 1), technology-intensive industries have the highest average technological sophistication index (1.125), which keeps on increasing at the rate of 0.024 in the sample period. The average technological sophistication index in capital intensive industries is 0.788, with the annual increase rate of 0.01. The average technological sophistication index in labor-intensive industries is as low as 0.767, with the annual increase rate of 0.003.

As shown in Fig. 1, the three classifications of industries all suffered from the international financial crisis in 2009 to different extents. The average technological sophistication index of exports in capital intensive industries was affected the most, with the fluctuation range of the average index larger than that in labor-intensive industries and technology-intensive industries, which is possibly due to the vulnerability of the capital intensive industries to the financial crisis. Nevertheless, the three classifications of industries regain the increase in 2010.

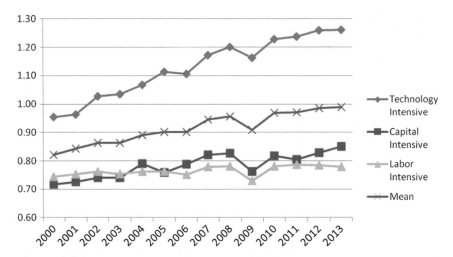

Fig. 1 The change tendency of the average technological sophistication indexes of exports in china's manufacturing industries in 2000–2013

3 Empirical Analysis of the Influencing Factors of the Technological Sophistication of Exports

3.1 Selection of Empirical Variables and Data Sources

To further analyze the influencing factors of the technological sophistication of exports in China's manufacturing industries, in this paper we select the following factors as the main explanatory variables:

(1) Foreign Direct Investment (FDI)

The technology spillover effect of FDI has a positive effect on the technical progress in the host country. In recent years, the strategy of "market for technology" was implemented in China to attract more foreign investment, which helped to promote technical progress and technical efficiency improvement, and contributed to the increase in the technological sophistication of exports in China's manufacturing industries. Therefore, this paper selects FDI as an important analytical variable of the technological sophistication of exports in China's manufacturing industries. As it is difficult to obtain directly the FDI data in China's manufacturing industries, instead we use the ratio of the foreign investment actually employed in enterprises above designated size in an industry to the annual average of net value of fixed assets in the industry. The data are collected from *China Statistical Yearbook* and are processed using the same methods as that mentioned in part 2.2.

(2) Fixed Capital Stock (FCS)

Fixed capital stock in an industry reflects its current scale of production and management and its technical level of equipment. To some extent, FCS determines the product quality and technology sophistication in this industry. In this paper, the net value of fixed assets per capita in the industries is taken as one of the important analytical variables in the technological sophistication of exports in China's manufacturing industries, which is defined as the ratio of the annual average of net value of fixed assets in enterprises above designated size in an industry to the number of employees in the industry. The data are collected from *China Statistical Yearbook* and are processed using the same methods as that mentioned in part 2.2.

(3) Research and Development Input (R&D)

R&D input plays an important role in technological innovation and its application in an industry. To a great extent, it determines the development of new products and the technical content of export products in this industry. In this paper, R&D input is defined as the ratio of the R&D expenses in an industry to the added value in the industry. The data are collected from *China Statistical Yearbook* and are processed using the same methods as that mentioned in part 2.2. *China Statistical Yearbook on Science and Technology* can also provide the reference data.

(4) Human Capital (HC)

Human capital, as the advanced elements of R&D, production and management, plays an important role in improving the product quality and increasing the technological sophistication in the manufacturing industry. In this paper, HC is defined as the ratio of the number of employees with Bachelor degree or above in an industry to the total number of employees. The data are collected from the *China National Bureau of Statistics* (paid service). The official provides the micro-data of the export enterprises in China's manufacturing industry. The relevant data in 2000–2013 are available by classifying the industries according to their main business.

In addition, this paper introduces three controlled variables in the model, namely the variable of total scale (TS) in industries reflects the industry's overall level of development and the relative position in the manufacturing industry, which is defined as the ratio of the added value in enterprises above designated size to the added value in the manufacturing industry; the variable of average scale (AS) in industries reflects the influence effect of the enterprise scale in industries, which is defined as the ratio of the added value in enterprises above designated size to the total number of enterprises in the industry; the variable of profitability capacity (PC) in the industries is defined as the ratio of the total profit (total profit + total taxes) in an industry to the total amount of the industrial capital input (the annual average net value of fixed assets + the annual average net value of current assets). It reflects the profitability of all the capital invested. Data of the three variables are collected from the *China Statistical Yearbook* and are processed using the same methods as that mentioned in part 2.2. The import variable (IMP) in industries is measured by the ratio of the value of imports to the added value of industry in the industry, which reflects, to some extent, the effect of import penetration on the exports' technical level. The relevant data are collected from the UNcomtrade database and are processed using the same methods as that mentioned in part 3.2.

In this paper, we make the technological sophistication of exports in industries for the explained variable. The data are computed according to the Model (6), the Model (7) and the relevant data (Table 4).

3.2 Selection and Construction of the Empirical Model

Firstly, based on the main explanatory variables and sample data, we construct the basic panel data model as follows,

$$\text{EXTS}_{it} = \alpha_{it} + \beta_{it}\text{FDI}_{it} + \gamma_{it}\text{R\&D}_{it} + \delta_{it}\text{FCS}_{it} + \mu_{it}\text{HC}_{it} + \varepsilon_{it} \qquad (8)$$

In Model (8), the subscript i refers to the industry; t refers to the period; EXTS_{it} refers to the export's technological sophistication in the industries calculated in part 3, FDI_{it} refers to the proportion of FDI, $\text{R\&D}_{it}$ refers to the proportion of input of research and development in industries, FCS_{it} refers to the proportion of fixed capital stock in the industries, and HC_{it} refers to the proportion of human capital in the industries; α_{it}, β_{it}, γ_{it}, δ_{it}, μ_{it} represent, respectively, the regression parameters corresponding to the main explanatory variables, ε_{it} refers to the random error term.

Using the software of EViews 6.0, we conduct Stationary Test, Covariance Examination and Hausmann Examination. According to the results, considering all these factors, it is more reasonable to choose the Entity Fixed-effect Variable-coefficient model in our paper. Stationary Test and Covariance Examination show the relevant data is significantly integrated of order one, so we take the logarithms of the variables on both sides of the model. Thus Formulation (8) can be restated as

$$\text{LnEXTS}_{it} = \alpha_{it} + \beta_{it}\text{LnFDI}_{it-1} + \gamma_{it}\text{LnR\&D}_{it-2} + \delta_{it}\text{LnFCS}_{it} + \mu_{it}\text{LnHC}_{it} + \varepsilon_{it} \qquad (9)$$

Since it takes time to transform the factors, such as FDI, input of fixed assets and input of research and development, into actual productivity (namely it takes time for those factors to have effects on increasing the technological sophistication of exports), we take the form of first-order time order or second-order time order for the variables in the regression model, indicated, respectively, with the subscript $t-1$ and the subscript $t-2$.

The empirical model has brought into analysis of regression the main influencing factors of the technological sophistication of exports in the industries, despite of the influence of the controlled variables on the technological sophistication of exports in the industries. Therefore, we try to make the following adjustments to Model (9):

Table 2 Test result of the main parameters in model (10)

Cross-section fixed (dummy variables)			
R-squared	0.976233	Mean dependent var	−0.211383
Adjusted R-squared	0.961478	S.D. dependent var	0.54664
S.E. of regression	0.107289	Akaike info criterion	−1.342867
Sum squared resid	2.762599	Schwarz criterion	0.182574
Log likelihood	411.859	Hannan-Quinn criter	−0.738174
F-statistic	66.16273	Durbin-Watson stat	2.249916
Prob (F-statistic)	0		

$$\text{EXTS}_{it} = \alpha_{it} + \beta_{it}\text{FDI}_{it-1} + \gamma_{it}\text{R\&D}_{it-2} + \delta_{it}\text{FCS}_{it} + \mu_{it}\text{HC}_{it} + B_1 Z_{i,t} + \lambda_i + \varepsilon_{it} \tag{10}$$

In Model (10), $Z_{i,t}$ refers to the controlled variable vector that influence the technological sophistication of exports in the industries, including the variable of total scale (TS) in industries, the variable of average scale (AS) in industries, the variable of profitability capacity in industries, and the import variable (IMP) in industries; B_1 refers to the regression coefficient vector; and λ_i refers to the unobserved heterogeneity in industries, which controls the omitted influencing factors at the level of industries.

Taking the controlled variable into consideration, Model (10) adopts the form of logarithms that are delayed for a period in the estimation process.

3.3 Empirical Test and Result Analysis

To avoid the possible heteroscedasticity, by adopting the method of weighted least square (wls) and EView software, based on variable-coefficient fixed-effects, we conduct regression analysis of Model (10) and the estimation results are shown in Table 2.

As seen in Table 2, the overall goodness-of-fit is desirable, with the adjusted coefficient of determination being 0.961478, which shows the four main factors account for 96.1478% of the variation in the exports' technological sophistication in industries. The statistics of F is 66.16273, with the concomitant probability of 0.000000, indicating a significant influence of the independent variable on the dependent variable at the significance level of 1%. The value of D.W is 2.249916 (close to 2), which proves no autocorrelation in the model.

Since the great heterogeneity among the industries can hardly be observed and will undermine the valid calculation, we estimate the rest models with fixed effect (see Table 3 for the result).

Table 3 Regression result of the main parameters of factors influencing the exports' technological sophistication in industries

Variable	(1) $t-1$	(2) $t-1$	(3) $t-1$	(4) $t-1$	(5) $t-1$	(6) $t-1$	(7) $t-1$	(8) $t-1$
FDI	−0.0003	0.056* (0.0321)	0.057* (0.0381)	0.079* (0.0477)	0.075* (0.0391)	0.099** (0.0475)	0.089* (0.0463)	0.089* (0.0464)
FCS	0.332*** (0.0276)	0.251*** (0.0714)	0.286*** (0.0695)	0.248*** (0.0867)	0.316*** (0.0711)	0.274*** (0.0861)	0.287*** (0.0839)	0.287*** (0.0839)
R&D	−0.193*** (0.0451)	−0.146*** (0.05161)	−0.134*** (0.0503)	−0.152*** (0.0561)	−0.124** (0.0516)	−0.130** (0.0521)	−0.145** (0.0447)	0.037** (0.0508)
HC	0.275* (0.1974)	0.455* (0.2320)	0.466** (0.2258)	0.399** (0.2406)	0.493*** (0.2257)	0.436** (0.2382)	0.488** (0.2118)	0.472** (0.2323)
TS			0.453*** (0.0999)				0.452*** (0.0996)	0.442*** (0.0993)
AS				0.049* (0.0569)			−0.00335	0.027* (0.0551)
PC					−0.00639		−0.163*** (0.0467)	−0.138** (0.0439)
IMP						−0.143*** (0.0451)	−0.00563	−0.00502
Constant	−0.891*** (0.1405)	−0.673*** (0.1515)	0.097* (0.2250)	0.600*** (0.1739)	−0.728*** (0.1507)	−0.653*** (0.1727)	−0.06262	0.065* (0.2334)
Samples	420	420	420	420	420	420	420	252
R-squared	0.9266	0.9332	0.9369	0.9333	0.9351	0.9352	0.9323	0.9386

Note 1. Numbers in brackets below the regression coefficients refer to the standard errors of the regression coefficients; ***,** and * indicate, respectively, the regression coefficients are statistically significant at the significance levels of 1, 5 and 10%; 2. $t-1$ and $t-2$, respectively, refer to first-order time order or second-order time order of the three variables of FDI, R&D, and FCS; 3. we take the logarithms of all variables in the models except those in model (1); 4. data in Model (8) are collected from the capital intensive industries and technology-intensive industries, excluding data from labor-intensive industries

According to the regression result, HC shows a significant regression coefficient (averages 0.3455) in every model, which has passed the significance test of 1%. This indicates that human capital has a significant positive effect on increasing the technological sophistication of exports in the industries, possibly under the effects of technical research and innovation. TS shows a less significant regression coefficient (averages 0.4455), which has passed the significance test of 1%. This indicates that the scale and the development level of the industries also have a significant positive effect on increasing the technological sophistication of exports in the industries. FCS shows a relatively significant regression coefficient (averages 0.314), which has passed the significance test of 5%. This indicates that the stock of fixed assets in industries also has a positive effect on increasing the technological sophistication of exports in the industries. Besides, the influence of R&D expense and FDI is not consistent. They have an insignificant positive effect on increasing the technological sophistication of exports in the industries in second-order time order, but have a negative effect under other conditions. This indicates that it requires further study on how to adjust the strategy of attracting foreign investment and how to maximize the positive effects of R&D input.

Import trade in the industries shows a significant minus regression coefficient (−0.192). Thus import trade has an obvious negative effect on increasing the technological sophistication of exports in the industries, which confirms, to certain extent, the statement that processing trade will possibly turn China into a "World Processing Factory" and lead to the "Low-Locked" phenomenon. Out of our expectation, enterprise scale shows a relatively significant minus regression coefficient (−0.057). In fact, scale economy is not well realized in most of the manufacturing industries and needs in the slow and arduous procession toward the market structural adjustment.

To figure out whether the negative variables will have different effects on increasing the technological sophistication of exports in different industries, we construct the panel Model (8) based on classification of the industries, aiming to examine the influence effects of the variables, such as R&D expense, FDI and AS on the technological sophistication of exports in the capital intensive industries and the technology-intensive industries. As the result shows, compared with the labor-intensive industries, R&D and FDI have more positive effects on increasing the technological sophistication of exports in the capital intensive industries and the technology-intensive industries, with the regression coefficient significant at the significance level of 10% (0.114, 0.072), even though R&D and FDI are not more significant and consistent than other factors.

4 Conclusion and Suggestions

Using the Malmquist index based on DEA, this paper estimates the TFP in China's 30 manufacturing industries, and then proposes a new method based on TFP for calculating the technological sophistication of exports in China's manufacturing industry in an all-around way, and then constructs a panel data model to analyze empirically the influencing factors of the technological sophistication of exports in China's manufacturing industry. The results are as follows: (1) In the sample period, technical progress has an obvious positive effect on increasing the production and the technological sophistication of exports in China's manufacturing industries. In general, in the sample period, the technological sophistication of exports in China's manufacturing industries illustrates a consistent increasing tendency. To be more specific, the technological sophistication of exports in the technology-intensive industries increases at the fastest rate, the technological sophistication of exports in the capital intensive industries increases at a fast rate, while the technological sophistication of exports in the labor-intensive industries increases at a low rate. (2) Human capital, the scale, and the development level of the industries and fixed capital stock have a significant effect on increasing the technological sophistication of exports in China's manufacturing industry. R&D expense and FDI have certain positive effects on increasing the technological sophistication of exports in the technology-intensive industries and the capital intensive industries in an inconsistent and insignificant way, and on the contrary, import trade, scale economy and profitability capacity of the enterprises may have negative effects on increasing the technological sophistication of exports in manufacturing industries.

Based on the results of the research, we propose the following suggestions: While maintaining the traditional comparative advantage, China should give full play to the positive role of human capital, strive to increase the technological sophistication of exports in manufacturing industries, and accelerate the process of cultivating our new competitive advantages of export that is focused on technology, quality and international brand; while optimizing the industrial structure, it is necessary to raise the level of technology innovation capacity and accelerate the process of scales expansion in manufacturing industry and promote the sustainable development in the industries, aiming to cultivate comprehensive competitive advantages in the industry and industries; we should further optimize FDI structure and import structure, maximize the technology spillover effects of FDI and import and minimize their negative effects, in the hope of increasing the technological added value of our exports and thus avoiding being locked in the low-end of the global value production chain; we should not only increase R&D input but also increase the efficiency of research and development of new technology and new products; besides, we should deepen the reform of systems and institutions of economics and trade, stimulate the motivation of independent innovation and strengthen the protection of intellectual property and interests from innovation, and make every endeavor to occupy the high-end of the global value production chain through continuous innovation.

Acknowledgements This work was supported by the project of National Natural Science Foundation of China (Grant No. 71573082) and the Ministry of Education Humanities and Social Sciences Planning Project (Grant No. 14YJA790056).

References

1. Alcalá, F.: Specialization across goods and export quality. J. Int. Econ. **98**, 216–232 (2016)
2. Amiti, M., Freund, C.: The anatomy of China's export growth, University of Chicago Press, pp. 35–56 (2010)
3. Anwar, S., Sun, S.: Foreign direct investment and export quality upgrading in China's manufacturing industry. Int. Rev. Econ. Financ. **54**, 289–298 (2018)
4. Ding, Y., Li, J.: Product space, potential comparative advantages and export technological complexity. Aust. Econ. Pap. **57**(3), 218–237 (2018)
5. Fang, M., Yang, R.: Research on effects of input servitization on export technological complexity of manufacturing industry of China. Des. Autom. Embedded Syst. **22**(3), 279–291 (2018)
6. Fare, R., Grosskopf, S., Norris, M., et al.: Productivity growth, technical progress, and efficiency change in industrialized countries. Am. Econ. Rev. 66–83 (1994)
7. Farrell, M.J.: The measurement of productive efficiency. J. Roy. Stat. Soc. Series A (General) **120**(3), 253–290 (1957)
8. Forsund, F.R., Hjalmarsson, L.: Frontier production functions and technical progress: a study of general milk processing in swedish dairy plants. Econometrica **47**(4), 883–900 (1979)
9. Gandenberger, C., Bodenheimer, M., Schleich, J.: Factors driving international technology transfer: empirical insights from a CDM project survey. Clim. Policy **16**(8), 1065–1084 (2016)
10. Hausmann, R., Hwang, J., Rodrik, D.: What you export matters. J. Econ. Growth **12**(1), 1–25 (2007)
11. Koopman, R., Wang, Z., Wei, S.J.: How much of Chinese exports is really made in China? Assessing domestic value-added when processing trade is pervasive. NBER Working Paper, 14109, (2008)
12. Li, B.: Export expansion, skill acquisition and industry specialization: evidence from china. J. Int. Econ. **114**, 346–361 (2018)
13. Malmquist, S.: Index numbers and indifference surfaces. Trabajos de Estadisticay de Investigacion Operativa **4**(2), 209–242 (1953)
14. Petralia, S., Balland, P.A., Morrison, A.: Climbing the ladder of technological development. Res. Policy **46**(5), 956–969 (2017)
15. Rodrik, D.: What's so special about china's exports? China & World Econ. **14**(5), 1–19 (2006)
16. Schott, P.K.: The relative sophistication of chinese exports. Econ. Policy **23**(53), 5–49 (2008)
17. Thorbecke W., Pai H.K.: The sophistication of East Asian exports. J. Asia. Pac. Econ. **20**(4), 658–678 (2015)

ICT and Innovative Applications

Universal Layers of IoT Architecture and Its Security Analysis

Amir Abdullah, Harleen Kaur and Ranjeet Biswas

Abstract Internet of Things (IoT)-based intelligent devices have been actively used in clouds to provide various data ranging from personal healthcare to disaster response. High acceptance rate of IoT devices resulting day-by-day increment in the number of Internet's connected devices. The enormous growth of IoT devices and their reliability on wireless technologies make it vulnerable to cyberattacks which posing challenges for digital forensic experts. Security concern such as secure communication, access control, secure storage of information and privacy is major concern in IoT surroundings. The rapid growth of IoT services and devices resulted in the deployment of various insecure and vulnerable system. Furthermore, every single device including sensors that deployed in IoT network and each byte that is transmitted within the IoT network may come under scrutiny at some layer. This chapter identifies universal layers within the IoT architecture and discussed security vulnerability in all layers.

Keywords Internet of things · IoT architecture · IoT layers · IoT security

1 Introduction

The phrase "Internet of things" was first used by Kevin Ashton in 1999 [1] as the title of a presentation at Procter & Gamble (P&G) during implementation of radio frequency identification (RFID) for application in supply chain management. After that uses of this phrase "Internet of things" (IoT) becomes common to define any possible system that can be connected to the Internet for knowledge formation, data

A. Abdullah (✉) · H. Kaur · R. Biswas
Department of Computer Science and Engineering, Jamia Hamdard, New Delhi, India
e-mail: amirabdullah93@gmail.com

H. Kaur
e-mail: harleen.unu@gmail.com

R. Biswas
e-mail: rbiswas@jamiahamdard.ac.in

© Springer Nature Singapore Pte Ltd. 2020
S. Patnaik et al. (eds.), *New Paradigm in Decision Science and Management*,
Advances in Intelligent Systems and Computing 1030,
https://doi.org/10.1007/978-981-13-9330-3_30

collection and automation. It acquires information of objects through automatic identification technologies and integrates the information of that object into the information network through communications technologies where intelligent computing technologies get utilized to generate managed information. Such computer-generated managed information further gets utilized by the application layer as per need of end user. IoT system consists of sensors, gateways, processors and applications and each of these have their unique characteristics to form a useful IoT system. IoT integrates heterogenic objects, sensors and smart nodes that can communicate autonomously with each other without human involvement. IoT devices may provide physical actuation such as controlling thermometers, door locks, light bulbs, along with virtual actuation such as alarming a sensed event to users [2] through email or texting. Collection of sensor data through the web is also possible, such as air quality and weather, using Application Programming Interface (API), also known as virtual sensors [2]. Tremendous advancement in computation and communication technologies enables rapid development in IoT and its deployment for various purposes. IoT devices in the past few years expanded from wearable devices and body sensors to industrial monitoring sensors and home appliances [3, 4]. Enormous business opportunities which exist within the IoT domain have increased number of autonomous intelligent devices and services. A massive and real-time data flow can be produced from the huge number of Internet-connected objects through sensors [5, 6]. It will become important to collect and analyse appropriate raw data in the efficient way for more valuable information extraction such as associations of things and services to provide Internet of services or web of things.

IoT drives the rapid evolution along diverse applications and services such as smart cities, healthcare and smart transportation systems. The rise in computing and networking have expedited the big data era which significantly influenced our day-to-day life. Learning the habits of the end users to customise the audio services by the smart audio system of amazon echo and google home is recent example of IoT [7]. Many more intelligent devices continuously appearing including humanoid robots [8]. IoT brings serious security and piracy concerns along with the technological and societal advantages which hurdle the further deployment and development of IoT devices [9–13]. Huge number of devices involved in the IoT with intrinsic complexity offers unprecedentedly challenging and costly cyber protection from attackers. Gaining access to a single IoT device eventually lead to large-scale attacks. Traditional security mechanisms may not be effective and challenging to apply in IoT enabled devices. Many IoT researchers including service providers, device manufacturers, and consumers are focusing on security of IoT architecture as well as communication protocols and applications. Researchers are struggling to achieve end-to-end privacy and security by ensuring confidentiality, integrity and authenticity of the collected data from various IoT smart devices. Secure integration, transmission and aggregation at edge devices and middleware with privacy preservation of data within the IoT infrastructure are other challenges which cannot be efficiently copped with existing cryptosystems and protocol. Developing IoT context-aware, energy-saving and service-oriented lightweight cryptography and protocols are expected as an efficient solution.

Recently several architectures of IoT have been proposed because a single agreement of an IoT architecture layer cannot satisfy the need of heterogenic devices used in the layer which demands inherent security [14–18]. Data storage reliability, interoperability, scalability and service quality are the major concern to be considered for determining an acceptable IoT architecture. In this chapter, we explore IoT's various architecture layer and major security concern in each layer by identifying four universal architecture layers of IoT, such as perception layer, network layer, middleware layer and application layer. In this chapter, we discuss the functionality of application layer, middle layer, network layer and perception layer with their unique security needs. We also identified and highlighted several essential research challenges existed in each architecture layers of IoT.

2 IoT Architecture's Identification

Recently IoT has attracted a lot of interest due to its endless possibility across various applications. Working of IoT is possible through the integration of various technologies and related securities. Numerous applications resulted in a variety of requirements that IoT systems should comply with. The requirements varied significantly in the targeted application realm which demands complex technical systems with various performance expectations. Such requirement affects the architecture design which results in a range of IoT architectures with not only different set of components and functionalities but also various used terminologies. IoT architecture design can be considered as a concept of several hierarchical layers which generally can be categorized into three basic layers known as perception layer, network layer and application layer. The perception layer basically collects information from the surroundings and converts it into digital signal [17]. The network layer collects digital signal from the perception layer and transmits these signals through network and subsequently the application layer transferred digital signals into different contexts for the end user application [18]. Different technology involved in each IoT layer, even the technology of the very same layer might vary depending upon used devices and targeted application. Combination of the technology and devices is used as IoT to offer a range of services and applications, each with its own specific requirements and limitations. These devices and technologies themselves are highly heterogeneous which makes a complex and difficult IoT management. Sometimes a middle layer is also added to address this challenge and manage different types of service and end application need [17]. This layer is called middleware which collects data from the previous network layer to store them into the database and cloud. Data from this middleware layer can be further pulled through various API for desired application output without compromising with device's privacy. The middleware layer of the IoT architecture also provides ability to process data. The four-layer IoT architecture constituted by the above factors has been discussed in this chapter which can be utilized in the development of actual application. The four-layered architecture of IoT along with their corresponding technologies is shown in Fig. 1.

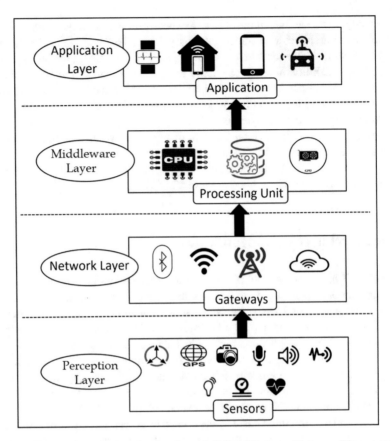

Fig. 1 Schematic representation of various layers in IoT architecture with its possible components

2.1 Perception Layer and Its Security Need

The perception layer of IoT architecture is considered as lowest physical layer which is known by many other names/terminologies, e.g. sensor's layer [19], recognition layer [15]. This layer consists of various sensors to acquire data from the surroundings as well as identifies other smart objects in the environment. The major technology of this bottom layer is sensors, RFID tags, cameras, wireless sensor network (WSN), etc. which is affected by computing power and energy [20]. This perception layer detects and collects data from the surroundings to process and then transmits it in digital format to the next network layer. The perception layer can be further divided into two sections, namely perception network which interconnects the network layer and the perception node (sensors, actuators, etc.). Sensor device of this IoT architecture layer can be easily damaged due to chances of working in a hostile environment. This can directly affect the efficiency of entire IoT system. There are significant security issues [16] in various technical domain of IoT perception layer as outlined below:

- The strength of digital data as most of such data is transmitted between IoT and sensor devices using various wireless network whose efficiency can be easily compromised.
- The sensors in IoT networks can be captured by the attackers as the sensors usually installed in outdoor environments. It also makes vulnerable to physical attacks on IoT network by tampering the hardware components of the device. Another confidentiality threatening attack is known as node capture attack, in which the attacker takes over the sensors to intercept all the sensitive data as well as addition of another node to send malicious data which threatens the integrity of this layer. This can consume the energy in the IoT devices leading to a DoS attack by disrupting it from sleep mode that always used to save energy [6].
- Third is due to the often movement of IoT nodes around different places resulting in intrinsic dynamic nature of network topology. The computation capability, power consumption and storage capacity of sensor devices in the IoT perception layer are very limited which further attract vulnerability to many kinds of attacks and threats. By altering or spoofing the identity information of one or more devices within the IoT network, confidentiality of the layers can be challenged. Timing attack is another threat in which the attacker can obtain the encryption key by analysing the time elapsed to perform the encryption.

The above-stated security concern at the perception layer can be solved by using end-to-end encryption, authentication and access control [9]. However, nodes at perception layer are short of compute power and storage capacity and hence unable to apply existing public key encryption algorithm for end-to-end protection. It is hard to set up end-to-end protection system at perception layer on the other hand attacks from the external network including deny of services attract new security issues. IoT nodes data still need the protection for authenticity, integrity and confidentiality. First of all, node authentication is crucial to prevent illegal and unwanted access; secondly, the confidentiality of information needs protection during transmission within the networks. End-to-end encryption of data is absolute need along with key process. However, stronger safety measures result in higher consumption of resources. To solve such requirement of more resources for stronger security, lightweight encryption technology has been suggested [21–23], which includes lightweight algorithm and protocol. Due to these factors, authenticity and integrity of sensor data become research priority [15].

2.2 Network Layer and Its Security Concern

The network layer is the most developed layer which process data and considered the brain of the IoT architecture [18]. It performs secure processing of data from the perception layer to the application layer via middleware layer. The network layer basically consists of mobile devices, cloud computing and the Internet [24]. This layer collects information from the perception layer and delivers to multiple applications at servers for initial processing of information, classification and polymerization which can be further processed and utilize. Internet gateway devices operate at this layer by using latest communication technologies such as LTE, WiFi, 4G, Zigbee and Bluetooth to provide various network-connected services. More precisely, this network layer serve as the intermediary between various IoT smart devices by filtering, aggregating and transmitting information from and to different sensor devices [25]. Data transmission depends on different basic networks, which are mobile communication network, Internet, wireless network, satellites, network infrastructure and other essential communication systems for the information exchange between devices.

The network layer is also vulnerable through various attacks including confidentiality, privacy breach, DoS attacks and passive monitoring [19]. Typically, attacks on the network layer affect information sharing among devices [26] and coordination of the works. These attacks have chances of recurrence due to data exchange and remote access mechanisms of smart devices. The communication in IoT devices is not restricted to only machine to human but also involves machine to machine communication which introduces many security issues of compatibility [27]. The major exchange mechanism in the IoT system should be secure enough for preventing any burglar from possibility of eavesdropping and identity theft. The heterogeneity of network components makes it tough to use the existing network protocols as is to produce efficient protection system. Protection of the network is very important in IoT devices along with protection of the objects within the network. IoT objects should have the capability to recognize the state of network as well as ability to defend themselves from such cyberattacks. Although the core network use in IoT has relatively safety protection capability, the counterfeit attack and middle attack still exist, meanwhile computer virus and junk mail cannot be underestimated. Hence, security system in the network layer is very important to the IoT.

Distributed denial-of-service attack (DDoS) is also known as a common attack method in the IoT network layer which is considered as severe in the IoT. Existing communication security mechanisms are difficult to apply for preventing the DDoS attack within the vulnerable device in this layer. IETF has made a great effort for assuring unique address of each individual connected device in the IoT network to forward IPv6 traffic within IoT architecture [28, 29] through implementation of 6LoWPAN protocol. Unique addressing and routing capability ensure unified integration of numerous IoT devices into a large network.

2.3 Middleware Layer and Required Security

This is third level layer in the IoT architecture which sometimes also referred as support layer [15] or processing layer [16]. The middleware layer processes and stores data obtained from the network layer and connect the IoT system to the database and cloud [26, 30] for further utilization by the application layer. There are many middleware solutions available to support interoperability and abstraction, which is the foremost requirement of middleware. This layer analyses, stores and processes an enormous amount of data using intelligent computing powers of cloud computing and network grids. It combines miscellaneous services of the downward network layer and upward application layer. Middleware layer is capable of automatically computation and processing information by employing latest technologies like cloud computing and big data processing [15]. Intelligent processing is limited for malicious information and hence improving the malicious information recognition ability is a big challenge [15].

Middleware layer is able to provide more powerful computing and storage capabilities to IoT by utilizing cloud computing which is continuously developing [31]. This layer also provides Application Programming Interface (API) to fulfil the demands of the various application which enables the utilization of same data by different personalized applications. Cloud security and database security are the main challenges in the middleware layer which may affect the service quality within the application layer. Middleware layer requires many application security architecture like strong encryption protocol and encryption algorithm, antivirus and stronger system security mechanism for secure multiparty cloud computation [32]. Mukherjee et al. [33] show implementation of the flexible security middleware in the IoT architecture for end-to-end encryption involving intelligent devices and applications. Managing irregular network connectivity along with device constraints in terms of energy, computational power, network bandwidth and memory was highlighted [33].

2.4 Application Layer and Security Challenges

Application layer is the topmost layer and known as termination level of the IoT architecture where the main purpose which is the creation of intelligent devices get accomplished. The application layer bridges the gap between the applications and the users by maintaining confidentiality, integrity and authenticity of the data [19]. This layer uses data from the middleware layer for implementation in required applications in various scenarios to provide excellent service to the end user. This layer is combined with industry expertise to accomplish an extensive set of smart applications [3, 4]. Personalized services as per the needs of end users achieved at application layer where users can access the IoT using application interface through Internet-enabled devices like, computer, mobile, tablet, television and many other wearable equipment [34]. Application layer integrates IoT network to build end applications, such as natural disaster monitoring, building condition monitoring for cultural and heritage conservation, intelligent transportation, health and medical monitoring, ecological environment monitoring, intelligent transportation protocols, smart grid, smart homes, smart cities and smart health [14, 35]. Many challenges related to the security in IoT devices arises due to unavailability of global policies to govern the development of such applications [19]. Different application environment requires different security measures [36]; however, information sharing is one of the major characteristics in the application layer which invites various data privacy issues such as disclosure of information and access control. Different authentication mechanisms used by different software in the application layer make difficult integration to ensure identity authentication and data privacy. The huge number of connected smart IoT devices that share data requires a control mechanism as well as awareness that what data will get discloses and how that data can be used, by whom and when. Further, IoT developers require considering responsibility for managing these applications in connected devices and the amount of data that will be revealed during different user's interaction. Application layer security concern can be resolved by arranging key agreement across various network and authentication as well as user's privacy encryption.

3 Conclusions

IoT has been introduced as the things which interconnect different types of devices anywhere anytime using the IP. The enormous growth of IoT devices and their reliance on wireless networks increase the vulnerability to cyberattacks in all four layers of the universal IoT architecture. Major security issues at perception layer appear due to the inability of existing encryption and authentication technology application for efficient operation. Lightweight cryptography is suggested solution to overcome such security issue as well as efficiency concern at perception layer. Routing and unique addressing capability to the integrated network of uncountable connected devices in a single accessible network is major concern in the network layer. These devices must be uniquely identifiable over a large network having unique IP address to reduce attack vulnerability which can be obtained by implementing 6LoWPAN protocol. The middleware provides an API for abstraction, data management, communication, computation which require specific techniques and tools for securing other IoT layers and nodes within the networks. The issue in application layer arises during the processing of sensitive information, such as malicious modification of the information, illegal access to the information and lifetime of access authorization. Attackers can inject code vulnerabilities to attack IoT network and gain sensitive information access for manipulation. Experts implement various forensic techniques by aiming to track internal and external attacks during investigation of such attacks. These techniques emphasize on the IoT's architectural vulnerabilities and communication mechanisms. IoT developers require to consider arrangement of key agreement and authentication across the heterogeneous network as well as user's privacy encryption by considering responsibility of managing these applications in connected devices.

References

1. Ashton, K.: In the Real World, Things Matter more than Ideas. That 'Internet of Things' Thing (2009)
2. Lee, U., et al.: Intelligent positive computing with mobile, wearable, and IoT devices: literature review and research directions. Ad Hoc Netw. **83**, 8–24 (2019)
3. Mohiuddin, I., Almogren, A.: Workload aware VM consolidation method in edge/cloud computing for IoT applications. J. Parallel Distrib. Comput. **123**, 204–214 (2019)
4. Alam, M.G.R., Hassan, M.M., Zi, M., Uddin, A.Almogren, Fortino, G.: Autonomic computation offloading in mobile edge for IoT applications. Futur. Gener. Comput. Syst. **90**, 149–157 (2019)
5. Sun, X., Ansari, N.: Dynamic resource caching in the IoT application layer for smart cities. IEEE Internet Things J. **5**(2), 606–613 (2018)
6. Karatas, F., Korpeoglu, I.: Fog-based data distribution service (F-DAD) for internet of things (IoT) applications. Futur. Gener. Comput. Syst. **93**, 156–169 (2019)
7. Zhang, N., Mi, X., Feng, X., Wang, X., Tian, Y., Qian, F.: Understanding and Mitigating the Security Risks of Voice-Controlled Third-Party Skills on Amazon Alexa and Google Home (2018)
8. Bahishti, A.A.: Humanoid robots and human society. Adv. J. Soc. Sci. **1**(1), 60–63 (2017)
9. Yaqoob, I., Hashem, I.A.T., Ahmed, A., Kazmi, S.M.A., Hong, C.S.: Internet of things forensics: recent advances, taxonomy, requirements, and open challenges. Futur. Gener. Comput. Syst. **92**, 265–275 (2019)
10. Qian, Y., et al.: Towards decentralized IoT security enhancement: a blockchain approach. Comput. Electr. Eng. **72**, 266–273 (2018)
11. Beecher, P.: Enterprise-grade networks: the answer to IoT security challenges. Netw. Secur. **2018**(7), 6–9 (2018)
12. Sadique, K.M., Rahmani, R., Johannesson, P.: Towards security on internet of things: applications and challenges in technology. Procedia Comput. Sci. **141**, 199–206 (2018)
13. Riahi Sfar, A., Natalizio, E., Challal, Y., Chtourou, Z.: A roadmap for security challenges in the internet of things. Digit. Commun. Networks **4**(2), 118–137 (2018)
14. Sethi, P., Sarangi, S.R.: Internet of things: architectures, protocols, and applications. J. Electr. Comput. Eng. **2017**, 1–25 (2017)
15. Suo, H., Wan, J., Zou, C., Liu, J.: Security in the internet of things: a review. Int. Conf. Comput. Sci. Electr. Eng. **2012**, 648–651 (2012)
16. Aziz, T., Haq, E.: Security challenges facing IoT layers and its protective measures. Int. J. Comput. Appl. **179**(27), 31–35 (2018)
17. Chen, K., et al.: Internet-of-Things security and vulnerabilities: taxonomy, challenges, and practice. J. Hardw. Syst. Secur. **2**(2), 97–110 (2018)
18. Silva, B.N., Khan, M., Han, K.: Internet of things: a comprehensive review of enabling technologies, architecture, and challenges. IETE Tech. Rev. **35**(2), 205–220 (2018)
19. Mahmoud, R., Yousuf, T., Aloul, F., Zualkernan, I.: Internet of things (IoT) security: current status, challenges and prospective measures. In: 2015 10th International Conference for Internet Technology and Secured Transactions (ICITST), pp. 336–341 (2015)
20. Li, L.: Study on security architecture in the internet of things. In: Proceedings of 2012 International Conference on Measurement, Information and Control, pp. 374–377 (2012)
21. Toshihiko, O.: Special Issue on IoT That Supports Digital Businesses Lightweight Cryptography Applicable to Various IoT Devices
22. Biryukov, A., Perrin, L.: State of the Art in Lightweight Symmetric Cryptography
23. Katagi, M., Moriai, S.: Lightweight Cryptography for the Internet of Things
24. Alaba, F.A., Othman, M., Hashem, I.A.T., Alotaibi, F.: Internet of things security: a survey. J. Netw. Comput. Appl. **88**, 10–28 (2017)
25. Leo, M., Battisti, F., Carli, M., Neri, A.: A federated architecture approach for Internet of Things security. In: 2014 Euro Med Telco Conference (EMTC), pp. 1–5 (2014)

26. Zhang, W., Qu, B.: Security Architecture of the Internet of Things Oriented to Perceptual Layer (2013)
27. Yaqoob, I., Hashem, I.A.T., Mehmood, Y., Gani, A., Mokhtar, S., Guizani, S.: Enabling communication technologies for smart cities. IEEE Commun. Mag. **55**(1), 112–120 (2017)
28. Naidu, G.A., Kumar, J., Garudachedu, V., Ramesh, P.R.: 6LoWPAN border router implementation for IoT devices on RaspberryPi. SSRN Electron. J. (2018)
29. Kavyashree, E.D.: 6LoWPAN network using Contiki operating system. In: NCICCNDA, pp. 300–310 (2018)
30. Farooq, M.U., Waseem, M., Khairi, A., Mazhar, S.: A critical analysis on the security concerns of internet of things (IoT). Int. J. Comput. Appl. **111**(7), 1–6 (2015)
31. Stergiou, C., Psannis, K.E., Kim, B.-G., Gupta, B.: Secure integration of IoT and cloud computing. Futur. Gener. Comput. Syst. **78**, 964–975 (2018)
32. Bittencourt, L., et al.: The internet of things, fog and cloud continuum: integration and challenges. Internet of Things **3–4**, 134–155 (2018)
33. Mukherjee, B., et al.: Flexible IoT security middleware for end-to-end cloud–fog communication. Futur. Gener. Comput. Syst. **87**, 688–703 (2018)
34. Yildirim, H., Ali-Eldin, A.M.T.: A model for predicting user intention to use wearable IoT devices at the workplace. J. King Saud Univ. Comput. Inf. Sci. (2018)
35. Jing, Q., Vasilakos, A.V., Wan, J., Lu, J., Qiu, D.: Security of the internet of things: perspectives and challenges. Wirel. Netw. **20**(8), 2481–2501 (2014)
36. Valmohammadi, C.: Examining the perception of Iranian organizations on internet of things solutions and applications. Ind. Commer. Train. **48**(2), 104–108 (2016)

Development of Model for Sustainable Development in Agriculture Using IoT-Based Smart Farming

Vinita Kumari and Mohd Iqbal

Abstract Agriculture is an important occupation which provides food to all over world. However, agriculture requires attention regarding farmland methane a greenhouse gas, and monitoring of different environmental parameters for controlled irrigation. Greenhouse gases are primarily responsible for global warming. Major contributor of farmland methane are ruminants like cows and buffalos which are often inherent part of farmland ecosystem. The solution could be achieved by using Internet of things (IoT) based smart farming. We have developed a model for IoT-based smart farming for sustainable development in agriculture which is capable of real-time monitoring of various farmland parameters. This model could be further used to reduce farmland methane emission and precise irrigation for reduced global warming as well as reduced water wastage in future. The monitoring and reduction of greenhouse gases and controlled irrigation can lead toward sustainable development.

Keywords Smart farming · IoT · Agriculture · Greenhouse gas · Precise irrigation

1 Introduction

In countries like India, agriculture is one of the main occupations and ruminant animals like cow, buffalo are integral part of our farmlands. These animals are source of milk and meat, and their products are helpful in providing manure to improve quality of soil in farm. However, enteric methane emission from these ruminant farmland animals is one of the major concerns for global warming [1, 2]. Methane is one of the greenhouse gases with the capability to trap four times more heat in comparison to carbon dioxide. Globally, ruminant animals are responsible for 30% methane emission. However, enteric methane emission can be modified by feed management for ruminants, particularly cow and buffalo [3]. Hence there is

V. Kumari (✉) · M. Iqbal
Department of Computer Science and Engineering, School of Engineering Sciences and
Technology, Jamia Hamdard, New Delhi, India
e-mail: vinita_bmi@yahoo.co.uk

© Springer Nature Singapore Pte Ltd. 2020
S. Patnaik et al. (eds.), *New Paradigm in Decision Science and Management*,
Advances in Intelligent Systems and Computing 1030,
https://doi.org/10.1007/978-981-13-9330-3_31

requirement for continuous monitoring of environmental methane level in order to enhance the productivity and reduction in methane emission for decrease in global warming. In addition to that monitoring of various environmental factors of farmland like soil moisture, temperature and humidity are also required.

It is also well known that annual yield of agriculture products relies heavily on environmental factors like temperature, humidity, soil moisture content, etc. If the environmental conditions of farmland are well maintained, productivity can be much enhanced [4, 5].

Internet of things (IoT) is a concept in which network of various physical devices, sensors and embedded system are capable of connecting, collecting and exchanging various data [6]. IoT-based smart farming including an embedded system and various sensors could be solution of these farming problems which can lead to sustainable development of agriculture [7].

2 Development of Model for IoT-Based Smart Farming for Sustainable Development

Sustainable development of agriculture can be achieved by enhancing the productivity of farmland and reducing greenhouse gas like methane from farmland which is responsible for global warming [2]. To achieve these aims, we have developed a model (Fig. 1) for providing smart farmland monitoring using IoT for different environmental parameters like temperature, humidity, soil moisture and local methane content. This model is capable of acquiring soil moisture, temperature, humidity and environmental methane content data using various sensors which are interfaced with the Arduino Uno board. In our model, the acquired data from farmland could be stored in cloud-based server in real time using IoT for further analysis and applications. To implement this model, Arduino platform is chosen as both the hardware and software are open source and freely available.

2.1 Embedded Control Unit

The embedded main unit is based on Arduino Uno microcontroller board that can be placed in farmland. It consists of a soil moisture sensor, an environmental temperature and humidity sensor and a gas sensor. It also includes a small DC motor water pump controlled by a 12 V relay. BC547 transistor connected to digital pin 11 of Arduino board has been used to drive this water pump. This pump can be triggered if soil moisture content is low or temperature is very high. A Nokia 5110 LCD module is also used for displaying various parameters under study.

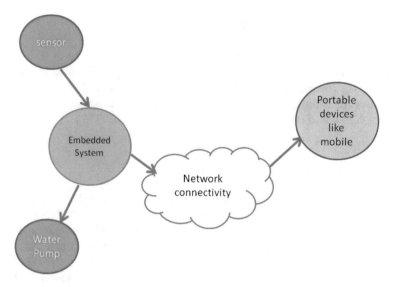

Fig. 1 Model for IoT-based smart farming for sustainable development

2.2 IoT Platform

The various parameters of farmland like temperature, humidity, soil moisture and local methane content can be monitored from remote devices with the help of IoT. The information about these parameters could be captured from respective sensors and sent to cloud using an IoT platform like Ubidots from Arduino Uno. Ubidots provides real-time dashboard for analyzing data and controlling devices [8].

2.3 Hardware Requirements

2.3.1 Arduino Uno Microcontroller Board

This board is designed to develop the applications and interactive controls using sensors, motors, actuators and other products. The hardware includes a high-performance 8-bit microcontroller ATmega328P. Arduino Uno board has the feature like: a USB interface, 14 digital I/0 and 6 analog inputs primarily for reading analog sensors but can also be used as GPIO pins, which allow interfacing with various other boards. Each of the digital pin can be used for input or output, using functions like pin Mode, digital Read and digital Write. Arduino Uno microcontroller board can be programmed using simple integrated development environment (Arduino IDE) using C or C++ [9].

2.3.2 Arduino Ethernet Shield

The Arduino Ethernet shield enables Arduino Uno board to connect with Internet to store data on cloud. The shield has a power connector, RJ45 Ethernet connector, connector for external USB board and micro SD card socket. For its working, the shield must be assigned a IP address and MAC address using the Ethernet.begin() function. The shield comes with pre-assigned MAC address [10].

2.3.3 Digital Humidity and Temperature Sensor (DHT11)

Term humidity represents water vapor present in the atmosphere while relative humidity is the function of both temperature and moisture content. Relative humidity is important as there is a change in relative humidity with temperature. DHT11 sensor is used for measuring both relative humidity and temperature. This sensor includes a humidity sensor and a NTC temperature sensor which generates digital output and can be interface with microcontroller like Arduino. It has attractive features like high reliability, good stability, low cost, low power consumption, tiny size and larger range of signal transmission [11].

2.3.4 Soil Moisture Sensor

The soil moisture sensor includes two probes for measuring soil moisture in terms of volumetric water content by using electrical property of soil. Through these probes, current passes through soil and then moisture value is measured by measuring soil resistance value. Higher the content of water in soil lesser is the resistance. So, dry soil represents higher resistance as the moisture level is lower. Sensitivity adjustment of the sensor can be made after inserting in soil. Relationship of the measured property and soil moisture level may dependent on the factors like soil type, temperature or electrical conductivity, so calibration is required [12].

2.3.5 Gas Sensor (MQ-6)

Farmland methane and other greenhouse gases can be monitored using gas sensors and storing the data on cloud using IOT or further machine learning and decision making regarding cattle feed management. MQ-6 is a gas sensor with six pins out of which four are used to fetch signals and two are used to provide heating current. This sensor is used in work to determine enteric methane emission in farmland which is one of the major concerns for global warming which can be modified by feed management for ruminants particularly cattle like cow and buffalo. This sensor has features like high sensitivity, fast response, stable and long life. The sensor includes tin dioxide sensing layer and a heater which provides necessary conditions for sensing [13].

3 Working Explanation

Working of our IoT-based smart farming model (Fig. 2) for sustainable development in agriculture is very simple. Level of methane in field can also be monitored the help of gas sensor (MQ-6) for management of cattle feed to achieve minimum methane emission which is a greenhouse gas. In this model, control of the entire process is done by Arduino Uno platform. In addition to environmental methane data, data of soil moisture, temperature and humidity level could also be acquired from soil moisture sensor and digital humidity and temperature sensor (DHT11). The captured data could be stored in real time on cloud-based IoT platform like Ubidots from Arduino Uno. The IoT platform Ubidots could provide real-time dashboard which can be accessed from portable devices like smartphone, computer, laptop or tablets from anywhere. The captured data on IoT platform could be used for further analysis and control. Thus, our model is also capable of detecting humidity, soil moisture level and temperature level which can be helpful for controlled irrigation.

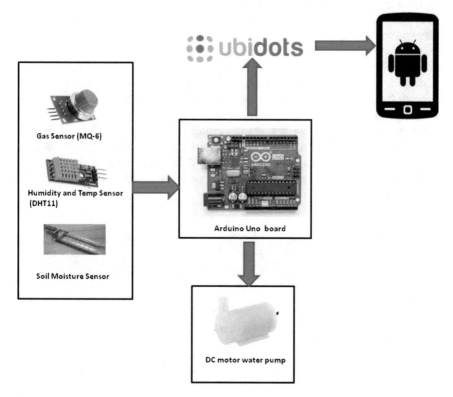

Fig. 2 Schematic diagram of the developed model

4   Results and Discussion

Keeping in view of issue of global warming, enteric methane emission is a major concern all over the world and could be controlled by feed management for ruminants like cows and buffalos. Our model provides a solution of this problem by continuous monitoring of farmland methane emission with the help of IoT using Arduino Uno microcontroller board interfaced with gas sensor MQ-6 and IoT platform Ubidots which could provide real-time dashboard access from portable devices like smart-phone, laptop or tablets from anywhere. In farmland, major source of methane are ruminant animals. The methane emission data can be helpful to diet management of ruminant to control methane level. In addition to that, our embedded system based model is capable of providing smart farming solution using IoT where differ-ent parameters like temperature, humidity and soil moisture could be monitored in real time using mobile devices. In developing countries like India, annual yield of agriculture products relies heavily on above environmental factors. Continuous mon-itoring of these environmental parameters of farmland is important for their better understanding and controlled irrigation [13].

The main embedded unit includes Arduino Uno microcontroller board interfaced with soil moisture, humidity, temperature and gas sensor and a small DC motor water pump which can be placed in farmland. The sensors have been used to sense the respective environmental parameters and the resulting data stored in cloud-based platform in real time for further study and analysis. This data can further used to estimate the requirement of farm irrigation and cattle feed management to reduce enteric methane emission.

Figures 3 and 4 show the reading of sensors at the cloud platform in dry soil condition and wet soil condition. In these figures, home page of Ubidots with the live data of farmland gas, soil moisture sensor is shown. It can be seen clearly that level of farmland gas is constant in both the conditions while soil moisture level is enhanced in Fig. 4.

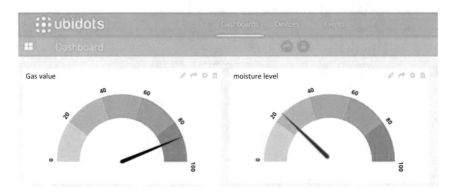

Fig. 3 Live farmland gas and moisture monitoring on smartphone using Arduino Uno system and Ubidots in dry soil

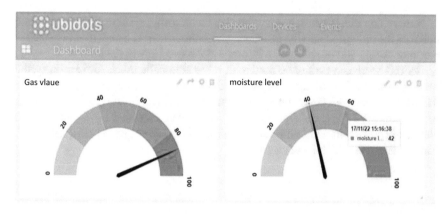

Fig. 4 Live farmland gas and moisture monitoring on smartphone using Arduino Uno system and Ubidots in moist soil

This work is done to provide IoT-based smart farming solution to address issue of global warming in order to achieve sustainable development goal in agriculture. In addition to that, our model is also capable of real-time monitoring of temperature, humidity and soil moisture in farmland to provide controlled irrigation and efficient water management.

5 Conclusion

In this work, a model for IoT-based smart farming system has been developed using Arduino Uno board to monitor farmland methane emission which major contributors are ruminants and various other environmental parameters like soil moisture, humidity and temperature which affect farmland productivity. In order to develop this model, Arduino Uno board, various sensors and Arduino Ethernet shield have been used for acquiring and storing the sensor data in the cloud-based server. Arduino Ethernet shield is used to connect our embedded device with Internet. The main source of methane is farmland animals like cows and buffalos which diet can be modified for reduced methane emission after studying data related to it.

Environmental parameter data are important to understand precise requirement of irrigation in farms which could be helpful in water conservation. Thus, our model can contribute significantly to sustainable development goals, i.e., reduction of global warming and water conservation in farmland.

Further, the acquired and stored data in cloud could be processed and analyzed to achieve the threshold values which could be set for controlling the motor and other devices in future.

Thus, using this IoT-based smart farmland monitoring system could be helpful to enhance the crop productivity, decreasing greenhouse effect by environment parameter monitoring and providing the information in real time to the farmer.

So, this work can further be used to develop automatic precise irrigation system leading to automation of farm, saving of water and reduced greenhouse gas emission. This can lead to reduction in physical effort and making cultivation accurate leading to cost-effective farming. This work can be further extended to test the soil for chemical constituent, nutrient level and salinity from remote and remedial measures can be taken with the help of experts.

References

1. Knapp, J.R., Laur, G.L., Vadas, P.A., Weiss, W.P., Tricarico, J.M.: Enteric methane in dairy cattle production: quantifying the opportunities and impact of reducing emissions. J. Dairy Sci. **97**, 3231–3261 (2014)
2. Eckert, M., Bell, M., Potterton, S., Craigon, J., Saunders, N., Wilcox, R., Garnsworthy, P.: Effect of feeding system on enteric methane emissions from individual dairy cows on commercial farms. Land **7**, 26 (2018)
3. Xia, F., Yang, L.T., Wang, L., Vinel, A.: Internet of things. Int. J. Commun Syst **25**, 1101–1102 (2012)
4. Brodt, S., Six, J., Feenstra, G., Ingels, C., Campbell, D.: Sustainable agriculture. Nat. Educ. Knowl. **3**, 1 (2011)
5. https://sswm.info/water-nutrient-cycle/water-use/hardwares/conservation-soil-moisture/crop-selection
6. Bamigboye, F.O., Ademola, E.O.: Internet of things (Iot): it's application for sustainable agricultural productivity in Nigeria. In: Proceedings of the 6th iSTEAMS Multidisciplinary Cross_Border Conference University of Professional Studies, Accra Ghana, pp. 621–628 (2016)
7. https://ubidots.com/platform/
8. https://datasheet.octopart.com/A000066-Arduino-datasheet-38879526.pdf
9. https://www.arduino.cc/en/Guide/ArduinoEthernetShield
10. https://www.mouser.com/ds/2/758/DHT11-Technical-Data-Sheet-Translated-Version-1143054.pdf
11. http://www.allianceforwaterefficiency.org/soil_moisture_sensor-intro.aspx
12. https://www.sgbotic.com/products/datasheets/sensors/MQ6-datasheet.pdf
13. Thessler, S., Kooistra, L., Teye, F., Huitu, H., Bregt, A.K.: Geosensors to support crop production: current applications and user requirements. Sensors **11**, 6656–6684 (2011)

Quality of Service Based Multimedia Data Transmission in Multi-constraint Environment

Krishan Kumar, Rajender Kumar and Amit Kant Pandit

Abstract Nowadays fast-growing multimedia data transfer applications based on quality of service. The quality of service depends upon the multiple constraints conditions. These multiple constraints provided the different level of quality of service for variety of user demands. To fulfill the quality of service demand, H.264/AVC video coder is used for multimedia data transfer. In the present work under multi-constraints environment multiple paths generated by the integration of modified A* prune algorithm and optimized technique. In this work rate and distortion are considered multi-constraints to achieve desire level of quality of service for mobile communication. Present work is validated with the software simulation using different constraints condition. In the result, k-number of multiple paths was generated with different bit budget of rate, distortion, and time complexity. With this novelty, present work is used for different application of video compression where different quality of service is required.

Keywords Video coding · Rate · Distortion · Multi-constraints shortest path · Motion estimation

K. Kumar (✉) · R. Kumar · A. K. Pandit
Department of Electronics and Communication Engineering, SMVD University, Katra, (J&K), India
e-mail: krishan.bpsmv@gmail.com

R. Kumar
e-mail: rajender.mtech@gmail.com

A. K. Pandit
e-mail: amitkantpandit@gmail.com

K. Kumar · R. Kumar
Department of Electronics and Communication Engineering, BPS Women University, Sonepat, (Haryana), India

© Springer Nature Singapore Pte Ltd. 2020   311
S. Patnaik et al. (eds.), *New Paradigm in Decision Science and Management*,
Advances in Intelligent Systems and Computing 1030,
https://doi.org/10.1007/978-981-13-9330-3_32

1 Introduction

Growing advancement in the wireless communication technology, data transmission is challenging task in respect of maintaining the quality and storage capacity of image or video. Error minimization and high compression ratio are two key parameters used for maintaining this task [1]. To find out most appropriate value of these parameters are the main topics of research in the field of video compression. Different video coding standards with different video compression techniques are developed over the last thirty years for finding out most appropriate value of these parameters. Video coding standards are the leading force to serve many applications of multimedia such as video conferencing, digital versatile disc (DVD), video on demand (VOD), digital television broadcasting, Internet video streaming, and digital camcorders [2]. Due to versatile nature of video compression in various video processing fields, different quality of service (QoS)-based compression will be required for different application. Increasing huge demand of video data transmission and also increase the user in multiplication nature there is requirement of verity of quality of service option for different variety of user in same application of video compression. To fulfill the above said requirement purposed algorithm is developed with the collective effect of A* prune algorithm and Lagrangian-based optimization with multi-constraint environment. In this work, multiple paths are generated with different bit budget for a particular application of video compression in constraint-based environment.

2 Related Works

The video storage and transmission mechanisms rely heavily on lesser error and compression technology in video coding standards. There are various international video coding standards [2, 3] as shown in Fig. 1.

The video sequence contains two types of redundancies, i.e., spatial and temporal redundancies. The compression of video frame can be done by removing these types of redundancies by intraframe coding or interframe coding. Removing these redundancies play vital role to achieve high video compression with acceptable degrade in the image quality. Quality of service of multimedia data transmission depends upon the different factor such as transform method, quantization, block partitioning method, motion estimation and compensation strategy, optimization strategy.

In spite of above parameters to increasing the performance of the video coding standards/encoder most prominently depends upon the size of block and the number of blocks which takes part in the encoding process. If number of blocks increases, motion vector increases and quality of the reconstructed frame is enhanced as well as complexity is increased and vice versa. To established tradeoff between complexity and quality a suitable block partitioning technique is required to control the motion vector by controlling the number of blocks as well as quality [4–6]. Further to improve video codec/encoder in terms of quality and compression a novel

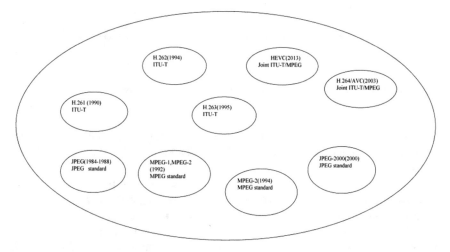

Fig. 1 Development of various video coding standards [2]

motion estimation and compensation technique is used by using efficient fast block matching algorithm. The main target of block matching algorithm is to minimizing the block matching error by using different error matching technique [7–10]. Quantization parameter is another most prominent factor used for the video compression. Level of compression depends upon the value of quantization parameter, high value of quantization parameter produced high compression and low quality of coded video stream and vice versa. It is necessary during designing the encoder, selection of appropriate quantization factor is mandatory step, and this is possible with the help of suitable optimization technique. Right selection of quantization parameter gives tradeoff between compression and quality [11, 12]. Optimization techniques also provided the tradeoff between the rate and distortion of encoded bit stream [13, 14]. Using A* prune algorithm and Lagrangian optimization techniques with multi-constraint-based environment methodology obtained k multi-constraint shortest path with different bit budget and quality of service parameter [15–17]. In this study found that multiple shortest paths are calculated for a given condition of constraints for a single application of multimedia such as video conferencing and mobile phones.

3 Framework for K-Multi-constraints Shortest Paths

Video compression techniques proposed in last few decades, and it can be divided into three phases as shown in Fig. 2 [18]. In the initial phase, problem definition for K-Multiple-Constrained-Shortest-Paths (K-MCSP) in video compression is presented in last section of work. The second phase is related to development of concept with the

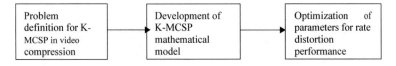

Fig. 2 Basic steps for development of k-multi-constraint shortest paths [18]

help of mathematical modeling. In last phase, different optimization techniques are used for finding out optimum value of rate and distortion for multi-constraint-based paths.

To understand the concept of the k multi-constraint shortest paths, it is necessary to understand the concept of network theory. Dijkstra's algorithm is used for finding out the distance of paths from source to destination in the given graph of present work [19, 20]. As bandwidth available for transmission is dynamic parameters, whenever, requirements changes, it is required to return multi-constrained-based shortest path (CSP) algorithm [21, 22]. A* prune algorithm fulfills the requirement of finding out the multi-constrained-based k-shortest paths with the integration of Lagrangian-based optimization algorithm. Concept of present work in this paper is explained with the help of data flow diagram shown in Fig. 3.

In this section, validation of present work had done with the help of mathematical equation. Let us consider a current frame (Fc) and reference frame (FRef) of given video sequences are represented with Eqs. 1 and 2. Total bit budget for a particular application of video compression, distributed among the motion vector (MVF) and residual error (DFD) express with the help of Eqs. 3 and 5.

$$\text{Current frame (Fc)} = \sum_{K<L=0}^{M,N} S(K, L) \tag{1}$$

$$\text{Referance frame} = \sum_{K<L=0}^{M,N} S^r(K + u, L + v) \tag{2}$$

where u and v are motion vector field as per given equation

$$\text{Motion vector field (MVF)} = \sum_{u,v=0}^{M,N} M(u, v) \tag{3}$$

Motion compensated reconstructed frame is given as

$$F(\text{MC}) = \sum_{K,L=0}^{M,N} S^r(K + u, L + v) - \sum_{u,v=0}^{M,N} M(u, v) \tag{4}$$

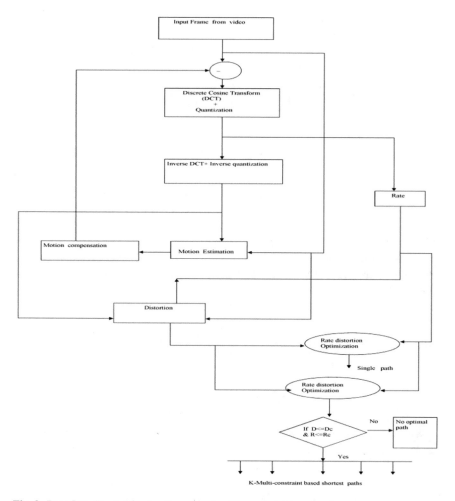

Fig. 3 Data flow diagram for development of multi-constraint-based paths

Than displaced field difference is defined as

$$F(\text{DFD}) = \sum_{K,L=0}^{M,N} S(K,L) - \left[\sum_{K,L=0}^{M,N} S^r(K+u, L+v) - \sum_{u,v=0}^{M,N} M(u,v) \right] \quad (5)$$

According to the problem definition, displaced field difference (DFD) is passed through quantization process and become quantized displaced field difference shown in Eq. 6. Finally reconstructed video sequence is represented by the combination of motion compensated reconstructed frame and quantized displaced field difference frame, shown in Eq. 7.

$$F(Q_{\text{DFD}}) = Q \left[\sum_{K,L=0}^{M,N} S(K,L) - \left[\sum_{K,L=0}^{M,N} S^r(K+u, L+v - \sum_{u,v=0}^{M,N} M(u,v) \right] \right] \quad (6)$$

$$F_{\text{reconst}} = F(MC) + F(\text{QDFD})$$

$$F_{\text{reconst}} = \sum_{K,L=0}^{M,N} S^r(K+u, L+v) - \sum_{u,v=0}^{M,N} M(u,v)$$

$$+ Q \left[\sum_{K,L=0}^{M,N} S(K,L) - \left[\sum_{K,L=0}^{M,N} S^r(K+u, L+v - \sum_{u,v=0}^{M,N} M(u,v) \right] \right] \quad (7)$$

To minimize the distortion in video compression algorithm, accurate motion estimation is requiring among the current frame and reference frame. This motion estimation is possible with the help of suitable matching criteria such as SAD, MAD MSE, and SSD. MSE is most appropriate method for motion estimation in the current work and represented in Eq. 8. Another most prominent parameter for quality of service measurement is PSNR derived with the use of MSE and represent in Eq. 9. Quality of service of encoded video sequence depends on the PSNR value. Good quality of service is achieved only with high level of PSNR value.

$$\text{MSE} = \frac{1}{M \times N} \left| \sum_{K,L}^{M,N} S(K,L) - \sum_{K,L=0}^{M,N} S^r(K+u, L+v) \right|^2 \quad (8)$$

$$\text{PSNR} = 10 \log_{10} \left(\frac{K^2}{\text{MSE}} \right) \quad (9)$$

With the help of mathematical equation, multiple paths are derived having different value of rate and distortion. Now according to Fig. 2, optimum paths are found among the set of paths with the help of multi-constraint-based optimization algorithm. In this paper, multi-constraint-based Lagrangian optimization is used for finding multi-constraints-based shortest paths. Below equation is used to find K-Multi-constraint-based Shortest Path.

$$\text{Cos} t(\text{Min}) = D(\text{Min}) + \lambda R(\text{Min})$$

$$\text{subject to} = D(\text{Min}) \leq D_C, R(\text{Min}) \leq R_C \quad (10)$$

where · is a Lagrange parameter provide the optimal bits allocation among the rate and distortion. In this algorithm by changing the value of multi-constraint condition, a set of different quality of service-based paths is obtained.

4   Results and Conclusion

In this paper, a modified multi-constraint-based algorithm is implemented with the integration of A* prune and Lagrangian optimization technique. The proposed work

Table 1 Rate-Distortion performance for reconstruction of Mother-Daughter Sequence with three frames skipping (frame 5 from 2) with multi-constraint (R = 1000, D = 2000) condition

Paths	Rate bits	Distortion bits	Total bits	PSNR	Time
1	668	1858	2526	33.88	133
2	722	1804	2526	33.88	137
3	776	1759	2535	33.88	140
4	848	1711	2559	33.88	144
5	932	1669	2601	33.88	148

Table 2 Rate-Distortion performance for reconstruction of Mother-Daughter Sequence with three frames skipping (frame 5 from 2) with multi-constraint (R = 950, D = 1800) condition

Paths	Rate bits	Distortion bits	Total bits	PSNR	Time
1	776	1759	2535	33.88	99
2	848	1711	2559	33.88	103
3	932	1669	2601	33.88	106

is simulated using MATLAB version 18.0 platform with i3 configured machine. In Table 1 k paths found under the multi-constraints condition (R = 1000, D = 2000) parameters for a specific application of video compression, these k paths having different bit budget as per the user requirement but within bit budget of rate and distortion. Similarly Table 2 shows that k paths under the different multi-constraint condition (R = 950, D = 1800) used for different video compression application these k paths also having different bit budget as per user requirement. So that novelty of work is to provide quality of service-based multiple paths obtained according to applied constraints condition for variety of user. Using purposed algorithm no need to develop a separate algorithm for generation of different quality of service-based paths, there is only need to change the constraint value for different application of video compression.

References

1. Wigand, T., Sullivan, G.J., Luthra, A.: "Draft ITU-T Rec. H.264/ISO/IEC 14496-10 AVC," JVT of ISO/IEC MPEG and ITU-T VCEG, Doc. JVT-G050r1 (2003)
2. Wigand, T., Sullivan, G.J., Bjøntegaard, G., Luthra, A.: Overview of the H.264/AVC Video Coding Standard. In: IEEE Transactions on Circuits and Systems for Video Technology, vol. 13, no. 7, pp. 30–37 (July 2003)
3. G.J. Sullivan, T. Wiegand, Video compression—from concepts to the H.264/AVC standard. Proc. IEEE **93**, 18–31 (2005)
4. Vaisey, D.J., Gersho, A.: Variable block-size image coding. In: Proceeding IEEE International Conference Acoustics, Speech, and Signal Processing (ICASSP), pp. 25.1.1–25.1.4 (April 1987)

5. Sullivan, G.J., Baker, R.L.: Efficient quadtree coding of image and video. IEEE Trans. Image Process. **3**(4), 327–331 (1994)
6. Sullivan, G.J., Baker, R.L.: Efficient quadtree coding of images and video. In: Proceedings of the IEEE International Conference on Acoustics, Speech and Singal Processing, Toronto, Canada (1991)
7. Jie-Bin, Xu, Po, Lai-Man, Cheung, Chok-Kwan: Adaptive motion tracking block matching algorithms for video coding. IEEE Trans. Circ. Syst. Video Technol. **9**(7), 1025–1029 (1999)
8. Zhu, Shan, Ma, Kai-Kuang: A new diamond search algorithm for fast block-matching motion estimation. IEEE Trans. Image Process. **9**(2), 287–290 (2000)
9. Cheung, Chun-Ho, Po, Lai-Man: Novel cross-diamond-hexagonal search algorithms for fast block motion estimation. IEEE Trans. Multimedia **7**(1), 16–22 (2005)
10. Rhee, I., Martin, G.R., Muthukrishnan, S., Packwood, R.A.: Quadtree-structured variable-size block-matching motion estimation with minimal error. In: IEEE Transactions on Circuits and Systems for Video Technology, vol. 10, no. 1 (February 2000)
11. Wang, H., Kwong, S., Kok, C.W.: Efficient predictive modal of zero quantized DCT coefficients for fast video encoding. Image and Vis. Comput. **25**, 922–933 (February 2007)
12. Viraktamath, S.V., Attimarad, G.V.: Impact of quantization matrix on the performance of JPEG. Int. J. Future Gener. Commun. Networking **4**(3) (September 2011)
13. Wiegand, T., Girod, B.: Lagrange multiplier selection in hybrid video coder control. In: Proceeding of International Conference on Image Processing, pp. 542–545, Thessaloniki, Greece (October 2001)
14. Everett 111, H., Generalized Lagrange multiplier method for solving problems of optimum allocation of resources. Oper. Res. **11**, 399–417 (1963)
15. Korkmaz, T., Krunz, M.: Multi-constrained optimal path selection. In: Proceeding of IEEE INFOCOM 2001, vol. 2, Anchorage, AK, pp. 834–43 (April 2001)
16. Kuipers, F., Mieghem, P.V., Korkmaz, T., Krunz, M.: An overview of constraint-based path selection algorithm for QoS routing. IEEE Commun. Mag. 50–55 (2002)
17. Liu, G., Ramakrishnan, R.: A*prune: an algorithm for finding K shortest paths subject to multiple constraints. INFOCOM (2001)
18. Schuster, G.M., Katsaggelos, A.K.: Rate Distortion based Video Compression. Kluwer Academic Publishers, Norwell, MA (1997)
19. Korkmaz, Turgay, Krunz, Marwan, Tragoudas, Spyros: An efficient algorithm for finding a path subject to two additive constraints. Comput. Commun. J. **25**(3), 225–238 (2002)
20. Sun, J., Sun, G.: SPLZ: An Efficient Algorithm for Single Source Shortest Path Problem Using Compression Method. Natural Science Foundation of China (No. 61033009 and No. 61303047) and Anhui Provincial Natural Science Foundation (No. 1208085QF106)
21. Chen, S., Song, M., Sahni, S.: IEEE, Two Techniques for Fast Computation of Constrained Shortest Paths, vol. 16, no. 1 (February 2008)
22. Santos, L., Coutinho-Rodrigues, J., Current, J.R.: An improved solution algorithm for the constrained shortest path problem, 756–771 (2007)

Optimization of PI Coefficients of PMSM Drive Using Particle Swarm Optimization

Sangeeta Sahu and Byamakesh Nayak

Abstract Permanent magnet synchronous motor has wide application in industries due to its various advantages. In this research work, the performance of the PMSM drive is boosted up with the help of particle swarm optimization technique by optimizing the coefficients of the PI controller. The same is achieved in the MATLAB environment using the optimization toolbox. ITAE criterion is implemented to control the speed of PMSM. The output torque, current and speed responses are compared with random and optimized PI coefficients.

Keywords PI control design · Permanent magnet synchronous motor · Particle swarm optimization · MATLAB environment · ITAE-based objective function

1 Introduction

Permanent magnet synchronous motors in comparison with induction motors have wide application in industries due to its various advantages like high efficiency, ease of control, low-maintenance cost, high-power factor to name a few. PMSM may be operated in constant power mode or variable power mode. In both modes with four quadrants' operation, it requires easier control algorithms, especially field-oriented control for generating between the switching signals for three-phase voltage source inverter. Control design plays a crucial part in control engineering for control of torque, speed and position in various industrial and domestic fields. The journey from the classical control method to the modern control methods during a span of 50 years has developed a lot of control algorithms starting from PI controller to LQ, robust, predictive, adaptive control or artificial control to list a few. In drive applications, though there are number of control algorithms, the industry communities prefer the simplest PI control for controlling the drives because of cost-effective and easy to

S. Sahu (✉) · B. Nayak
School of Electrical Engineering, KIIT University, Bhubaneswar, India
e-mail: sahu.sangeeta@gmail.com

© Springer Nature Singapore Pte Ltd. 2020
S. Patnaik et al. (eds.), *New Paradigm in Decision Science and Management*,
Advances in Intelligent Systems and Computing 1030,
https://doi.org/10.1007/978-981-13-9330-3_33

handle. The dynamic responses such as overshoot, settling time, good tracking and above all the stability of the system mainly depend on coefficients of the PI controller.

The response of PMSM drive is improved using a PI controller that is designed using Nelder-Mead optimization [1]. The advantage of fuzzy logic and PI controller is achieved using hybrid fuzzy PI speed controller that is faster and less computational [2]. The adaptive PID controller proposed in [3] contains three control terms which is quite straightforward and easily implementable with greater accuracy. It comprises tuning laws for online adjustment of control gains. PSO is applied to vector control PMSM for optimizing the efficiency of the motor. The value of d-axis stator current that is used to control and optimize the electrical loss is optimized using PSO. A comparative study is done to analyse the performance of multi-phase PMSM with several rotor topologies using finite element analysis [4]. An analysis is performed on the operation principle, motor design and control algorithms of a FOC-based PMSM using PID as a speed controller in a closed-loop operation [5]. A control strategy based on direct flux vector control is used for PMSM which does not require regulator tuning. Smooth speed control of PMSM using current and speed proportional-integral-resonant (PIR) control strategy is proposed [6]. The disadvantage of PID controller is overcome by a fractional order PID controller using fuzzy logic inference algorithm [7]. A fuzzy PI-type controller is proposed for PMSM which embeds dual control terms, i.e. a decoupling term and a fuzzy PI control term [8]. The speed step response of a PMSM is studied to derive a model-free self-tuning method for PI speed controller and the PI coefficient is tuned using the binary search algorithm [9]. The speed of a PMSM is controlled using rotor velocity that is calculated using rotor position tracking PI controller [10] by minimizing the rotor position error to zero.

2   PMSM Dynamic Model

The primary difference between the permanent magnet synchronous motor and the wound rotor synchronous motor lies with the flux production in the rotor. The main flux production in PMSM is constant but in the wound rotor synchronous motor it is variable, and it can be varied by changing of excitation voltage in the field coil. So, the back emf generated by the permanent magnet and that generated by an excited coil is alike. This result proofs that the mathematical model of a PMSM and a wound rotor synchronous motor is similar. The motor is inverter controlled, and the controlled output parameters are six switching signals of the semiconductor switches of the inverter. The PMSM dynamic model in the rotor reference frame can be written without considering the saturation and parameter variations effects. Neglecting eddy current and hysteresis loss in the PMSM, the stator d-q axes voltage equations in rotor reference frame is given as [11]:

Fig. 1 Phasor diagram of field-oriented control for PMSM

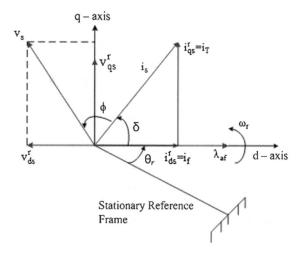

$$\begin{bmatrix} v_{qs}^r \\ v_{ds}^r \end{bmatrix} = \begin{bmatrix} R_s + L_q S & \omega_r L_d \\ -\omega_r L_q & R_s + L_d S \end{bmatrix} \begin{bmatrix} i_{qs}^r \\ i_{ds}^r \end{bmatrix} + \begin{bmatrix} \omega_r \lambda_{af} \\ 0 \end{bmatrix} \tag{1}$$

The torque developed in the machine is

$$T_e = \frac{3}{2}\frac{p}{2}\big[\lambda_{af} + (L_d - L_q)i_{ds}^r\big]i_{qs}^r \tag{2}$$

In order to decouple the flux component of stator and torque component of stator, the flux component of stator must be oriented along the rotor flux linkage line. The vector control, using field orientation, is achieved by orienting the flux component current along rotor flux linkage at every instant of time and it should be zero.

Figure 1 illustrates the rotor flux linkage that is rotated at a rotor speed ω_r and always located away from a stationary reference by the rotor angle θ_r. Therefore, the stator dq current should be rotated with the same speed ω_r so that d-axis component of stator current must be aligned to rotor flux position and rendering the stator flux current component zero by making $\delta = 90°$.

$$i_f = i_{ds}^r = i_s \cos \delta = 0 \tag{3}$$

Therefore

$$T_e = \frac{3}{2}\frac{P}{2}\lambda_{af}i_{qs}^{'r} = \frac{3}{2}\frac{P}{2}\lambda_{af}i_s = \frac{3}{2}\frac{P}{2}\lambda_{af}i_T \tag{4}$$

Since the machine is a permanent magnet, the λ_{af} is constant. Therefore, the torque can be written as

$$T_e = K_T i_T \tag{5}$$

The rotor dynamic equations in vector control mode are

$$K_T i_T = T_L + JS + B\omega_r \tag{6}$$

and

$$\theta_r = \int \omega_r dt \tag{7}$$

where i^r_{ds}, i^r_{qs}, i_f and i_T are stator direct axis, quadrature axis, field component and torque component currents, respectively. S is the Laplace's operator. δ is the torque angle. ω_r and ω_m are rotor electrical and mechanical speeds in rad/s.

3 Control Procedure for Vector Control of PMSM

Figure 2 is the block diagram of speed-controlled PMSM drive. It consists of the PMSM, speed sensor and current sensor, PI speed controller with limiter, hysteresis current controller and the three-phase voltage source inverter. The rotor speed that is noise polluted is sensed by the sensor and removed by the first-order low-pass filter. The error in speed between the reference speed and actual rotor speed is handled by the PI speed controller with a limiter to get the output of the torque reference. The limiter limits maximum torque production and PI controller nullifies the steady-state error in speed. The PI output is divided by the torque constant (K_T) to produce the rotor reference torque component current (i_T). The field component current must be zero in order to achieve the vector control. The above two currents are converted into three-phase stator abc current commands using Eqs. 8 and 9. The hysteresis current control (HCC) is used to generate six current switching pulses for inverter operation in feedback current mode by comparing each reference current commands generated and each actual sensed stator three-phase currents. The optimization of the coefficients of PI controller has been carried out by applying particle swarm optimization technique to improve the performances of responses.

$$\left[T_{dq} \right] = \begin{bmatrix} \cos\theta_r & \sin\theta_r \\ -\sin\theta_r & \cos\theta_r \end{bmatrix} \tag{8}$$

$$\left[T_{\alpha\beta} \right] = \frac{2}{3} \begin{bmatrix} 1 & -\frac{1}{2} & -\frac{1}{2} \\ 0 & \frac{\sqrt{3}}{2} & -\frac{\sqrt{3}}{2} \end{bmatrix} \tag{9}$$

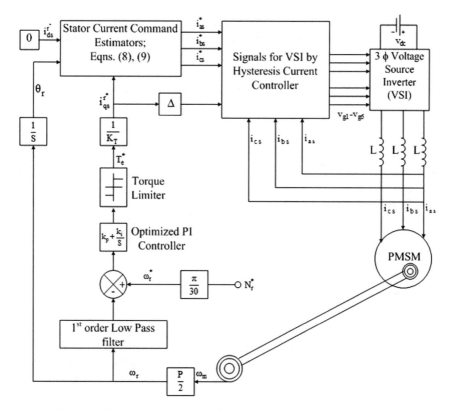

Fig. 2 Schematic diagram of the speed-controlled PMSM drive using FOC mode

4 Particle Swarm Optimization

In 1995, James Kennedy and Russell Eberhart introduced a nature-inspired meta-heuristic optimization technique known as particle swarm optimization (PSO) [12]. This technique replicates the societal behaviour in herd of birds. This algorithm tries to find the best possible solution by keeping a track of the potential solutions that are distributed randomly. This search is carried out by moving the set of particles in a multidimensional space which gives its own best position till then. The position and the velocity vectors determine each particle. Millonas, in the due course, found that the swarming birds pursue five principles [12] that are principle of proximity, principle of quality, principle of diverse response, principle of stability and principle of adaptability. Finally, the PSO technique is the movement of each particle in the N-dimensional space to its best position, and the best position is determined by their own performance.

Kennedy and Eberhart made a few minute changes and deletion of the variables and then ended up with the rule to calculate the next best position of the particle. The rule is:

$$v_{i,j}^{k+1} = wv_{i,j}^k + v_{i,j}^k c_1 r_1 \left(x\text{best}_{i,j}^k - x_{i,j}^k \right) + c_2 r_2 \left(x\text{gbest}_j^k - x_{i,j}^k \right) \tag{10}$$

$$x_{i,j}^{k+1} = x_{i,j}^k + v_{i,j}^{k+1} \tag{11}$$

where w is the inertia weight $x_{i,j}^k$ is the jth component of the position of ith particle in the kth iteration and $v_{i,j}^k$ is the jth component of the velocity vector of ith particle in the kth iteration; r_1 and r_2 are uniformly distributed random numbers in the range $(1, 0)$; $x\text{best}_i$ is the ith particle's best position so far and $x\text{gbest}$ is the whole swarm's best position so far. The particle's confidence in itself (cognition) is denoted as c_1 and the particle's confidence in the swarm (social behaviour) is denoted as c_2.

5 Simulation Results and Discussion

The procedure used in the industry for finding out the coefficients of the PI controller is to adopt the lower coefficient values and then progressively tune them till the best possible performance is achieved [1]. But, actually, it is difficult to ascertain the best coefficient values and also time-consuming. Therefore, the optimization of PI coefficients is the best solution.

As per procedures described above, the built-in function AC6 is simulated repetitively by modifying the coefficients of PI based on minimization of the objective function. The objective function considered here to nullify the steady-state speed error is based on integral of time-weighted absolute error (ITAE) criterion (Figs. 3, 4, 5, 6 and 7).

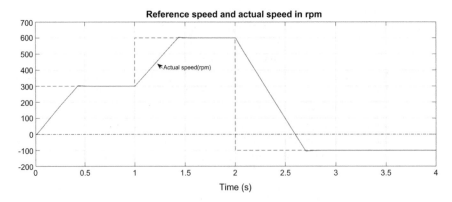

Fig. 3 Reference speed and actual speed for optimized values of PI controller

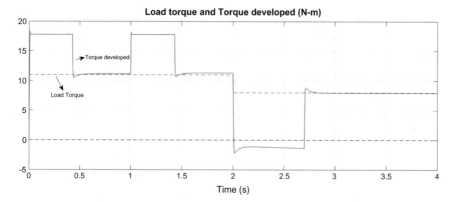

Fig. 4 Load torque and actual torque developed in PMSM drive

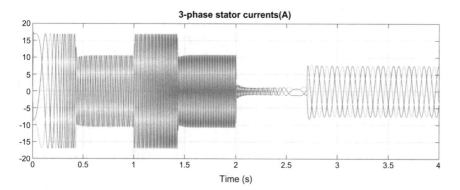

Fig. 5 Three-phase stator a-b-c currents in A

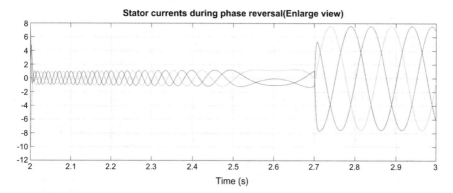

Fig. 6 Enlarge stator currents during phase reversal

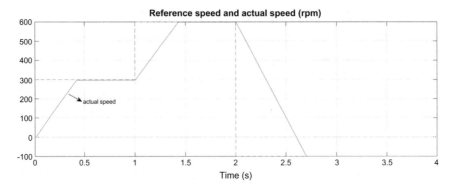

Fig. 7 Reference speed and actual speed for random chosen PI controller values

$$\text{ITAE} = \int_0^\infty t|e(t)|dt \tag{16}$$

where $|e(t)|$ is the difference of absolute time-dependent between actual speed and command speed and t is the time at that instant. The complete PMSM drive system is simulated, and the outputs are shown here for step speed inputs. The various responses of PMSM drive are compared with two different PI coefficients, one is randomly chosen values (initial values considered for optimization, i.e. $k_p = 2$ and $k_i = .2$) and other is estimated optimized values ($k_p = 1.8207$ and $k_i = 35.7655$). At the end of optimization, the values are found as:

$X = 1.8207, 35.7655$ (The coefficients of PI controller)
FVAL $= 1.8062e+06$
EXITFLAG $= 1$
OUTPUT $=$
iterations: 48
funcCount: 117
algorithm: 'Particle Swarm Optimization'

In order to verify the robustness of optimization values, the motor is operated at a speed command of 300 rpm at load torque of 11 N-m. At 1 s, the command speed is suddenly changed to 600 rpm without changing the load. At 2 s, a speed command of -100 is given with a load torque of 8 N-m. The comparisons are made by initial parameters of PI controller chosen and optimized value calculated offline by considering the initial values. It is observed in Fig. 3 that the rotor speed exactly tracks the reference speed without any steady-state error in optimized PI controller though there is slightly overshoot in the speed response. But there is an appreciable steady-state error in randomly chosen parameters (296 rpm instead of 300 rpm) as shown in enlarge view of Fig. 8. At starting, the electromagnetic torque developed in the machine is equal to the value of torque limiter, which is the maximum torque capability of the machine. This ensures that the motor accelerates very quickly and

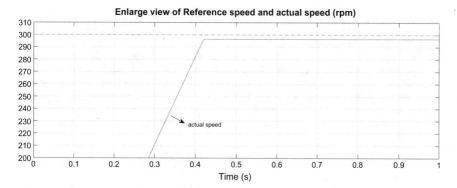

Fig. 8 Enlarge view to show the steady-state error

stabilizes the command speed. In order to change the direction of rotation, the phase sequence should be changed.

6 Conclusion

A PI with hysteresis band current controlled permanent magnet synchronous motor drive has been considered for optimization of the coefficients of PI controller. Normally, in industries the hit and trial methods starting from lower coefficient values and gradually tuning the values to get the best performances are used. But, this practice affects the system performances during the time of variations of PI controller values and also difficult to get the optimized values. Therefore, offline estimation of the PI controller is the best choice by building the replica of the industrial model and simulated it with different values of the PI controller. The different iterative values are chosen based on the objective function. In this paper, the gains of the PI controller of PMSM drive are optimized by using particle swarm optimization algorithms. The difference between the reference speed and actual speed is considered here as an objective function. The optimized values ($k_p = 1.8207$ and $k_i = 35.7655$) are found after 48 iterations.

References

1. Nayak, B., Choudhury, T.R.: Optimization of PI coeffecients of permanent magnet synchronous motor drive. Indian J. Sci. Technol. **10**(25), 1–11 (2017)
2. Sant, A.V., Rajagopal, K.R., Sheth, N.K.: Permanent magnet synchronous motor drive using hybrid PI speed controller with inherent and noninherent switching functions. IEEE Trans. Magn. **47**, 4088–4091 (2011)

3. Sreejeth, M., Singh, M., Kumar, P.: Particle swarm optimisation in efficiency improvement of vector controlled surface mounted permanent magnet synchronous motor drive. IET Power Electr. **8**, 760–769 (2015)
4. Okbuka, C.U., Nwosu, C., Agu, M.: A new high speed induction motor drive based on field orientation and hysteresis current comparison. J. Electr. Eng. **67**, 334–342 (2016)
5. Majhi, P., Panda, G.K., Saha, P.K.: Field oriented control of permanent magnet synchronous motor using PID controller: Int. J. Adv. Res. Electr. Electron. Instrum. Eng. **4**, 632–639 (2015)
6. Xia, C., Ji, B., Yan, Y.: Smooth speed control for low-speed high-torque permanent-magnet synchronous motor using proportional–integral–resonant controller. IEEE Trans. Ind. Electron. **62**, 2123–2134 (2014)
7. Zhang, B.T., Pi, Y.: Robust fractional order proportion-plus-differential controller based on fuzzy inference for permanent magnet synchronous motor. IEEE Trans. Control Theory Appl. **6**, 829–837 (2012)
8. Jung, J.W., Choi, Y.S., Leu, V.Q., Choi, H.H.: Fuzzy PI-type current controllers for permanent magnet synchronous motors. IEEE Trans. Electric Power Appl. **5**, 143–152 (2011)
9. Okbuka, C.U., Nwosu, C., Agu, M.: A high performance hysteresis current control of a permanent magnet synchronous motor drive. Turkish J. Electr. Eng. Comput. Sci. **25**, 1–14 (2017)
10. Seok, J.K., Lee, J.K., Lee, D.C.: Sensorless speed control of nonsalient permanent-magnet synchronous motor using rotor-position-tracking pi controller. IEEE Trans. Ind. Electron. **53**, 399–405 (2006)
11. Krause, P.C.: Analysis of Electric Machinery. McGraw Hill (1956)
12. Millonas, M.M.: Swarms, phase transitions and collective intelligence. In: Computational Intelligence: A Dynamic System Perspective (1994)

Implementation of SSA-Based Fuzzy FOPID Controller for AGC of Multi-area Interconnected Power System with Diverse Source of Generation

Tapas Kumar Mohapatra and Binod Kumar Sahu

Abstract The authors developed a fuzzy logic FOPID controller for a multi-area interrelated power system having several sustainable sources like wind and solar and conventional sources like thermal, hydro, redox flow batteries (RFBs), gas system, diesel, flywheel and ultra capacitors in the proposed article. RFB is implemented to reduce the oscillations that occur due to sudden disturbances in power system. Renewal energy sources (RES) are used to provide continuous power supply and quality of service to the users to fulfil their load demand. The best suitable gain of FLC and appropriate FOPID controller parameters are optimally developed using SSA optimisation technique. The results are also matched with recently used other optimisation techniques such as whale optimisation algorithm (WOA), particle swarm optimisation (PSO) and teaching learning-based optimisation (TLBO). The outcomes of fuzzy FOPID and FOPID are compared. Furthermore, performances of the planned system and robustness of the controller are inspected by varying load, RES input power and system parameters.

Keywords Fractional order PID (FOPID) controller · Automatic generation control (AGC) · Redox flow batteries (RFB) · Renewal energy sources (RES) · Fuzzy logic controller (FLC) · Salp swarm algorithm (SSA)

1 Introduction

The change in power in tie line (P_{tie}) and the frequency of the power system (f_s) are the elementary parameters which decide the stability of an interrelated multi-area power system. As stated by the control theory of power system, equilibrium between generation power (PG) and instantaneous fluctuation of load (PL) affects the nominal

T. K. Mohapatra (✉) · B. K. Sahu
Siksha 'O' Anusandhan Deemed to be University, Bhubaneswar, Odisha, India
e-mail: tapasmohapatra@soa.ac.in

B. K. Sahu
e-mail: binoditer@gmail.com

© Springer Nature Singapore Pte Ltd. 2020
S. Patnaik et al. (eds.), *New Paradigm in Decision Science and Management*,
Advances in Intelligent Systems and Computing 1030,
https://doi.org/10.1007/978-981-13-9330-3_34

frequency of system. If $P_L > P_G$, the speed of generator as well as f_s reduces and vice versa. To hold f_s and P_{tie} in power system at their scheduled value, AGC plays a very substantial role. The AGC maintains the fs at a specified minimal value and preserves P_{tie} flow between numerous control areas.

Stable frequency and good power quality are fundamental for an energy system. Power quality and reliability at peak load period can improve with the energy storage facilities like battery storage system (BSS). Several types of BSS like lithium ion, lead–acid, sodium sulphur and redox flow batteries (RFBs) are implemented in power system. For high power and large extent storage, the RFB is the best one among all. RFB significantly improves the dynamic performances and the stability of interconnected power system for peak saving loads. RFB can likewise rapidly remunerate dynamic power and improve the execution of AGC. A dynamic power source like RFB is valuable for voltage and frequency regulation and stabilisation. Active power of AGC can also be quickly compensated by RFB. RFB units in AGC have been broadly studied in the literature [1].

The expansion in energy demand and the ascending of an unnatural weather change have required the combination of sustainable power source with the existing sources in the power grid. This has offered a boost to hybrid distributed energy generation and storage system [2, 3]. The wind and solar-based power plants are stochastic in nature and changes according to the climate conditions. So at some point the heap request might be more than the supply. Energy storing appliances can be fixed with such system to modest the unbalance. Power quality can be enhanced, and variation in frequency can be reduced by these storage gadgets [4–6]. In the event that there is additional power accessible from these non-conventional energy sources over the load demand, these storing gadgets store for a brief span and discharge to the grid later on, if the load request is advanced than the supply. Storing gadgets are not ready to control totally the change in system frequency and power of tie line. So it is particularly fundamental to build up a solid controller to limit the change of system frequency and power of tie line. We can employ PID and FOPID controller. Here, we utilise FOPID for preferred execution over PID.

The derivative and integral order is not a whole number in FOPID. In papers [7, 8], a few fractional order controller (FOC) issues and tuning techniques are exhibited and replicated. Paper [9] manages the tuning of FOC for different industries. The paper displays the additional things.

1. The FOC is better control of dynamical system.
2. The FOC is less delicate to parameter variety because of extra degree of freedom to more readily modify the dynamic properties of FOPID.

So we have implemented FOPID controller to improve the dynamic behaviour and robustness of the system.

Literature survey reveals that a number of control strategies have been presented in various research articles for AGC study in power system. Control techniques like classical, optimal, fuzzy logic based, ANFIS based, etc. and many metaheuristic algorithms like GA, PSO, DE, TLBO, etc. have been projected for controller design of AGC of interrelated multi-area power system. So, nowadays researcher are paying more attention in proposing innovative optimisation and control techniques to tackle problems associated with AGC of power system. In this article, the swarm intelligence technique is implemented. The SSA is implemented to regulate the appropriate gain of FLC and order of FOPID parameters, like order of integrator (λ) and differentiator (μ).

2 Details of the System for Analysis

There are two interconnected systems consisting of diverse sources of generation such as thermal unit, hydro unit with mechanical governor and gas power unit, storage units like batteries, flywheels or ultra-capacitors, fuel cells and some renewal sources like solar, wind is revealed in Fig. 1. The complexity of thermal and hydro units is increased with the insertion of GDB and GRC. $\pm 3\%$/min, 320%/min of GRC is considered for thermal and hydro units, respectively. To provide a sufficient damping to electromechanical oscillation in both the areas, RFB units are introduced. RFB units convert and store electrical energy into chemical energy and release electrical energy as and when required in a controlled fashion. RFB units are extremely reliable and help in improving the systems' dynamic stability because they have very long reliable charging/discharging process and rapid response to damp out systems' oscillations due to sudden disturbances. The system frequency is controlled by fuzzy FOPID controller that is introduced in both areas. Area control error (ACE) and rate of area control error (ΔACE) are taken as the inputs of fuzzy and ΔP_{C1} and ΔP_{C2}, respectively, outputs of fuzzy FOPID controllers. ACE is the linear combination of f_s and P_{tie}. Figure 1 shows mathematical model of ACEs. The mathematical expressions are given by:

$$\text{ACE}_1 = \Delta P_{\text{tie}12} + \beta_1 \Delta f_1 \tag{1}$$

$$\text{ACE}_2 = \alpha_{12} \Delta P_{\text{tie}12} + \beta_2 \Delta f_2 \tag{2}$$

where $\Delta P_{\text{tie}12}$ is the power change in tie-line connecting the both areas, β_1 and β_2 are bias factors of the frequency of both areas, and Δf_1 and Δf_2 are the deviation in frequency of both areas, respectively, due to occurrence of small disturbance, ACEs act as activating signals for the systems to diminish ΔP of tie line and Δf to minimum value when steady state is reached. Some external power supply is given to wind and solar power plant. The gain parameters of fuzzy FOPID controllers should be implemented properly to improve the system performance. The both area RFBs input is taken from the output of ACEs. The RFB linearised model is given in [4]. Both the area parameters are same. FAMCON toolbox is implemented in MATLAB for FOPID controller designing.

To examine the execution, the power plants are modelled as first-order transfer functions. The transfer functions of WTG, STPG, FC, AE, FESS, UC and DEG are explained in [10]. The parameters are shown in Table 1 in Appendix. The transfer functions are given in the following equations.

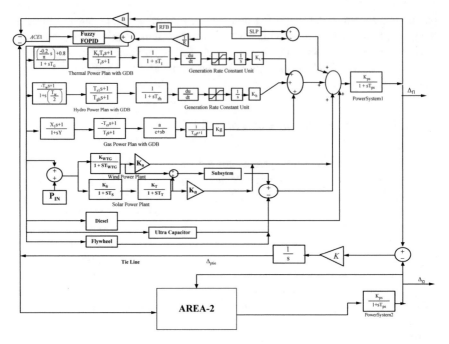

Fig. 1 Model of two area interrelated power system

Table 1 Fuzzy logic rules

Fuzzy rules /principles base					
ACE	Rate of ACE				
	BN	SN	Z	SP	BP
BN	BN	BN	BN	SN	Z
SN	BN	BN	SN	Z	SP
Z	BN	SN	Z	SP	BP
SP	SN	Z	SP	BP	BP
BP	Z	SP	BP	BP	BP

$$G_{\text{WTG}}(s) = \frac{K_{\text{WTG}}}{1 + sT_{\text{WTG}}} \tag{3}$$

$$G_{\text{FC}}(s) = \frac{K_{\text{FC}}}{1 + sT_{\text{FC}}} \tag{4}$$

$$G_{\text{DEG}}(s) = \frac{K_{\text{DEG}}}{1 + sT_{\text{DEG}}} \tag{5}$$

$$G_{\text{STPG}}(s) = \frac{K_{\text{S}}}{1 + sT_{\text{S}}} * \frac{K_{\text{T}}}{1 + sT_{\text{T}}} \tag{6}$$

$$G_{AE}(s) = \frac{K_{AE}}{1 + sT_{AE}} \tag{7}$$

$$G_{FESS}(s) = \frac{K_{FESS}}{1 + sT_{FESS}} \tag{8}$$

3 Controller Structure and Objective Function

FOPID is very renowned in industries as well as in research than the conventional. PID controller. FOPID was first introduced by Alomoush [11] in AGC application. In several other literatures [10, 12], FOPID is also applied in AGC. The FOC concept is dealing with differential equations through fractional calculus. In FOPID, derivative and integral controller orders are fractions instead of integer. Better flexibility and dynamic performance can be achieved by FOPID controller. The FOPID controller transfer function can be explained as:

$$G_C(s) = K_p + \frac{K_I}{s^\lambda} + K_d s^\mu \tag{9}$$

where K_p, K_I and K_d are proportional integral and derivative gains, and λ and μ are integrator and differentiator order, respectively.

$$K_I = \frac{K_P}{T_I}, \quad K_D = K_P * T_D$$

where T_I and T_D are time constants.

The conventional PID controller is one of the cases of the FOC, where $\lambda = \mu = 1$ (refer to Fig. 2).

The purpose of the optimisation techniques is to find most suitable solution by selecting proper objective function (OF). Different types of objective functions are explained in [10]. Out of these, ITAE diminishes the steady-state time and peak

Fig. 2 Fractional PID controller converge

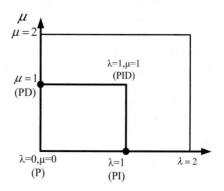

Fig. 3 Fuzzy FOPID
controller

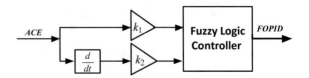

overshoot, but reduced steady-state time cannot be achieved by other tuning. ITAE
is not advantageous as it provides great controller output for an unexpected change
in fixed point. So, not suitable for controller design aspect. Hence, ITAE (ITAE =
$\int_0^t t|ACE|dt$) is preferred as OF in this paper. Five parameters will be tuned for each
FOPID controller and two parameters for each fuzzy by using SSA optimisation
technique. The main aim of the optimisation is to minimise the OF:

FOPID executes better performance in the accompanying conditions/applications:

1. Five unique conditions can be acknowledged, which is outlandish on account of
   orthodox PID.
2. The FOPID can undoubtedly accomplish the property of iso-damping, in contrast
   to ordinary PID.
3. The FOPID gives improved outcomes to upper-order system than traditional PID.
4. For substantial time postpone system a FOPID gives preferred outcomes over
   customary PID.
5. FOPID provides better robustness and stability over traditional PID.
6. The FOPID can achieve enhanced response for non-minimum phase system.

This controller is likewise less sensitive to parameter variety because of two
additional level of flexibility. However, in few cases this controller preparing ends
up complex. In MATLAB, FOPID controller is executed utilising FOMCON tool.

In this paper, fuzzy FOPID controller is also implemented for all the plants in both
the areas. The fundamental part of the article is to provide upgraded parameters of
fuzzy FOPID controller to get least error. Different optimisation algorithms like PSO,
TLBO, WOA and SSA are implemented, and the results are compared with out fuzzy
controller. Also, the results of FOPID and fuzzy FOPID controller are compared using
SSA optimisation techniques. The structure of fuzzy FOPID controller is shown in
Fig. 3. ACE and ΔACE are the two inputs of the fuzzy. The K_1 and K_2 are the input
scaling factors. In the article K_1 and K_2 of fuzzy, gain parameters and order of FOPID
are optimised by SSA technique.

Five membership functions (MF) such as big negative (BN), small negative (SN),
zero (Z), small positive (SP) and big positive (BP) are used. MFs are triangular in
nature to reduce computational time [13, 14]. Since MFs as shown in Fig. 4 are
involved, 25 rules/principles to be formed for fuzzy output. Table 1 shows the principle involved in fuzzy controller. These standards have basic impact on the execution
of FLC, and in this manner, in this paper these rules are inspected extensively by
examining the dynamic behaviour of the given system. For fuzzifying the inputs and
also to combine the fuzzified inputs with fuzzy principles, Mamdani interface system
is used. Here, centre of gravity (COG) method is used for defuzzification.

Fig. 4 Triangular
membership functions

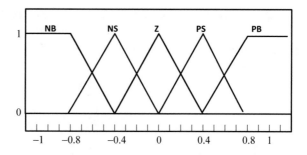

4 Overview of Optimisation Algorithm

In this projected paper, SSA optimisation technique is implemented for tuning of fuzzy FOPID controller. SSA is one kind of metaheuristic optimisation algorithm (MHOA) which has been presented by Mirjalili and Gandomi [10, 12].

The masses are first separated to two social occasions: pioneer and supporters. The pioneer is at the front of the chain, and the straggling leftovers of Salps are considered as supporters. As the name of these Salps derives, swarms are guided by the pioneer and the supporters take after each other. The circumstance of Salps is portrayed in n-dimensional intrigue space. Where n represents number of elements of the issue. Thusly, the Salps circumstance are anchored in two-dimensional structure named as x that there is a F is the sustenance source in the pursuit space which is the object of the swarm's. To resuscitate the circumstance of the pioneer the running with condition is proposed:

$$
x_j^1 = \begin{cases} F_j + C_1((ub_j - lb_j)C_2 + lb_j) & C_3 \geq 0. \\ F_j - C_1((ub_j - lb_j)C_2 + lb_j) & C_3 < 0. \end{cases} \tag{10}
$$

where

X_j^1 The position of pioneer (the first Salp) along the jth dimension.
F_j jth dimension food source position.
ub_j jth dimension upper bound.
lb_j jth dimension lower bound.

The coefficient C_1 balances exploration and is considered as the most important factor in SSA. It is defined as follows:

$$
C_1 = 2e^{-\left(\frac{4l}{L}\right)^2} \tag{11}
$$

where

l The current iteration.
L Maximum number of iterations.

The random number C_2 and C_3 consistently created in the interim of $[-1, 1]$. They direct whether the position of jth measurement will move along positive or negative endlessness and additionally estimate progression.

$$x_j^i = \frac{1}{2}at^2 + v_0 t \tag{12}$$

$$x_j^i = \frac{1}{2}\left(x_j^i + x_j^{i-1}\right) \tag{13}$$

where $i \geq 2$, x_j^i shows the ith follower Salp location in jth dimension, t is time, v_0 is the initial speed, and $a = \frac{v_{final}}{v_0}$

- SSA calculation refreshes main Salp position unless the best arrangement acquired, so the pioneer dependably investigates and misuses the space around it.
- SSA calculation refreshes the adherent slaps location concerning each other and moves step by step to the main Salp location.
- Adherent Slaps gradual developments effectively stagnant neighbourhood optima.
- Coefficient C_1 is reduced suddenly and completed the course cycle, so the pursuit initially investigated space and afterwards misuses it.

SSA calculation is basic and very simple to execute. The above comments ensure that the SSA calculation is hypothetical and ready to take care of single target advancement issues with unknown inquiry spaces. The versatile instrument of SSA enables this calculation to keep away from neighbourhood arrangements and in the end finds a precise estimation of the best arrangement acquired amid advancement. Thusly, it can be connected to both unimodal and multi-modular issues.

Virtual Code of SSA Algorithm:

```
Instate the Salp populace x_i (i = 1, 2, . . . , n) considering u_b and l_b
while(end condition is not fulfilled)
Compute the wellness of each hunt agent(Salp)
F = the best pursuit specialist Update C_1 by Eq. (45)
for every salp(x_i)
if(i == 1)
Refresh the situation of the main Salp by Eq. (44)
else
Refresh the situation of the devotee Salp by Eq. (47)
end
end
Change the salps in view of the upper and lower limits of factors
end
return F
```

5 Result and Discussion

It is very difficult to get quick response and excellent stability at the same time. So we have to make a negotiation among stability and quicker response. This negotiation can be accomplished by appropriately defining OF and choosing a suitable controller. Optimisation technique helps to achieve this objective function.

In the projected article, a fuzzy FOPID controller is introduced for AGC of the interrelated two-area diverse source of generation-based power system. The ITAE is taken as objective function (since it lessens both steady-state time and maximum undershoot/overshoot [13, 14]). The articulation for ITAE objective function is

$$\text{ITAE} = \int_{t=0}^{t_{\text{sim}}} (|\Delta f_1| + |\Delta f_2| + |\Delta P_{\text{tie}}|) \cdot t \cdot dt \tag{14}$$

where Δf_1, Δf_2 and ΔP_{tie} are area-1 frequency change, area-2 frequency change and change in tie-line power, respectively. 't_{sim}' = The simulation time.

The gain of the controller K_P is taken in the range of [0.1, 100], and the integrator and differentiator time constants are chosen in the range of [0.1, 1]. The fuzzy controller gain of both the area is considered in the range of [0.1, 2]. 100 numbers of population and maximum iteration are considered. The simulation is conducted by using a SLP of 0.01 pu in area 1 and the wind and solar power input of 0.001 pu. The dynamic performances such as undershoot, overshoot, steady-state error and settling time are compared for the different parameters obtained by optimisation techniques such as WOA, PSO, TLBO and SSA. The results are shown in Figs. 5, 6 and 7, respectively, without FLC. The comparison results of parameters and responses for different optimisation techniques are shown in Tables 2 and 3 with a tolerance band of ±0.000005. From the result, it is observed that the parameters obtained by SSA optimisation technique gives the best performance than others.

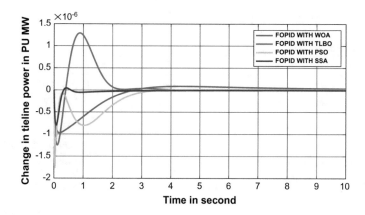

Fig. 5 Comparison of results of power deviation in tie-line

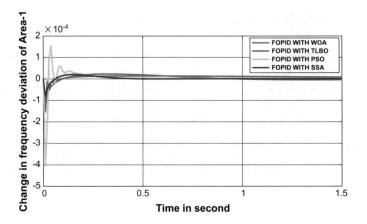

Fig. 6 Area-1 frequency deviation

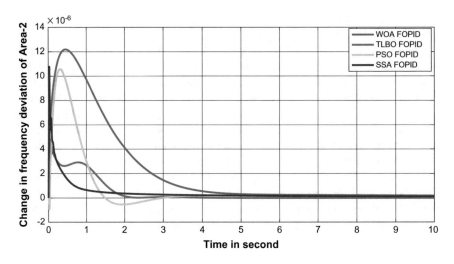

Fig. 7 Area-2 frequency deviation

Table 2 System parameters

Parameters	Symbols	Values
Area rated capacity	P_{rt}	2000 MW
Nominal area load	P_L^0	1740 MW
Nominal system frequency	f_s	60 Hz
Power system gain	K_{PS}	68.9655 Hz/pu MW
Power system time constant	T_{PS}	11.49 s
Steam turbine time constant	T_T	0.3 s

(continued)

Table 2 (continued)

Parameters	Symbols	Values
Steam turbine reheat time constant	T_r	10.2 s
Steam turbine reheat constant	K_r	0.3
Governor speed regulation parameters of thermal, hydro and gas units	R	2.4 Hz/pu MW
Frequency bias coefficient	β	0.4312 pu MW/Hz
Synchronisation coefficient	T_{12}	0.0433
Gain	α_{12}	-1
Governor time constant in steam plant	T_G	0.06 s
Hydro turbine governor time constant	T_{RH}	28.749 s
Resetting time	T_R	4.9 s
Governor time constant	T_{GH}	0.2 s
Water starting time	T_W	1.1 s
Lead time constant of gas turbine governor	X	0.6 s
Lag time constant of gas turbine governor	Y	1.1 s
Valve positioner gains	$a = c$	1
Valve positioner time constant	b	0.049 s
Gas turbine combustion reaction time delay	T_{CR}	0.01 s
Gas turbine fuel time constant	T_F	0.239 s
Compressor discharge volume time constant	T_{CD}	0.2 s
Participation factors of thermal, gas, and hydro units	$K_T, K_G, K_H,$	0.5747, 0.1380, 0.2873
Gain of RFB	K_{RFB}	1.8
Wind turbine generator (WTG) gain	K_{WTG}	1
WTG time constant	T_{WTG}	1.5
Aqua electrolyser (AE) gain	K_{AE}	0.002
AE time constant	T_{AE}	0.5
Fuel cell (FC) gain	K_{FC}	0.01
FC time constant	T_{FC}	4
Flywheel energy storage system (FESS) gain	K_{FESS}	-0.01
FESS time constant	T_{FESS}	0.1
Solar thermal power generation (STPG) gains	K_S, K_T	1.8, 1
STPG time constants	T_S, T_T	1.8, 0.3
Diesel engine generator (DEG) gain	$K_{DEG} = 0.003$	$T_{DEG} = 2$
Ultra-capacitor (UC) gain, time constant	K_{UC}, T_{UC}	$-0.7, 0.9$

Table 3 Optimized control parameters of several controllers for two-area multi-source power system

Controller type	Control parameters									
	K_{P1}	T_{I1}	T_{D1}	λ_1	μ_1	K_{P2}	T_{I2}	T_{D2}	λ_2	μ_2
PSO optimization	82.0001	0.6557	0.7595	0.8976	0.6020	79.9691	0.7580	0.5364	0.9612	0.5926
TLBO optimization	50.3284	0.1635	0.6900	1.0000	0.7020	100.0000	0.1000	0.7962	1.0000	0.3751
WOA optimization	66.7684	0.1000	0.4647	0.7768	0.4867	100.0000	0.4060	0.4305	0.9371	0.7497
SSA optimization	69.1857	0.06000	0.6091	0.7212	0.7167	92.576	0.1000	0.2723	0.7541	0.5000
SSA optimization with fuzzy FOPID	60.6898	0.07000	0.6111	0.9250	0.5538	100.0000	0.1646	0.6448	0.9340	0.7756

Fuzzy other parameters are $K_1 = 2.0000$, $K_2 = 2.0000$, $K_3 = 0.9869$, $K_4 = 0.1076$

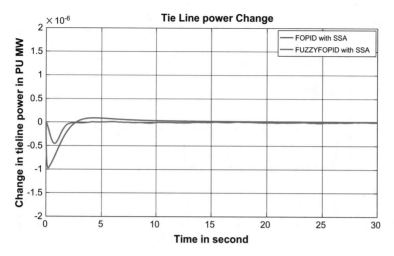

Fig. 8 Power deviation in tie-line

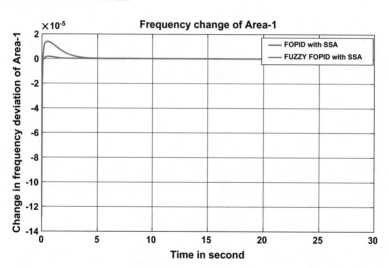

Fig. 9 Area-1 frequency deviation

The projected paper also observed the effect of FLC. The results of Δf_1 (area-1 frequency deviation), Δf_2 (area-2 frequency deviation) and ΔP_{tie} (power deviation in tie line) with fuzzy FOPID and FOPID are shown in Figs. 8, 9 and 10, respectively, using SSA optimisation technique. The parameters and the comparison results of fuzzy FOPID and FOPID are shown in Tables 2 and 3, respectively, with a tolerance band of ±0.000005. From the result, it is observed that fuzzy FOPID controller provides better performance than FOPID controller. The robustness of the fuzzy FOPID controller is tested for the variation of SLP from 10 to 100%, wind and solar

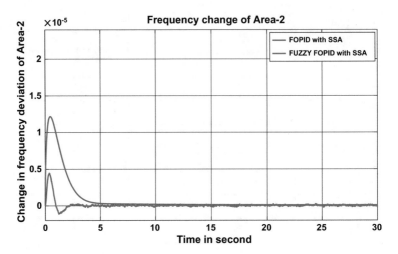

Fig. 10 Area-2 frequency deviation

input power from 10 to 100% and for both the component variation simultaneously from 10 to 100%. The comparison results are shown in Figs. 11, 12 and 13. From the result, it is also cleared that the controller is very robust in nature.

6 Conclusion

In the anticipated paper, SSA is utilised to optimise the control parameters of the proposed fuzzy FOPID controller and FOPID controller keeping in mind the end goal to handle the load frequency control issue of a two-region interconnected multi-source power system with non-conventional sources like wind, solar, battery storage and so on. For each power plant, fuzzy FOPID controllers are utilised in this investigation. At first, the gains of FOPID controllers for AGC of the multi-source power system with AC tie-lines just are optimised utilising PSO, TLBO, WOA and the proposed SSA algorithms. Accepting ITAE as objective function, a step load change of 0.01 PU is connected all of a sudden in area 1 with 0.001 pu input power to solar and wind to think about the execution of the system. The outcome acquired is contrasted and three other optimisation techniques like PSO, TLBO and WOA without FLC and it is seen that the SSA optimised FOPID controller is giving greater dynamic performance. At that point, the proposed algorithm is utilised to appropriately plan a fuzzy FOPID controller for AGC of the power system with AC tie-lines as it were. It is additionally demonstrated that SSA tuned fuzzy FOPID controller is performing better regarding undershoot, overshoot and settling time than only FOPID controller.

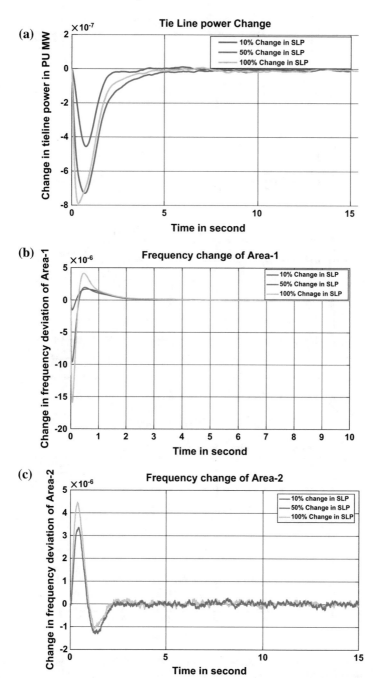

Fig. 11 **a** Power deviation in tie-line. **b** Area-1 frequency deviation. **c** Area-2 frequency deviation for SLP Variation

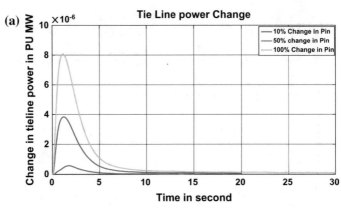

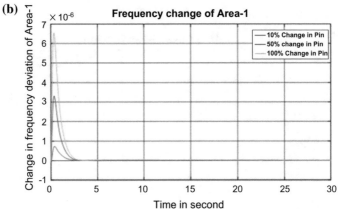

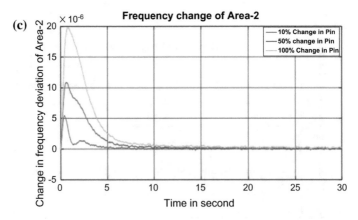

Fig. 12 **a** Power deviation in tie-line. **b** Area-1 frequency deviation. **c** Area-2 frequency deviation for variation of input power of renewal source

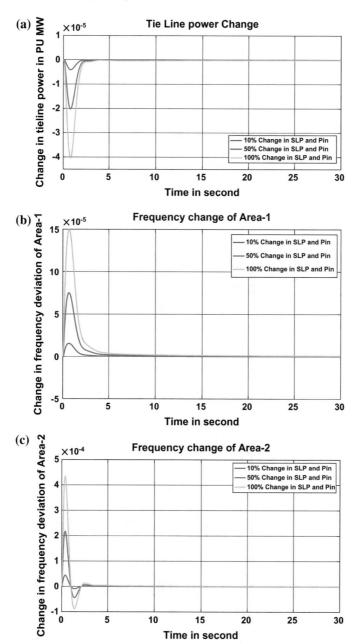

Fig. 13 **a** Tie-line power change. **b** Frequency change in area-1. **c** Frequency change in area-2 for variation of both input power of renewal source and SLP

Table 4 Numerical values of under shoot, over shoot and settling time (for tolerance band of ± 0.000005) of two areas with several controllers

Controller type	Undershoot			Overshoot			Settling time (s)		
	ΔP_{tie} (10^{-7})	Δf_1 (10^{-5})	Δf_2 (10^{-7})	ΔP_{tie} (10^{-7})	Δf_1 (10^{-5})	Δf_2 (10^{-5})	ΔP_{tie}	Δf_1	Δf_2
PSO optimization	−8.03	−40.39	−5.356	0.872	15.26	1.057	0	0.39	0.82
TLBO optimization	−12.47	−15.32	0	12.86	2.92	1.271	0	0.81	0.11
WOA optimization	−9.67	−12.85	0	0.869	1.385	1.218	0	2.05	1.8
SSA optimization	−7.95	−8.04	0	0.451	2.055	1.077	0	0.35	0.1
SSA optimization with fuzzy	−4.56	−0.155	0	0.115	0.168	0.441	0	0	0

Appendix

See Tables 2, 3 and 4.

References

1. Gorripotu, T.S., Sahu, R.K., Panda, S.: AGC of a multi-area power system under deregulated environment using redox flow batteries and interline power flow controller. Eng. Sci Technol. Int. J. **18**(4), 555–578 (2015)
2. Ponnusamy, M., Banakara, B., Dash, S.S., Veerasamy, M.: Design of integral controller for load frequency control of static synchronous series compensator and capacitive energy source based multi area system consisting of diverse sources of generation employing imperialistic competition algorithm. Int. J. Electr. Power Energy Syst. **73**, 863–871 (2015)
3. Shankar, R., Chatterjee, K., Bhushan, R.: Impact of energy storage system on load frequency control for diverse sources of interconnected power system in deregulated power environment. Int. J. Electr. Power Energy Syst. **79**, 11–26 (2016)
4. Senjyu, T., Nakaji, T., Uezato, K., Funabashi, T.: A hybrid power system using alternative energy facilities in isolated island. IEEE Trans. Energy Convers. **20**(2), 406–414 (2005)
5. Ganti, V.C., Singh, B., Aggarwal, S.K., Kandpal, T.C.: DFIG-based wind power conversion with grid power levelling for reduced gusts. IEEE Trans. Sustain. Energy **3**(1), 12–20 (2012)
6. Arya, Y.: AGC performance enrichment of multi-source hydrothermal gas energy systems using new optimized FOFPID controller and redox flow batteries. Energy **127**, 704–715 (2017)
7. Fractional order fuzzy control of hybrid power system with renewable generation using chaotic PSO
8. Panda, S., Sahu, B.K., Mohanty, P.K.: Design and performance analysis of PID controller for an automatic voltage regulator system using simplified particle swarm optimization. J. Franklin Inst. **349**(8), 2609–2625 (2012)
9. Mirjalili, S., Gandomi, A.H., Mirjalili, S.Z., Saremi, S., Faris, H., Mirjalili, S.M.: Salp swarm algorithm: a bio-inspired optimizer for engineering design problems. Adv. Eng. Softw., 1–29 (2017)
10. Pan, I., Das, S.: Fractional order fuzzy control of hybrid power system with renewable generation using chaotic PSO. ISA Trans. **62**, 19–29 (2016)
11. Alomoush, M.I.: Load frequency control and automatic generation control using fractional-order controllers. J. Electr. Eng. **91**(7), 357–368 (2010)
12. Mohapatra, T.K., Sahu, B.K.: Design and implementation of SSA based fractional order PID controller for automatic generation control of a multi-area, multi-source interconnected power system. In: Technologies for smart-city energy security and power (ICSESP), IEEE (2018)
13. Sahu, B.K., Pati, S., Panda, S.: Hybrid differential evolution particle swarm optimisation optimised fuzzy proportional-integral-derivative controller for automatic generation control of interconnected power system. IET Gener. Transm. Distrib. **8**(11), 1789–1800 (2014)
14. Martín, F., Monje, C.A., Moreno, L., Balaguer, C.: DE-based tuning of PID controllers. ISA Trans. **59**, 398–407 (2015)

Use of Fuzzy-Enabled FOPID Controller for AGC Investigation Utilizing Squirrel Search Algorithm

Manjit Bahadur Singh, Sushobhan Pal, Siddharth Mohanty
and Manoj Kumar Debnath

Abstract This article presents an innovative fuzzy-enabled fractional order proportional-integral-derivative (Fuzzy-FOPID) controller for automatic generation control (AGC) of a dual control area amalgamated power framework. Because of the convergence superiority, the parameters for the fuzzy-FOPID controllers are calibrated with the newly developed optimization strategy termed as squirrel search algorithm (SSA). A disturbance having an extent of 0.01 p.u. is applied to control area-1 to check the performance of the proposed system. Toward the finish of the investigation, the implementation of the fuzzy-FOPID controller was contrasted against fuzzy-enabled PID, traditional PID controllers regarding peak and least overshoots, settling time, relaxing time and steady-state error. The efficiency of the projected SSA tuned fuzzy-FOPID controller has been evidenced by intensifying the loading of the system.

Keywords Squirrel search algorithm · Fuzzy controller · Automatic generation control · Fractional controller · Controller tuning

1 Introduction

Electrical power is required to be provided within admissible breaking points of voltage and frequency. According to the load demand changes, the framework voltage and frequency veer off from the underlying working conditions. Hence, the system

M. B. Singh · S. Pal · S. Mohanty · M. K. Debnath (✉)
Siksha 'O' Anusandhan University, Bhubaneswar, Odisha, India
e-mail: mkd.odisha@gmail.com

M. B. Singh
e-mail: manjitbahadur.singh@gmail.com

S. Pal
e-mail: susobhan.pal96@gmail.com

S. Mohanty
e-mail: siddharthshankar@gmail.com

© Springer Nature Singapore Pte Ltd. 2020
S. Patnaik et al. (eds.), *New Paradigm in Decision Science and Management*,
Advances in Intelligent Systems and Computing 1030,
https://doi.org/10.1007/978-981-13-9330-3_35

is equipped with certain control techniques to minimize the deviation. The voltage dissimilarity at the load end is the signal of the unbalance between the reactive power produced and immersed. Balance between net generated power and subsequent loads with losses is essential. Voltage collapse occurs, if voltage departs excessively from the rated value. The intention of AGC is that the frequency deviations as well as steady-state error should give zero faults. Area control error (ACE) is a function of power differences and frequency deviations for accomplishing the power equilibrium in interconnected control region; ACE can be used as an effective means in LFC studies. The singular objective is to retain the frequency close to the nominal value and power to planned values. In ACE signal, frequency bias factor is taken into account which makes it a great tool for automatic generation control method [1].

Several approaches for AGC of interconnected power system network have been proposed by researchers over the past decades. The primary work on AGC was devised by Elgerd [2]. In paper [3], authors have used a P-controller along with feedback action to AGC of a power framework. An intermittent nonlinear control technique [4] called ramp following controller (RFC) was realized in automatic generation control in a multi-area power system framework. In [5], Fosha et al. suggested to implement AGC in a deregulated situation. New methods, for example, artificial neural network [6] and fuzzy logic [7] have been used for improving the performance of AGC controllers. Fuzzy controller and traditional controllers were combined, and an increase in efficiency was resulted [8]. Recent philosophies in various control characteristics [9] regarding the AGC problem have been discussed. Earlier, by utilizing manual methods, controller gain parameters were being tuned. After the discovery of soft computation techniques and nature-encouraged and evolutionary computational procedures, the regulation controllers became much more efficient as well as free of error. Fuzzy logic controller intended for AGC with SMES units was implemented [10]. Automatic generation control in a hydropower plant by using conventional controller was employed by Nanda [11]. In an interconnected power system, numerous evolutionary techniques such as ABC (artificial bee colony), hybrid DE-PSO, and Hybrid DE-GWO technique were implemented successfully for the AGC [12–14].

Thorough literature survey shows that most of the works pertains to the evolutionary methods and traditional control method in the province of automatic generation control. Here, the conniving of a different control method that utilizes a fuzzy logic controller through a fractional PID controller, which is later on, applied to the AGC of a power system. To optimize the gain parameter of fuzzy-enabled FOPID controller, our work additionally recommends a novel streamlining strategy named the SSA enhancer [15]. Finally, the suggested control method was contrasted against usual PID [12] and fuzzy-enabled PID controller [13].

2 System Inspected

Our framework is incorporated with thermal power plants equipped with two non-reheat-type steam turbine which is shown in Fig. 1. Fuzzy-enabled FOPID controller is responsible to handle the deviations in nominal frequency. Parameters such as frequency-biased constant β, regulation constant R, time constants of the various units such as hydraulic amplifier, steam turbine, etc. are extracted from Sahu et al. [13]. Area control error (ACE) acts as an input signal to fuzzy-enabled FOPID controller. u_1 and u_2 are nominated to be the outputs of the fuzzy-FOPID blocks. ACE employs to accomplish the primary targets of AGC, i.e., maintaining the power and frequency to their ostensible values.

For the given framework, the area control error signal can be communicated as:

$$\text{ACE}_1 = \Delta P_{\text{tie}1-2} + B_1 \Delta f_1 \text{ and } \text{ACE}_2 = \Delta P_{\text{tie}2-1} + B_2 \Delta f_2 \tag{1}$$

The parameters ΔP_{12}, ΔP_{21} are called "line power differences." Whereas B_1, B_2 are called frequency bias constants. Δf_1 and Δf_2 are frequency divergences in both areas. When a little disturbance is applied to the given system, ACE creates an activating error signal affirming that Δf and ΔP_{12} are made zero ideally. Fuzzy-

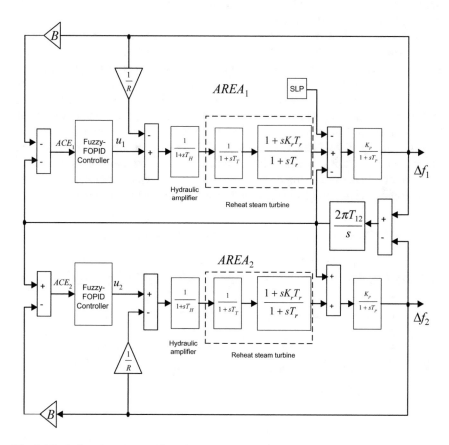

Fig. 1 Block diagram representation of two area unified model

enabled FOPID controller specified gain parameters should be allocated properly to improve the time response of the planned system.

3 Methodology

3.1 Controller Configuration

Traditional controllers have attracted many researchers because of its coherence, robustness with superior control action. The conventional PID controller was found to be less appropriate with increase in complexity in plants. One of the limitations of traditional PID controllers is that only one operating conditions can be handled a time. Modification in operating conditions will lead to change in controller parameters. Subsequently, the performance of the controller is subjected to operating point variations of the system. To beat this constraint, fuzzy-enabled fractional order PID controller can be a great advantage.

Linguistic method of computation such as soft computing tool-based fuzzy logic controller can take care of fuzziness. It was discovered to be an extraordinary controller to get an enhanced solution in contrast to traditional strategies. Figure 2 evidently illustrates the construction of the proposed controller comprising of seven constants such as $(K_1, K_2, \lambda, \mu, K_I, K_D, K_P)$. Two input parameters (ACE$_1$ and ACE$_2$) are given as inputs to the fuzzy-enabled FOPID controller and the proposed fuzzy controller consisted of explicitly five kinds of membership functions as depicted in Fig. 3. In Table 1, a rule base of 25 rules is created based on the five membership functions. Membership function which takes care of the proposed controller evaluates the required output and further the gain constants are calculated with recently acknowledged named squirrel search algorithm method.

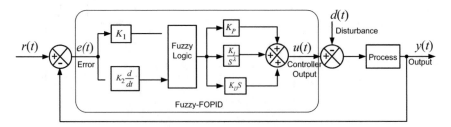

Fig. 2 Internal design of the projected controller (fuzzy-FOPID)

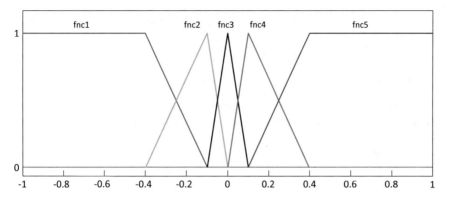

Fig. 3 Five fuzzy membership functions used for all fuzzy variables (fuzzy-FOPID controller)

Table 1 Rule base defining fuzzy IF THEN rules

ACE	$\dot{ACE}$				
	fnc1	fnc2	fnc3	fnc4	fnc5
fnc1	fnc1	fnc1	fnc2	fnc2	fnc3
fnc2	fnc1	fnc2	fnc2	fnc3	fnc4
fnc3	fnc2	fnc2	fnc3	fnc4	fnc4
fnc4	fnc2	fnc3	fnc4	fnc4	fnc5
fnc5	fnc3	fnc4	fnc4	fnc5	fnc5

3.2 SSA Optimization Technique

Squirrel search algorithm (SSA) is a recently progressed optimization method introduced by Jain et al. [15]. This method emulates the food searching actions with its hovering elements with its competent movement recognized as gliding technique, which is basically utilized by the animals for food collection in a large area.

In a forest, on each tree n numbers of squirrels are there. Each flying squirrel is assigned to search for food and preserve the food by using the dynamic foraging concept. In the forest, there are three groups of plants that are available for instance normal tree, oak tree, hickory tree. Hence, one hickory tree has been taken. The position of elements is denoted by the matrix below and initialized randomly. The entries obtained from the solution vector are put into the objective function to evaluate the fitness of each element. Estimation of fitness of every flying squirrel's position is arranged in a rising manner. The elements of irrelevant fitness esteem are relegated on the hickory nut tree. Following finest elements (3 squirrels) are considered on acorn trees. Residual elements are considered on normal trees.

Later by using the random selection procedure few squirrels approaches the optimal solution (hickory tree).

Conditions during updating place are enlisted below.

Situation 1: Elements on acorn trees possibly will change their position in the direction near hickory tree.
Situation 2: Elements sitting on normal trees may change their position to acorn nut trees.
Situation 3: Elements sitting on normal trees may approach to hickory tree.

Change of season knowingly disturbs the food searching capacity of elements. Elements which are trapped by local optima are avoided by a seasonal monitoring condition.

The elements sitting on hickory nut tree are the finest result.

4 Result and Discussion

In our work, SSA technique is executed to reduce some limitations of Fuzzy-enabled FOPID regulator. Unsteadiness results on selecting large value of gain. The controller gain parameters of the proposed controller are taken in between 0.01 and 4.0 and λ and μ are considered within [0, 1]. The initial population is taken as 50, and the iteration count was taken as 100. Integral time absolute error (ITAE) is measured as the objective function. A disturbance of magnitude 0.01 p.u. was applied in area 1, and simulation was run several times. The given fuzzy-enabled FOPID controller was tweaked by SSA, and the outcomes were compared with traditional controller. In Table 2, the parameters of the fuzzy-enabled FOPID are given.

$$\text{ITAE} = \int_0^t (|\Delta f_1| + |\Delta f_2| + |\Delta P_{\text{tie}}|) \cdot t \cdot \mathrm{d}t \tag{2}$$

Figures 4, 5, and 6 undoubtedly display the debauched response of the fuzzy-enabled FOPID controller as contrasted to the traditional fuzzy-PID [13] and PID

Table 2 The gain values of fuzzy-enabled FOPID controller achieved by SSA technique

Parameters	Area 1 (thermal unit)			Area 2 (thermal unit)		
	Fuzzy-FOPID	Fuzzy-PID [13]	PID [12]	Fuzzy-FOPID	Fuzzy-PID [13]	PID [12]
K_1	3.9457	1.9986	–	1.396	1.2074	–
K_2	2.6978	0.9232	–	0.998	1.8265	–
μ	0.98		–	0.97	–	–
λ	0.91		–	0.9	–	–
K_p	1.2987	1.9875	1.966	2.593	1.2204	0.710
K_i	3.8978	1.9992	9.590	0.424	1.3348	0.682
K_d	0.0985	–	3.932	1.019	–	0.741

Fig. 4 Frequency
abnormalities in area 1

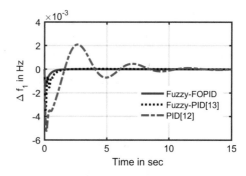

Fig. 5 Frequency
abnormalities in area 2

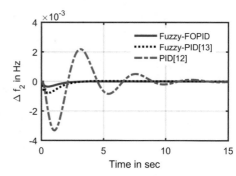

Fig. 6 Interline power
abnormalities

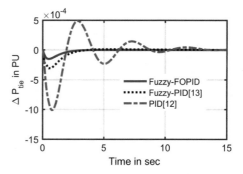

[12]. The consequences found in accordance with settling period, maximum and minimum exceed given in Table 3. In second case, the robustness is proved by augmenting the loading in control area 1 by an amount of 5%. Figures 7, 8, and 9 depict the abnormalities of the system response during these augmented loading conditions. Also Figs. 7, 8 and 9 verify the ascendancy of projected fuzzy-FOPID controller over other control methods [12, 13].

Table 3 Different performance specifications attained with several methods

Deviations	Time response evaluative factors	Fuzzy-FOPID	PID [12]	Fuzzy-PID [13]
Δf_1	Undershoots (Hz)	−0.0014	−0.0052	−0.0035
	Settling time (s)	1.1600	12.3700	3.9500
	Overshoots(Hz)	Nil	0.0021	0.00005986
Δf_2	Undershoots (Hz)	−0.0004	−0.0033	−0.0008
	Settling time (s)	2.3800	12.9300	2.8500
	Overshoots(Hz)	Nil	0.0022	0.00003291
ΔP_{tie}	Undershoots (PU)	−0.0002	−0.0010	−0.0003
	Settling time (s)	1.7100	8.3400	2.3600
	Overshoots (PU)	Nil	0.0005	0.00001176

Fig. 7 Frequency abnormalities in area 1 for case 2

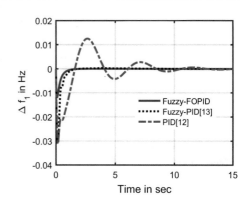

Fig. 8 Frequency abnormalities in area 2 for case 2

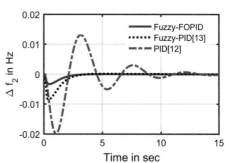

Fig. 9 Interline power
abnormalities for case 2

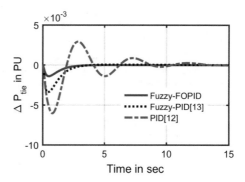

5 Conclusion

The innovative fuzzy-enabled FOPID controller was presented here for a two-area
AGC in an integrated power system model. Design variables are altered with the
scaling factor of the proposed controller. The gain constants are evaluated by using
the squirrel search algorithm. The projected fuzzy-FOPID controller minimizes the
frequency irregularities and power aberrations. Sturdiness scrutiny shows the domi-
nance of projected regulator. Parameters of the proposed controller were optimized
by SSA procedure. The results obtained were contrasted with previous journal results
like Fuzzy-PID and traditional PID controller. Further evaluations reinforce the fact
that fuzzy-enabled FOPID controller is more operative and improved in contrast to
earlier work.

References

1. Kundur, P., Balu, N.J., Lauby, M.G.: Power System Stability and Control, vol. 7. McGraw-Hill,
   New York (1994)
2. Elgerd, O.I., Fosha, C.E.: Optimum megawatt-frequency control of multiarea electric energy
   systems. IEEE Trans. Power Apparatus Syst. **4**, 556–563 (1970)
3. Kumar, A., Malik, O.P.: Automatic generation control of interconnected power systems using
   variable-structure controllers. In: IEE Proceedings C-Generation, Transmission and Distribution, vol. 130, no. 4. IET (1983)
4. Bakken, B.H., Grande, O.S.: Automatic generation control in a deregulated power system.
   IEEE Trans. Power Syst. **13**(4), 1401–1406 (1998)
5. Fosha, C.E., Elgerd, O.I.: The megawatt-frequency control problem: a new approach via optimal
   control theory. IEEE Trans. Power Apparatus Syst. **4**, 563–577 (1970)
6. Demiroren, A., Sengor, N.S., Lale Zeynelgil, H.: Automatic generation control by using ANN
   technique. Electr. Power Compon. Syst. **29**(10) (2001): 883–896
7. Chown, G.A., Hartman, R.C.: Design and experience with a fuzzy logic controller for automatic
   generation control (AGC). IEEE Trans. Power Syst. **13**(3), 965–970 (1998)
8. Nanda, J., Mangla, A.: Automatic generation control of an interconnected hydro-thermal system using conventional integral and fuzzy logic controller. In: Proceedings of the 2004 IEEE

International Conference on Electric Utility Deregulation, Restructuring and Power Technologies, 2004 (DRPT 2004), vol. 1. IEEE (2004)

9. Shayeghi, H., Shayanfar, H.A., Jalili, A.: Load frequency control strategies: a state-of-the-art survey for the researcher. Energy Convers. Manag. **50**(2), 344–353 (2009)

10. Sudha, K.R., Vijaya Santhi, R.: Load frequency control of an interconnected reheat thermal system using type-2 fuzzy system including SMES units. Int. J. Electr. Power Energy Syst. **43**(1), 1383–1392 (2012)

11. Nanda, J., Mangla, A., Suri, S.: Some new findings on automatic generation control of an interconnected hydrothermal system with conventional controllers. IEEE Trans. Energy Convers. **21**(1), 187–194 (2006)

12. Gozde, H., Cengiz Taplamacioglu, M., Kocaarslan, I.: Comparative performance analysis of artificial bee colony algorithm in automatic generation control for interconnected reheat thermal power system. Int. J. Electr. Power Energy Syst. **42**(1), 167–178 (2012)

13. Sahu, B.K., Pati, S., Panda, S.: Hybrid differential evolution particle swarm optimisation optimised fuzzy proportional–integral derivative controller for automatic generation control of interconnected power system. IET Gener. Transm. Distrib. **8**(11), 1789–1800 (2014)

14. Debnath, M.K., Mallick, R.K., Sahu, B.K.: Application of hybrid differential evolution—Grey Wolf optimization algorithm for automatic generation control of a multi-source interconnected power system using optimal fuzzy–PID controller. Electr. Power Compon. Syst. **45**(19), 2104–2117 (2017)

15. Jain, M., Singh, V., Rani, A.: A novel nature-inspired algorithm for optimization: squirrel search algorithm. Swarm Evol. Comput. (2018)

Security Issues in Internet of Things (IoT): A Comprehensive Review

Mohammad Reza Hosenkhan and Binod Kumar Pattanayak

Abstract Internet of things (IoT) as an extension of existing global Internet is supposed to make it possible for device-to-device communication beyond the human-to-human communication pattern of the global Internet. Billions of smart devices connected to IoT environment can communicate among themselves using sensors and actuators that lead to complex security provisioning for efficient communication across IoT globally. In this paper, we detail the major security as well as privacy issues and possible resolution strategies as extracted from the research work of various authors in this field.

Keywords IoT · Security · Smart devices · Communication

1 Introduction

Information technology (IT) as one of the biggest innovations of the twentieth century has dominated the processes involving varieties of application domains starting from business to scientific applications. It has left a significant global effect on humanity for several decades. IT has played a key role in simplification as well as amplification of outcomes in respective application domains thereby making the processes time as well as cost-effective. Cognitive computing as an integral part of IT has added intelligent behaviour in solving various real-world problems. Moving further, the emergence of global Internet has significantly amplified the effectiveness of IT-oriented processes thereby establishing a global communication environment.

M. R. Hosenkhan (✉)
Faculty of Information and Communication Technology, Université des Mascareignes,
Rose-Hill, Mauritius
e-mail: rhosenkhan@udm.ac.mu

B. K. Pattanayak
Department of Computer Science and Engineering, Siksha 'O' Anusandhan Deemed
to be University, Bhubaneswar, India
e-mail: binodpattanayak@soa.ac.in

© Springer Nature Singapore Pte Ltd. 2020
S. Patnaik et al. (eds.), *New Paradigm in Decision Science and Management*,
Advances in Intelligent Systems and Computing 1030,
https://doi.org/10.1007/978-981-13-9330-3_36

Operating on Internet Protocol Version 4 (IPV4) backbone, Internet has made it possible for two individuals from any corners of the globe to communicate among themselves in real time that has simplified the business processes thereby significantly improving the business outputs. A wide range of services are supported by Internet that has covered practically all fields of human life. However, Internet in present form can facilitate only human-to-human communications. Recent advancements brought into Internet that is regarded as Internet of things (IoT) are presumed to facilitate communications between anything, anywhere and anytime. The principal motivation behind the innovations of IoT is to facilitate communications between intelligent devices independent of human interventions. IoT is supposed to operate on IPV6 Internet backbone that can facilitate billions of devices around the globe to get connected. As statisticians predict, by the year 2025, as many as 70 billion of intelligent/smart devices can effectively communicate among themselves using the services provided by IoT on IPV6 backbone. A simplified architecture of IoT is depicted in Fig. 1. Devices in IoT referred to as "things" possess their identity belong embedded with sensors can be capable of collecting information from their surroundings and disseminate further to other devices effectively. These "things" can virtually be controlled and regulated remotely. An enormous amount of physical devices referred to as "things" incorporated with computer-based application software can significantly improve the efficiency, accuracy of results thereby providing economic benefits as well. Examples of such "things" can be electric clams in water, biochip transponders, automobiles with built-in sensors, heart monitoring implants and so on. IoT can also be effectively used in disaster recovery applications such as operational on-field devices that facilitate search as well as rescue operations. These devices are supposed to effectively collect relevant information from the surrounding environment through sensors and then share these information autonomously with other devices as and when needed. Alongside the provisioning of the necessary infrastructure along with necessary applications in order for communication among billions of smart intelligent devices in IoT environment, these devices are capable of collecting huge amount of data, aggregate them swiftly, store and disseminate as per requirement. Advantages from the exploration of IoT technology are immense as compared with the existing global Internet. However, for the reason that a large number of devices are supposed to communicate among themselves using the services of IoT autonomously independent of human interventions, there arises ample amount of security issues associated with this innovative method of communication. Devices in IoT environment are vulnerable to external attackers with a higher degree as compared to the current Internet. This necessitates optimal solutions to be innovated to protect the devices in IoT from such attackers and make IoT communication more reliable. In this paper, we outline the security issues related to IoT communication and perform an extensive survey of literature available in the context of IoT security as of today contributed by various authors. The rest of the paper is organized as follows. In Sect. 2, we detail the IoT security architecture and layerwise security issues as well as vulnerabilities. In Sect. 3, we discuss a set of security issues and challenges as reported in the literature. Section 4 outlines the related work in the

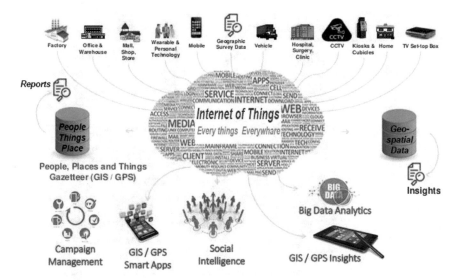

Fig. 1 IoT environment

field of IoT security. Finally, Sect. 5 concludes the paper along with probable future work in this direction.

2 IoT Security Architecture

Security issues related to IoT devices can be split into privacy, ethical and technological levels. Relating to security issues of IoT devices, various authors express their own perceptions. As claimed by authors in [1], the security requirements as required by the IoT devices are:

- Secure authentication;
- Secure bootstrapping;
- Security of IoT data;
- Secure access to data by authorized individuals.

Authors in [2] outline the security requirements for IoT devices as:

- Attack resiliency;
- Data authentication;
- Access control;
- Client privacy.

Authors in [3, 4] focus on:

- Data integrity;
- Data authentication.

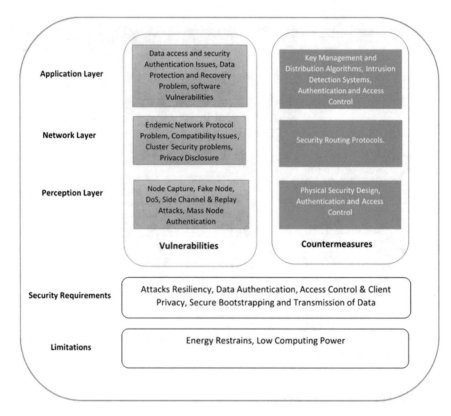

Fig. 2 IoT security architecture

In the present scenario, with highly diversified characteristics of IoT, ensuring security and privacy imposes a lot of challenges. Highly scalable and distributive properties of IoT need more flexible as well as innovative security architecture to be devised to meet such challenges in supporting the functionality of IoT devices in the long run. Here, we discuss a security architecture as proposed by authors in [5] that is depicted in Fig. 2.

(a) **Perception Layer**:

   (i) As claimed by authors in [6], the limitations of a Wireless Sensor Network (WSN) are power management, network discovery, control and routing. A WSN comprises of the modules hardware, communication stack, and middleware and secure data aggregation and, the basic components like sensor node, micro-controller, memory, radio transceiver and battery. A WSN node is associated with the security requirements like data confidentiality, integrity, freshness, availability, organization autonomy and authentication. The vulnerabilities in a WSN as suggested by authors in [7] are attacks on accuracy and authentication, silent attack on service integrity and attacks on network availability. Further, authors in [8] claim that the concerns of security in WSN can be resolved making use of authentication procedures using public key infrastructure (PKI).

(ii) Radio frequency identification (RFID): Radio frequency identification (RFID) technology can be used for unique identification by virtue of associating passive tags to the items they are attributed to. Authors in [9] propose various security standards for RFID technologies not only in order for addressing security concerns, rather the interaction issues as well.

(iii) IEEE 802.11 Wireless Protocol Standard: IEEE 802.11-based wireless networks are vulnerable to passive attacks such as eavesdropping [10]. Nevertheless, active attacks too can be viable by the virtue of exploitation of hardware vulnerabilities [11].

(iv) Long-term evolution (LTE): Long-term evolution (LTE) devices can be used to connect IoT devices to Internet in wireless mode in the absence of wired communication that guarantees efficient communication [12]. LTE devices are vulnerable to active as well as passive attacks. However, active attacks can be eliminated using cryptographic method. As claimed by authors in [13], passive attacks in such networks are almost impossible.

(v) WiMax (IEEE802.16): This technology can be extremely useful for connecting IoT devices in metropolitan areas and does not entertain the popularity in modern scenario though. It can support higher data rates with longer range of communication. Security specification in this standard basically exists in MAC layer which is also regarded as security or privacy layer [14]. That indicates that physical layer remains virtually unprotected. The security concerns related to this technology represent jamming at the level of physical layer that may very much lead to denial-of-service attack [15]. At the same time, the security/MAC layer can be susceptible to man-in-the-middle attack.

(vi) Near-field communication: Near-field communication (NFC) can be used for wide-range communication of IoT devices for operations such as payments, authentication and so on and has a short range of 20 cm though. NFC is also susceptible to security threats like denial-of-service (DOS) attacks, compromise of information and so on [16]. NFC suffers from the major security threat that sometimes it may not be encrypted that leads to the fact that the signal emerging from an IoT device may be captured by external antennas [17].

(vii) Bluetooth: Bluetooth technology has been proactively used for indoor systems in the context of iBeacons [18]. It can also be used for sensor networks meant for monitoring of earthquake. Security measures in this technology can be achieved in three different ways: (1) using pseudorandom frequency hopping; (2) restricted authentication; and (3) encryption [19].

(viii) Ultra-wideband (UWB): It represents a high-precision low power technology that can be used for IoT smart applications. It is considered to be more secured as compared to other technologies due to its high processing power and high range and most importantly its security aspects [20].

(b) **Middleware**:

IoT middleware is intended for interaction with the cloud technology and peer-to-peer systems. The list of services offered by middleware [21] follows.

 (i) Event-based: Authors in [21] claim that all the components of middleware communicate among themselves through events where the events operate on specific parameters that refer to changes in state. Some of the applications in the middleware do incorporate security features and some do not.
 (ii) Service-based: Service-based applications in the middleware are identical to that in a service-oriented computer (SOC) system that further relies on service-oriented architecture (SOA). These applications do encompass security attributes, and at the same time, they are vulnerable to security threats too [22].
(iii) Virtual machine (VM)-oriented: As assumed by authors in [21], the applications that rely on virtual infrastructure in order for safe execution are regarded as virtual machine (VM) applications in the middleware. Each application includes specific modules which interact with the VM running on each node in the network. Mate, a middleware application, incorporates a specific component that is dedicated to security provisioning which is responsible for blocking malicious programs propagating in the network [23].
(iv) Agent-based: Agent-based middleware includes mobile agents that are responsible for the security analysis of vulnerabilities [24].
 (v) Tuple spaces: Tuple space middleware comprises of components that incorporate a repository called as tuple space and those do not support any security mechanism as such.
(vi) Application-specific: This middleware focuses on the management of resources for various applications as when required by the application [25].

(c) **Application Layer**:

The application layer comprises of the following protocols.

 (i) Message Queue Telemetry Transport (MQTT): The IoT devices are incapable of dealing with the higher layer traditional application layer protocols like SMTP, SNMP and so on. Mostly lightweight protocols are devised for these devices by researchers in this field. Message Queue Telemetry Transport (MQTT) belongs to the class for such lightweight protocols especially devised for IoT devices that are constrained by factors such as high latency and low bandwidth operating in unreliable network environments [26]. The biggest advantage of MQTT is that it is compatible with any identification, authorization and authentication procedures in the context of network security. Identity of MQTT server can be achieved from its IP address along with the digital certificate associated with it. Its authorization too is obtained from MQTT server. Seekit that is a model-based security tool is responsible for privacy as well as protection of data in MQTT [27].

(ii) Extensible Messaging and Presence Protocol (XMPP): Extensible Messaging and Presence Protocol (XMPP) represents an extension of Extensible Mark-up Language (XML) that facilitates the exchange of real-time extensible data among network nodes. It does not possess any end-to-end security feature, rather protected with TLS alone. It relies upon Salted Challenge Response Authentication Mechanism (SCRAM) for security provisioning. SCRAM and TLS together provide authentication as well as confidentiality in MQTT [5].

(iii) Blockchain: Blockchain proposed in [28] is intended for solving double spending problem in cryptocurrency systems. It can be successfully applied without relying upon any third-party authentication mechanism. Blockchain can only provide pseudo-anonymity. In order for ensuring privacy in IoT systems, additional mechanisms must be implemented.

3 IoT Security Issues and Challenges

Security provisioning in such a huge heterogeneous environment with various devices like IoT becomes extremely challenging. In this section, a list of issues and challenges pertaining to security of an IoT system is detailed below [29].

(a) Privacy Context Awareness: This feature needs that the essential portion of the context of an object must be recognized effectively.

(b) Coupling of Digital Device and Physical Ambience: It needs that the processor must be coupled with the physical environment in order for the measurement of various information.

(c) Identification in IoT: For the reason that enormous number of applications may run on IoT environment, each of them must have an identification in all layers of the IoT protocol stack.

(d) Device Authentication: The smart devices in an IoT environment relying on sensors for data collection and aggregation are governed by a set of rules that provide the authentication for authorization of these sensors for sharing information.

(e) Data Combination: The heterogeneity of devices in IoT leads to different data formats supported by different devices. These data need to be combined to produce more meaningful information that further necessitates stringent security policies for its realization.

(f) IoT Scalability: With continuous innovations more and more devices can be added to IoT environment that needs the establishment of communication patterns for communication among such devices and it needs strict security measures to be incorporated in the environment.

(g) Secure Configuration and Setup: Resolving the scalability issue in IoT needs that a secure setup mechanism should also be provisioned which can be achieved on the basis of privacy.

(h) IoT and Critical Infrastructure (CI): The issues related to threats and privacy in IoT developed on critical infrastructure (CI) such as telecom, energy, etc. Need to be addressed optimally.

(i) Conflicting Market Interest: IoT becomes a more competitive environment that is capable of providing correlated data from different sources, and hence, the need arises for protection of personal correlated data.

(j) IoT with Evolution of Internet: Continuous evolution of global Internet has a direct impact on IoT, and it necessitates provisioning of data security and privacy of different elements constituting IoT environment.

(k) Trust Level Between IoT and Human: There must be guaranteed a specific level of trust between the human and IoT elements. Along with the trust level of machines belonging to IoT, human privacy must be ensured too.

(l) Data Management: Data protection on IoT can be achieved using cryptographic methods as well as respective protocols and for this purpose, exclusive policies need to be defined.

(m) IoT Devices' Durability: The fact that every entity on IoT has a limited and more specifically a short lifespan should be kept in mind during IoT security implementation.

4   Related Work

A limited spectrum of work has been reported in the literature pertaining to the security issues related to IoT. In this section, we discuss the contributions from different authors on IoT security as a whole. A general survey of various issues relating to IoT security as well as user privacy has been conducted by the authors in [30] wherein authors focus on the issues from end user's perspective with spreading of IoT technology. It should be noted that authors mostly pay attention to the security issues arising from information exchange among different entities on IoT. As a matter of fact, conventional security and privacy measures fail to provide the desirable level of security in IoT due to its decentralized topology and thus researchers now focus on blockchain (BC) technology that relies on cryptocurrency called Bitcoins. Again, BC considered to be computationally expensive is also incapable of providing desirable level of security for the reason that in incurs high bandwidth overhead. This problem is addressed by authors in [31] wherein authors propose a new secure IoT architecture using BC that is experimented on a smart home application that virtually addresses the limitation of bandwidth overhead successfully. This methodology of security provisioning in IoT using BC has been further explored by authors in [32] thereby proposing a lightweight scalable blockchain (LSB) scheme for this purpose that is again experimented on a smart home and results justified. A comprehensive analysis of security threats pertaining to IoT environment along with the probable resolution strategies is conducted by authors in [33] thereby putting emphasis on the significant role of encryption technology in security provisioning for IoT. Details of various IoT security issues along with their countermeasure with the help of some newly proposed algorithms have been addressed by authors in [34]. A matrix of security concerns has been innovated by the authors in [35] wherein focus is given on IoT middleware security provisioning with reference to the said matrix and probable measures are detailed as well. The underlying protocol architecture along with the security concerns of IoT is elaborated in a comprehensive survey work of authors in [36]. A discussion on IoT security issues related to access control to devices on IoT and their privacy concerns are addressed by authors in [37]. A layerwise security issues with respect to IoT protocol architecture along with resolution methods have been addressed by authors in their comprehensive survey work [38]. Security issues related to IoT devices specifically with low computing power and less memory capacity are addressed and measures outlined by the authors in [39]. Analysis of security as well as privacy concerns keeping in mind the heterogeneity of IoT environment is performed by authors in [40]. Taking into account the context-aware intelligent integrated services provided by IoT, the major security concerns in the context of the IoT protocol architecture are detailed in a comprehensive review work by authors in [41].

5   Conclusion and Future Work

Internet of things (IoT) as an innovation of the existing global Internet is supposed to connect billions of devices to the Internet worldwide. It is fascinated by its tremendous scalability features accommodating heterogeneous smart devices thereby provisioning an efficient communication among these devices using sensors and actuators. More than collection and aggregation of data, it is more important to make the communication environment more secure. With the dimensions of IoT and heterogeneity of devices connected to it, security provisioning becomes an extremely challenging task. Before security provisioning can be facilitated, it is vital to identify all security challenges and vulnerabilities related to this amazing technology. We have attempted to identify the security concerns with reference a proposed IoT security architecture and also conducted a comprehensive review of work available in the literature up to date. This paper may be quite useful for the researchers working in the area of IoT security. Moving further, we acknowledge that the amazing scalability feature of IoT that grows exponentially in terms of number of devices attached to it may lead to new security concerns that need to be resolved. We aim at extending our research work in this direction in future.

References

1. Borgia, E.: The internet of things vision: key features, applications and open issues. Comput. Commun. **54**, 1–31 (2014)

2. Weber, R.H.: Internet of things–new security and privacy challenges. Comput. Law Secur. Rev. **26**(1), 23–30 (2010)
3. Zafari, F., Papapanagiotou, I., Christidis, K.: Micro-location for internet of things equipped smart buildings. CoRR, vol. Abs/1501.01539 (2015)
4. Zafari, F., Papapanagiotou, I., Christidis, K.: Microlocation for internet-of-things-equipped smart buildings. IEEE Internet Things J. **3**, 96–112 (2016)
5. Mendez, D., Papapanagiotou, I., Yang, B.: Internet of things: survey on security and privacy. Inf. Secur. J. A Glob. Persp., 1–16 (2018)
6. Atamli, A.W., Martin, A.: Threat-based security analysis for the internet of things. In: 2014 International Workshop on Secure Internet of Things (SIoT), pp. 35–43, IEEE (2014)
7. Borgohain, T., Kumar, U., Sanyal, S.: Survey of security and privacy issues of internet of things (2015). arXiv preprint, arXiv:1501.02211
8. Medaglia, C.M., Serbanati, A.: An overview of privacy and security issues in the internet of things. In: The Internet of Things, pp. 389–395, Springer (2010)
9. Phillips, T., Karygiannis, T., Kuhn, R.: Security standards for the rfid market. IEEE Secur. Priv. **3**(6), 85–89 (2005)
10. Djenouri, D., Khelladi, L., Badache, N.: A survey of security issues in mobile ad hoc networks. IEEE commun. Surv. **7**(4), 2–28 (2005)
11. Naeem, T., Loo, K.-K.: Common Security Issues and Challenges in Wireless Sensor Networks and IEEE 802.11 Wireless Mesh Networks, vol. 3, no. 1 (2009)
12. Costantino, L., Buonaccorsi, N., Cicconetti, C., Mambrini, R.: Performance analysis of an LTE gateway for the IoT. In: 2012 IEEE International Symposium on a World of Wireless, Mobile and Multimedia Networks (WoWMoM), pp. 1–6, IEEE (2012)
13. Bilogrevic, I., Jadliwala, M., Hubaux, J.-P.: Security issues in next generation mobile networks: LTE and femtocells. In: 2nd International Femtocell Workshop, No. EPFL-POSTER-149153 (2010)
14. Papapanagiotou, I., Toumpakaris, D., Lee, J., Devetsikiotis, M.: A survey on next genera-tion mobile wimax networks: objectives, features and technical challenges. Commun. Surv. Tutorials IEEE **11**(4), 3–18 (2009)
15. Hasan, S.S., Qadeer, M.A.: Security concerns in wimax. In: First Asian Himalayas International Conference on Internet, 2009. AH-ICI 2009, pp. 1–5, IEEE (2009)
16. Madlmayr, G., Langer, J., Kantner, C., Scharinger, J.: NFC devices: security and privacy. In: Third International Conference on Availability, Reliability and Security, 2008. ARES 08, pp. 642–647, IEEE (2008)
17. Curran, K., Millar, A., Mc Garvey, C.: Near field communication. Int. J. Electr. Comput. Eng. **2**(3), 371 (2012)
18. Estimote, Estimote Real World Context for Your Apps. http://www.estimote.com. Online; Accessed 26 Sept 2014
19. Bouhenguel, R., Mahgoub, I., Ilyas, M.: Bluetooth security in wearable computing applica-tions. In: 2008 International Symposium on High Capacity Optical Networks and Enabling Technologies, pp. 182–186, IEEE (2008)
20. Ullah, S., Ali, M., Hussain, A., Kwak, K.S.: Applications of UWB Technology (2009). arXiv preprint, arXiv:0911.1681
21. Razzaque, M.A., Milojevic-Jevric, M., Palade, A., Clarke, S.: Middleware for internet of things: a survey. IEEE Internet Things J. **3**(1), 70–95 (2016)
22. Eisenhauer, M., Rosengren, P., Antolin, P.: Hydra: a development platform for integrating wireless devices and sensors into ambient intelligence systems. In: The Internet of Things, pp. 367–373, Springer (2010)
23. Costa, N., Pereira, A., Serodio, C.: Virtual machines applied to WSN's: the state-of-the-art and classification. In: 2007 Second International Conference on Systems and Networks Commu-nications (ICSNC2007), pp. 50–50, IEEE (2007)
24. Nagy, M., Katasonov, A., Szydlowski, M., Khriyenko, O., Nikitin, S., Terziyan, V.: Challenges of Middleware for the Internet of Things. INTECH Open Access Publisher (2009)

25. Murphy, A.L., Picco, G.P., Roman, G.-C.: Lime: a middleware for physical and logical mobility. In: 21st International Conference on Distributed Computing Systems, 2001, pp. 524–533, IEEE, 2001
26. Stanford-Clark, A.N.A.: MQTT version 3.1.1. OASIS Std., Oct 2014
27. Neisse, R., Steri, G., Baldini, G.: Enforcement of security policy rules for the internet of things. In: 2014 IEEE 10th International Conference on Wireless and Mobile Computing, Networking and Communications (WiMob), pp. 165–172, IEEE (2014)
28. Christidis, K., Devetsikiotis, M.: Blockchains and smart contracts for the internet of things. IEEE Access **4**, 2292–2303 (2016)
29. Zolanvari, M.: IoT security: a survey. (2015). http://www.cse.wustl.edu/~jain/cse570-15/ftp/iot_sec/index.html
30. Borgohain, T., Kumar, U., Sanyal, S.: Survey of security and privacy issues of internet of things. Int. J. Adv. Netw. Appl. **6**(4), 2372–2378 (2015)
31. Dorri, A., Kanhere, S.S., Jurdak, R.: Blockchain in Internet of Things: Challenges and Solutions. arXiv preprint arXiv:1608.05187 (2016)
32. Dorri, A., Kanhere, S.S., Jurdak, R., Gauravaram, P.: LSB: A Lightweight Scalable Blockchain for IoT Security and Privacy, pp. 2–17 (2017) arXiv:1712.02969v1cs.CR
33. Ahmed, A.W., Ahmed, M.M., Khan, O.A., Shah, M.A.: A comprehensive analysis on the security threats and their countermeasures of IoT. Int. J. Adv. Comput. Sci. Appl. (IJCSA) **8**(7), 489–501 (2017)
34. Pandey, E., Gupta, V.: An analysis of security issues of internet of things (IoT). Int. J. Adv. Res. Comput. Sci. Softw. Eng. (IJARCSSE) **5**(11), 1768–1773 (2015)
35. Fremantle, P., Scott, P.: A survey of secure middleware for the Internet of Things. PeerJ Comput. Sci. **3**, e114 (2017)
36. Chetan, C., Tejaswini, N.P., Guruprasad, Y.K.: A survey on applications, privacy and security issues in internet of things. Int. J. Adv. Res. Comput. Sci. (IJARCS) **8**(5), 2433–2436 (2017)
37. Satish, K.J., Patel, D.R.: A survey on internet of things: security and privacy issues. Int. J. Comput. Appl. (IJCA) **90**(11), 20–26 (2014)
38. Jaychand, Behar N.: A survey on IoT security threats and solutions. Int. J. Innov. Res. Comput. Commun. Eng. (IJIRCCE) **5**(3), 5187–5193 (2017)
39. Fernandes P., Monteiro, A., Lasrado, S.A.: Evolution of internet of things (IoT): security challenges and future scope. Int. J. Latest Trends Eng. Technol. (IJLTET) (Special Issue), 164–169 (2016)
40. Suchitra, C., Vandana, C.P.: Internet of things and security issues. Int. J. Comput. Sci. Mob. Comput. (IJCSMC) **5**(1), 133–139 (2016)
41. Choudhury, A., Godara, S.: Internet of things: a survey paper on architecture and challenges. Int. J. Eng. Technol. Sci. Res. (IJETSR) **4**(6), 442–447 (2017)

Design of Fuzzy Logic-Based PID Controller for Heat Exchanger Used in Chemical Industry

Mohan Debarchan Mohanty and Mihir Narayan Mohanty

Abstract In industries, the use of heat exchanger is badly necessary to transfer heat through fluid. For such purpose, control of temperature has a great role, so that PID controller is used vastly. This paper basically focuses upon the feedforward–feedback control strategy involved along with the PID controller. This control strategy cancels the effect of the measured disturbances and load changes in the system. The use of fuzzy controller tuning of PID controller makes the system intelligent and robust. The simulation results endorse the effectiveness of fuzzy controller-based tuning along with the feed-forward–feedback control strategy.

Keywords PID controller · Feed-forward–feedback · Fuzzy controller · Robust · Disturbances · Load changes

1 Introduction

In all chemical industries, the energy production occurs in the form of heat. Heat exchanger is utilized for transferring heat from hot fluid to cool fluid. They are widely used in space heating, refrigeration, petrochemical industries, air-conditioning and sewage treatment. The low-temperature fluid outlet works as the control variable. Many types of heat exchanger are used in the chemical industries to control the temperature of the outlet fluid. Most common type of it is shell and tube [1]. This work provides a PID controller based on fuzzy logic which is designed to control the heat exchanger. This method becomes advantageous here because it has the capability to

M. D. Mohanty
Department of Electronics and Instrumentation Engineering, College of Engineering and Technology, Bhubaneswar, India
e-mail: mohan.debarchan97@gmail.com

M. N. Mohanty (✉)
Department of Electronics and Communication Engineering, ITER, Siksha 'O' Anusandhan Deemed University, Bhubaneswar, Odisha, India
e-mail: mihirmohanty@soa.ac.in

© Springer Nature Singapore Pte Ltd. 2020
S. Patnaik et al. (eds.), *New Paradigm in Decision Science and Management*,
Advances in Intelligent Systems and Computing 1030,
https://doi.org/10.1007/978-981-13-9330-3_37

model the uncertainties themselves. Hence, it becomes an effective and intelligent method for tuning of PID controllers. This PID controller gives the better solution to many real-world problems. Most of the industries have implemented the controller based on PID algorithm because of its simple and clear functionality [2, 3]. *Several other works have also been done with the design of heat exchangers. Some of them are sited here.* In [4], the heat exchanger was designed by adaptive technique and experimentally estimated the usefulness of tuning methods, as internal model control (IMC) and relay auto-tuning. They have seen that the controller based on IMC shows a good result. For the purpose of regular tuning of the PID procedure control constraints, typically called 'auto-tuning' was introduced in [5]. The process of applying the technique consists of (a) input signal that can be tested using sampling procedure, (b) handling data for evaluating characteristic values of the process and (c) ascertaining the best value of control parameters. *The authors in [6] have designed a fuzzy-based PID controller which was found to be an energy efficiency approach. In [7], they have designed a heat exchanger for a nonlinear system and interval parametric study has done.* In [8], a new technique of predictive control is presented based on fuzzy logic. Also, multiple model approach for control of tubular heat exchanger system was presented in [9]. In [10], a predictive controller based on fuzzy model was used for controlling temperature of a heat exchanger. A new approach for water supply in the modern healthcare industry was analyzed by the authors in [11]. That work shows the application of rule-based fuzzy logic to organize the process in water supply section. To refine fuzzy rules' initial approximate set automatically, a self-organizing fuzzy controller has been used. A two-level control system for fossil fuel power units was planned in [12]. For improving the control, partial differential equation was introduced for optimization of the proportional–integral fuzzy controller was designed in [13].

Whatever is left of the paper is sorted out as following way: The proposed method is portrayed in Sect. 2. The obtained results are presented in Sect. 3. Section 4 concludes the work, and the future scopes are also presented.

2 Materials and Methods

The objective of the proposed system is to enhance the control performance of a heat exchanger system in the industries. As compared to the tuning methods of the conventional controllers, tuning of PID controllers based on FL controller serves as the most efficient methods of tuning. The FL controllers can be customized and of easy implementation [11]. Many have been proposed with the accommodation of industry and academic for complex systems [14, 15].

The use of controller based on fuzzy logic in control system gives a good active response, discards interruption, also permits low constraint responsive disparity and eliminates external power. This control algorithm is robust, effortless and proficient. The proposed control system has exposed significant tracking capability and

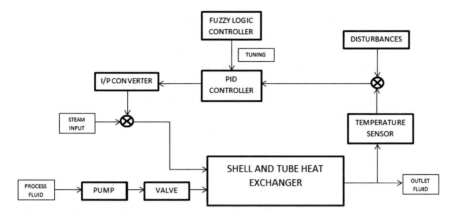

Fig. 1 Proposed system block diagram

displays the effectiveness and power of hybrid fuzzy controller credibly with elevated performance in constraint dissimilarities and load uncertainties.

Feedback control strategy is used for the compensation of the deviation of output temperature. For improvement and practical strategies, this control includes an interruption and therefore it is not the perfect result. The combination of feed-forward control and feedback control strategy maintains the constancy of the method and delivers products at objective values [16]. The combinational strategy also reduces the effect of the disturbances and the load changes on the controller. The proposed work is shown in Fig. 1.

3 Design of Heat Exchanger System

Chemical reactors and heat exchanger are used by a chemical heating process. The processed fluid is provided to the heat exchanger system by a storage tank. The exchanger system heats up the process to an optimized point using superheated stream to 180 °C. Subsequently, the steam warms the fluid and reduces to 100 °C [8, 17]. Another path is there to leave out the shell and heat exchanger. It keeps away from blocking of the heat exchanger. There are certain hypotheses taken for the situation; first, the rate of inflow and outflow of the fluid is equal; secondly, insulating wall has a negligible heat storing capability. The feed-forward loop contains the flow rate and the minor temperature disturbances. The PID controller acts reverse in the feedback process control loop. In this work, the value is of air to open to close type. One temperature sensor is employed as the sensing element, and it is applied in feedback path of the controller. Temperature sensor is used for measuring the temperature of the outlet fluids, and the output of the temperature sensor is kept in the form of voltage, which is ultimately fed to the controller by converting the output of the sensor to standard current signal in the range between 4 and 20 mA. The control algorithm works for controller with the fixed point by which the control element uses through the actuator. In this, the actuator works as an input converter and the final control element is the air to open–close valve. The actuator is fed with the controller output ranging from 4 to 20 mA and converts that output to a standardized pressure signal ranging from 3 to 15 psig [18]. Gradually, the valve performs according to the controller decisions.

4   Mathematical Modeling

The transfer function of a generalized shell and tube heat exchanger is represented as

$$Gc = \frac{Ke^{s\tau}}{1 + sT} z \qquad (1)$$

where τ is the delay time of the transportation lag, T represents the time constant of the process and K is the gain factor.

Based on [19], the experimental values for different parameters are substituted in (1) that results as

$$Gc = \frac{50}{1 + 30s} \quad \text{and} \quad Gv = \frac{0.13}{1 + 30s}$$

where Gv is the transfer function of the final control element. Hence, the transfer function of the heat exchanger plant becomes

$$Gp = Gc * Gv = \frac{6.5K}{90s^2 + 33s + 1} \qquad (2)$$

The characteristic equation of the liberalized system becomes

$$1 + G(s) \cdot H(s) = 0$$

Or,

$$900s^3 + 420s^2 + 43s + (1 + 1.04K) = 0 \qquad (3)$$

where $Gs = \frac{0.16}{1+10s}$ is the transfer function of the desired set point.

Applying Routh's stability criteria for stability, we obtain $K = 18.33$ and $T = 28.53$ s and used in this design.

5 PID Controller Tuning Using FL Controller

The obtained values of ultimate gain and ultimate time are tuned with the help of an FL controller that makes the system intelligent and robust. The controller has four stages of processing that are mentioned below.

1. Fuzzification: Process of transforming the preprocessed inputs to the fuzzy domain.
2. Knowledge base: It depends upon two strategies:

   a. Database: This strategy provides the necessary data for running the FL module and their individual rule base.
   b. Rule base: Structural representation of the control policies.
   c. Fuzzy Inference System: It has an uncomplicated input–output relationship. Input is managed with the help of certain algorithms specifically named as fuzzy inference algorithm. Mamdani fuzzy algorithm is one of the systems that is widely used.
   d. Defuzzification: Process of transformation of fuzzy sets to processed value is called defuzzification. There are different methods used for defuzzification. The method used here is the center of sum (COS) method. The method focuses on double calculation of the overlapping area. The defuzzified value $X*$ is hence given by:

$$X* = \frac{\sum_{i=1}^{N} xi}{\sum_{i=1}^{n} 1} \cdot \frac{\sum_{k=1}^{n} mu(xi)}{\sum_{k=1}^{n} mu} \tag{4}$$

Here, N is a total number of fuzzy variables and $mu(xi)$ is the membership function for the kth fuzzy set (Figs. 2, 3 and Table 1).

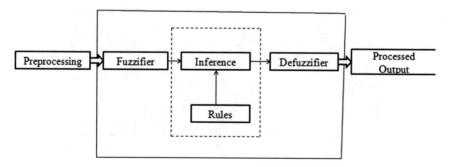

Fig. 2 Fuzzy inference system

(a)

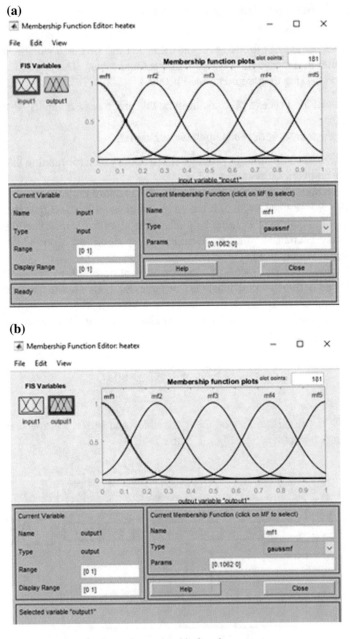

(b)

Fig. 3 **a** Membership function input, **b** membership function output

Table 1 Fuzzy rule base with 25 rules

E\dE	(+) Large	(+) Small	0	(−) Small	(−) Large
(+) Big	(−) Large	(−) Large	(−) Small	(−) Small	0
(+) Small	(−) Large	(−) Small	(−) Small	0	(+) Small
0	(−) Small	(−) Small	0	(+) Small	(+) Small
(−) Small	(−) Small	0	(+) Small	(+) Small	(+) Large
(−) Big	0	(+) Small	(+) Small	(+) Large	(+) Large

6 Results

Experimental results are as follows:

1. The heat exchanger designed without FL tuning results in an output aperiodic overshoots and more oscillations and is shown in Fig. 4.
2. The PID controller tuning based on FL controller gives an output of minimized error; the effect of disturbances and load changes is almost negligible.
3. The use of FL controller makes the system robust.

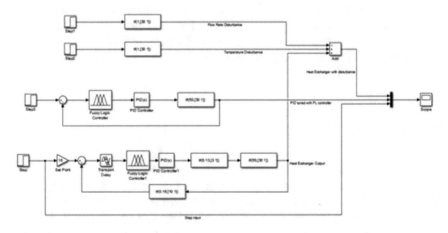

Fig. 4 Simulation model for the proposed system

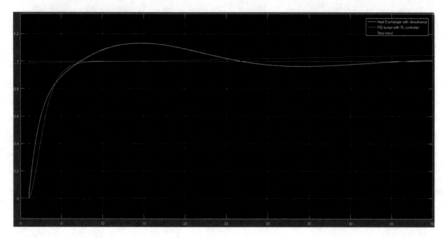

Fig. 5 Response of the Simulink proposed model

4. The overshoot decreases with the inclusion of disturbances, final control element and the sensor in the system.
5. The FL controller does not rely upon the mathematical model, however just the structure's semantic clarification.
6. Combination of feed-forward and feedback control strategy provides stability to the system.

It is a necessary requirement that the inference rules should be customized while performing the experiments on the equipment either by the reduction of rules or by the addition of complementary rules. The installation is model dependant. An FL controller makes the required arrangements for the transformation of the linguistic control strategy into automated systems. Figure 4 shows the Simulink model of the proposed system (Fig. 5).

7 Conclusion

This paper evaluates the effectiveness of combinational strategies of feed-forward–feedback control system. The outlet fluid temperature is controlled with the help of a PID controller. The PID controller used is tuned with the help of an FL controller. This tuning provides a robustness to the system, minimizes the error and reduces the effect of disturbances of flow rate and temperature variations. The simulation results obtained from MATLAB and Simulink show the efficiency of the control strategy and the superiority of the FL controller-based tuning method.

References

1. Padhee, S., Khare, Y.B., Singh, Y.: Internal model based PID control of shell and tube heat exchanger system. In: Students' Technology Symposium (TechSym), pp. 297–302. IEEE. (2011)
2. Ang, K.H., Chong, G., Li, Y.: PID control system analysis, design, and technology. IEEE Trans. Control Syst. Technol. **13**(4), 559–576 (2005)
3. Wang, L., Barnes, T.J.D., Cluett, W.R.: New frequency-domain design method for PID controllers. IEE Proc. Control Theor. Appl. **142**(4), 265–271 (1995)
4. Sahoo, A., Radhakrishnan, T.K., Rao, C.S.: Modeling and control of a real time shell and tube heat exchanger. Resour. Efficient Technol. **3**(1), 124–132 (2017)
5. Nishikawa, Y., Sannomiya, N., Ohta, T., Tanaka, H.: A method for auto-tuning of PID control parameters. Automatica **20**(3), 321–332 (1984)
6. Tabatabaee, S., Roosta, P., Sadeghi, M.S., Barzegar, A.: Fuzzy PID controller design for a heat exchanger system: the energy efficiency approach. In: IEEE International Conference on Computer Applications and Industrial Electronics (ICCAIE), pp. 511–515 (2010)
7. Vasickaninova, A., Bakosova, M., Meszaros, A., Oravec, J.: Fuzzy controller design for a heat exchanger. In: IEEE 19th International Conference on Intelligent Engineering Systems (INES). IEEE, pp. 225–230, Sept 2015
8. Skrjanc, I., Matko, D.: Predictive functional control based on fuzzy model for heat-exchanger pilot plant. IEEE Trans. Fuzzy Syst. **8**(6), 705–712 (2000)
9. Mazinan, A.H., Sadati, N.: Fuzzy multiple models predictive control of tubular heat exchanger. In: IEEE International Conference on Fuzzy Systems. FUZZ-IEEE World Congress on Computational Intelligence. IEEE, pp. 1845–1852, June 2008
10. Fischer, M., Nelles, O., Isermann, R.: Adaptive predictive control of a heat exchanger based on a fuzzy model. Control Eng. Practice **6**(2), 259–269 (1998)
11. Mohanty, M.D., Mohanty, M.N.: Design of intelligent PD controller for water supply in healthcare systems. In: Proceedings of the First International Conference on Information Technology and Knowledge Management, Annals of Computer Science and Information Systems, vol. 14, pp. 95–98 (2017)
12. Mohanty, M.D., Mohanty, M.N.: Design of an intelligent controller for 6 degrees of freedom quad rotor UAV. In: International Conference on Information Technology (ICIT) (2018) (Accepted)
13. Garduno-Ramirez, R., Lee, K.Y.: Wide range operation of a power unit via feedforward fuzzy control [thermal power plants]. IEEE Trans. Energy Convers. **15**(4), 421–426 (2000)
14. Maidi, A., Diaf, M., Corriou, J.P.: Optimal linear PI fuzzy controller design of a heat exchanger. Chem. Eng. Process. **47**(5), 938–945 (2008)
15. Torshizi, A.D., Zarandi, M.H.F., Zakeri, H.: On type-reduction of type-2 fuzzy sets: A review. Appl. Soft Comput. **27**, 614–627 (2015)
16. Stylios, C.D., Groumpos, P.P.: Modeling complex systems using fuzzy cognitive maps. IEEE Trans. Syst. Man Cybern. Part A Syst. Hum. **34**(1), 155–162 (2004)
17. Shi, L., Kapur, K.C.: A synthesis of feedback and feedforward control for process improvement under stationary and nonstationary time series disturbance models. Qual. Reliab. Eng. Int. **31**(3), 343–354 (2015)
18. Khare, Y.B., Singh, Y.: PID control of heat exchanger system. Int. J. Comput. Appl. **8**(6), 0975–8887 (2010)
19. Ismail, E.E., Abdel Rassoul, R., Abdelbary, A., El Said, O.M.: Improving the performance of heat exchanger system by better control circuits. Int. J. Comput. Appl. **121**(11) (2015)
20. Padhee, S.: Controller design for temperature control of heat exchanger system: Simulation Studies. WSEAS Trans. Syst. Control **9**, 485–491 (2014)

On the Introduction of the Teaching Model of "Research Learning"

Qing Zhang

Abstract "Research-based Learning" is aimed at stimulating innovative thinking, taking the basic structure of the subject as its content, taking independent thinking as the learning essentials, and fully performing the teaching principles of students as the main body, in order to train students to learn knowledge and to create for the purpose of improving students' independent thinking ability, inductive application ability, mind-imagining, and problem-solving skills.

Keywords Research learning · Teaching innovation · Innovative thinking

The introduction of "research Learning" method is a kind of exploration of classroom teaching innovation; it is a kind of teacher to establish the situation, guide the students to enrich join, automatically explore the coordinated and unified course, situational teaching mode. It is characterized by the full performance of students as the main principle of teaching, in order to train students to learn knowledge, learning to create for the purpose of improving students' ability to think independently, inductive application ability, mental imagination ability, and problem-solving ability.

1 The Idea of "Research Learning"

The essence of "research learning" is to promote a great change of learning concept (revolution), that is, the mastery of knowledge and skills from the traditional emphasis on the learner, and the change to the self-experience of the learners' inner mind and emotion in the course of learning (thoughts process), the passive, one-way, fixed learning, which is controlled by the traditional teachers in the classroom, transforms into the automatic, divergent and interactive learning controlled by the students in the situation of mutual cooperation and coordination.

Q. Zhang (✉)
Basic Department of Armed Police University of Engineering, Xi'an, China
e-mail: ama412@163.com

© Springer Nature Singapore Pte Ltd. 2020
S. Patnaik et al. (eds.), *New Paradigm in Decision Science and Management*,
Advances in Intelligent Systems and Computing 1030,
https://doi.org/10.1007/978-981-13-9330-3_38

In the course of this change, the teacher will be the dominant decision-maker and protagonist in the course of learning, and the teaching effect will gradually be diluted (back to the background, the configuration and completeness of the learning situation), and the students will become the real masters and decision-makers of the learning activities. In this model, both teachers and students can describe and evaluate the knowledge, break through the absolute concept of scientific knowledge, break through the superstition of scientific knowledge, and form the idea of exploration and asking questions, and discuss the equal exchange between teachers and students, It is helpful for students to overcome physical and linguistic stagnation, stop students from attending classes such as listening to lullabies, so that students' physiology is relaxed and the potential of each student's brain activity is easily aroused.

2   Connotation of the Concept of "Research Learning"

The important characteristics of this teaching mode are: To stimulate the creative thinking, take the basic structure of the subject as the content, take the independent thinking as the learning essentials. It assumes that the student is the decision-maker, causes the student to the topic own interest, under the teacher guidance like the scientist invents the truth, emphasizes the student to explore and study, advocates the student's practice activity, encourages the student to discover the law, grasps the basic principle method, then solves the problem. In the teaching structure sequence, the important performance is: The teacher first must establish a self-study situation for the student, then under the guidance of the teacher will enlighten, the reading, the inquiry, the review, summarizes the organic union, thus lets the student to feel completely a "raises the topic (research topic)-explores the discussion-summarizes the progress" the course.

3   Application of the Concept of "Research Learning"

3.1   Establishing the Situation of Autonomous and Exploratory Learning

1. Re-establish a new type of teacher–student relationship, set up an autonomous learning atmosphere. To create democracy and the same teacher–student relationship, teachers should respect and trust each student, especially with a sincere love, to close to those backward students. Well-known educator Xingzhi Tao teacher has a famous saying: "Your pointer under the watt, your cool-eyed Newton, your ridicule has Edison." Students experience equality, democracy, respect, trust, friendliness, clarity, affection, and friendliness while being encour-

aged, influenced, summoned, and guided to form a positive attitude toward life and emotional experience and enrich their autonomy and creativity.

2. Establish the best learning situation and trigger the learning motivation system. The teaching of each strategy, first of all to consider how students learn, in-depth study of students' physiological characteristics and the rules of mind, to establish autonomous learning situation, triggering students learning motivation, the best way to promote students' self-study, promote students' physical and mental growth. ① set up an objective inquiry situation, including various experimental situations, intuitive situations, natural situations, and social situations. In this situation, students learn to be lively, to learn to be active, so that potential can be further developed. ② establish intelligence situations, including various problems and situations. Teachers should study the recent knowledge structure of the students, find the best combination of new and old knowledge, determine the best difficulty of the students' research problems, set up the problem situation, and promote the new knowledge to be mixed in the original knowledge structure quickly.

3. Establish the situation of enrichment and realize the students' self-study. We want to allow students to join, all the way to join the whole. ① fully give students the time to study independently, so that students have enough time to explore, to think, and to exchange. ② fully gives students the right to question. Teachers should encourage students to question questions, allow students to argue, support students to express independent views, and exert students' innovative spirit. ③ guide students to master the principles and methods of dealing with problems.

3.2 The Elements of Practical Autonomy and Exploratory Learning

1. Raise a question. Firstly, teachers should find materials that are closely related to the teaching content, which can cause the students' interest, establish a specific situation, put forward the field or direction to be studied, and guide the students to set up questions and ask questions that need to be explored. For example, when it comes to global environmental governance issues, especially the water pollution, the following topics can be planned: What are the causes of water pollution? Is the level of pollution in each area similar? What are the important pollutants? What are the effects of weather conditions on water pollution and what are the hazards caused by water pollution? What are the governance measures adopted by all governments? With the latest developments in current affairs, what policies have we adopted to protect the environment?

2. Exploration and discussion. To solve the problem, put forward the beginning of the problem planning. The contents of the plan include: What inquiry essentials (hands-on research), what tectonic essentials (groups), when to complete, etc., according to the information available to the proposed topic. Students will book,

network collation of the relevant information, combined with their own practice to get the experimental signs, data, to hold the exposition, inference, and thus draw conclusions.
3. Summarize forward. Teachers guide students to the beginning of the conclusion of exchange, supplement, and completeness, so as to summarize the more rigorous and accurate presentations. For example, through research study, students know that water pollution treatment can be used a chemical treatment, physical treatment, biological treatment, as well as saving water to eliminate emissions, mediate industrial structure, innovative product structure, enhance water resources management, and other essentials. So that students can further realize that environmental protection is a good thing for the nation.

3.3 The Requirements of Research Learning Teaching to Students

The training of students' ability and the cultivation of quality are the starting point and destination of teaching courses. Therefore, the course teaching should focus on the progress of students' overall quality, for the formation of students' sound character and the formation and improvement of ability, knowledge, and so on, the research-oriented learning teaching should cherish the cultivation of students' comprehensive ability especially: ① should have keen observation and thinking ability, ② have the ability to collect and accumulate information, ③ should have the ability to comprehensively apply the knowledge of various subjects to solve realistic problems, and ④ should have good interpersonal skills and solidarity spirit. In the teaching of research-based learning, the students get not only knowledge, but also a learning character and ability, which laid a solid foundation for their lifelong learning and future growth.

3.4 The Requirements of Research Learning Teaching to Teachers

In the teaching of research learning, the teacher is the constructor, the person who joins and the guide. Teachers' plans for teaching purposes, the structure of teaching activities, the teaching methods, and the use of skills should be beneficial to each student's growth. Teacher's teaching is a creative activity, each teacher has the responsibility to guide and nurture the students' spirit of exploration, the spirit of innovation, to create an atmosphere of advocating knowledge, seeking truth, promote students' self-study, independent thinking, for students talent and potential free, enrich the growth of the situation to create loose.

It is important to have creative high-quality teachers to practice research-based learning teaching and cultivate students' creative ability. As teachers, we must also have: (1) modern educational philosophy, (2) absorbing the ability to master new knowledge, (3) a variety of professional teaching skills: ① can give full play to the students' main role ability, ② ability to skillfully use modern teaching methods, ③ skilled moral skills, and (4) with pioneering and innovative spirit and strong scientific research ability.

Acknowledgements I would like to express my gratitude to all those who helped me during the writing of this thesis.
Thanks to my family, they helped me quite a lot through advice, comments, and criticism. Their encouragement put me through frustration and frustration. It would not have been possible to finish the paper without their promotion. In addition, I am very grateful to my friends and colleagues for their contributions to this paper in various ways.

Reference

1. Huang, X., et al.: Curriculum Leadership and School-Based Curriculum Development, 29. Educational Science Press, Beijing, China (2005) (in Chinese)

Digital Learning: The Booster for Smart India

Jui Pattnayak and Sabyasachi Pattnaik

Abstract Digital India, the dream of the Government of India, is nothing but the inclusion of information and communication technology (ICT) to turn India into the digitization of knowledge to achieve knowledge economy. Successful integration of ICT in learning, i.e. digital learning or e-learning, is the most essential requirement now to accomplish Digital India. India is far behind to reach the milestone of Digital India. It requires a lot of efforts and dedication from all the government sectors as well as the private sector. This paper proposes an e-learning framework for the adoption of digital learning in different spheres of the country. To achieve collaboration, e-learning must be embedded in knowledge management (KM) environment. With successful collaboration process, knowledge becomes globalized to form knowledge society (KS) by which India will move towards Smart India.

Keywords Digital learning/e-learning · Digital India · ICT · KM · KS · PLE

1 Introduction

Digital revolution commences from Digital India. Digital learning now-a-days is the most challenging teaching/learning methodology for the fulfilment of Digital India, the dream of our Hon'ble Prime Minister. The twenty-first century is termed as the digital era where the Internet brought a significant change in people. The growth of information and communication technology (ICT) is moving exponentially. E-education has certainly boosted the sector [1]. "Digital learning is the learning pedagogy alleviated by the technology or by the instructional practice which uses

J. Pattnayak (✉)
Department of CSE, DRIEMS (Autonomous), Cuttack, Odisha, India
e-mail: juipattnayak@gmail.com

S. Pattnaik
Department of I & CT, Fakir Mohan University, Vyasa Vihar, Balasore, Odisha, India

© Springer Nature Singapore Pte Ltd. 2020
S. Patnaik et al. (eds.), *New Paradigm in Decision Science and Management*,
Advances in Intelligent Systems and Computing 1030,
https://doi.org/10.1007/978-981-13-9330-3_39

technology effectively". It includes the application of a wide range of practices that includes blended and virtual learning. Learners are allowed to understand concepts within no time in a more efficient manner. With the evolution of mobile phones, computers, etc., learning is achieved rapidly. Digital learning, virtual lectures, game-based learning, learning portals become the principal components for the development of digital learning. Digital learning permits learners and instructors to associate and communicate with other people around the world. This provides a richer learning expertise and nurtures in-depth knowledge to the learners. The Internet enables learners to access a huge quantity of data from different sources of the world at a very low cost [2]. The tools and technologies of digital learning enable instructors to share information very rapidly with other instructors in real time. By widening digital devices and connected learning, classrooms around the country and around the globe can not only coordinate with each other to share ideas but also encourage learning, experience and communication skills [3].

Digital learning is being accepted as the step up to make students smarter. The self-directed nature of digital learning enhances the critical thinking power of students so as to improve analytic reasoning. The Internet tools and technologies allow students to learn collaboratively in groups where knowledge is globalized through various knowledge-related processes. These processes form the foundation of knowledge management (KM) which is discussed in Sect. 4. KM is the software environment to achieve collaboration. The collaborative groups gradually converge to form learning society where the primary production is knowledge. The society is becoming a warehouse of knowledge termed as knowledge society (KS). But digital learning is not confined to academic purpose only. Rather, it is used in every field right now such as medical, defence, engineering, agriculture and business. It is a day-to-day requirement in the current technological world. In this regard, the concept of Digital India came into picture, i.e. a programme to turn India into a digitally empowered society and knowledge economy which is a march towards knowledge society (KS).

2   Vision of Digital India

The key aspects of Digital India are the following:

I.   Infrastructure as a Utility to Every Citizen

- High-speed Internet facility for quick delivery of public services.
- Formation of digital identity for every Indian.
- Shareable private space on public demand.
- Safe and secure cyberspace.

II. Governance and Services on Demand Integrated across Departments or Juris-
dictions

- Availability of the government service through online in a real-time environment.
- To improve the simplicity of business process through transformed electronic
services.
- Cashless transaction.

III. Digital Empowerment of Citizens

- Globalization of digital literacy.
- Global access to digital content.
- Cloud computing.
- Availability of multilingual content.
- Collaborative learning.

2.1 Initiatives of Digital India

1. *Broadband Connectivity*: The objective is to connect 250,000 village panchayats
   under National Optical Fibre Network (NOFN) by December 2016. Cloud service
   will provide high-speed Internet.
2. *Universal Mobile Internet Connectivity*: The objective is to enhance network
   penetration and to connect 44,000 villages through the mobile network by 2018
   with an investment of RS. 16,000.
3. *Internet Access Programme for Public*: Each gram panchayat would be provided
   with one Common Service Centre (CSC), and the proposal has been made to
   convert 150,000 post offices into multi-service centres.
4. *E-Governance*: The government services can be used in a more efficient way
   with the help of information technology. All information may be available in
   electronic form.
5. *e-Kranti*: The focus is on electronic delivery of services to people; it may be
   education, health, financial inclusion or justice.
6. *Information for All*: The government launched MyGov.in website to provide
   a bidirectional communication between the citizens and the government to
   exchange ideas or suggestion with the government.
7. *Electronics Manufacturing*: The government is concentrating not to import electronics by 2020 through manufacturing of items locally such as smart energy
   metres, micro-ATMs, mobile, consumer and medical electronics.
8. *IT for Jobs*: The focus is to train 10 million people in towns and villages in
   five years for jobs in IT sectors. Another focal point is to train five lakh rural
   workforces on telecom and telecom-related services and to set BPOs in each
   north-eastern state.

9. *Early Harvest Programmes*: As per the planning of the government, Wi-Fi facilities are to be installed in all universities across the country. All books will be transformed into digital form. To achieve paperless environment, Email is used as primary mode of communication and Biometric Attendance System is used to maintain online attendance.

3 Digital Learning Framework

The overall framework of digital learning is given in Fig. 1 which consists of three logical divisions, the client, the server and the database part known as repository.

The design of client-side components has already been suggested by Pattnayak et al. [4]. The design of server-side components and database is left for the later part. The overall idea is that the server will serve the requests of the learners and the user data, and contents are stored in the database for future use.

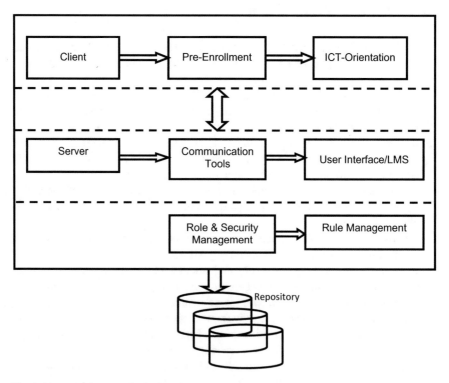

Fig. 1 Proposed framework of e-learning system

4 Role of Knowledge Management in Digital Learning

Collaborative learning is a significant part of digital learning without which societal learning is impossible. Collaboration is possible through knowledge management (KM). Digital learning focuses on individual learning where KM focuses on group learning. The objective of KM is to create learning community for collaboration. KM is used for quick processing of knowledge acquisition and knowledge organization and also for delivering large amounts of organizational knowledge. Sammour [5] stated that online education has already been replaced as the approach of tomorrow; knowledge could also be distributed across time and place. In World Wide Web (WWW), new knowledge can be acquired from the existing knowledge by the knowledge worker at the time of learning. Digital learning focuses on individual intelligence. KM-based learning supports the development of intellectual assets, workplace collaboration and innovation [6]. On the other hand, today, the benefits offered by knowledge sharing are increasing and this fact leads to widespread initiatives of collaboration [7]. Depending upon the various knowledge processes required for collaboration, the following KM process framework (Fig. 2) has been suggested.

- *Knowledge creation and acquisition*: Knowledge creation and acquisition involve the growth of people with knowledge either individually or in groups or communities of practice to acquire knowledge from intangible tacit knowledge.
- *Knowledge sharing*: Knowledge sharing involves in creating learning process when people are interested to develop new knowledge by helping each other.
- *Knowledge capture*: The process of converting tacit knowledge to explicit knowledge and also vice versa is called knowledge capture. It is done through externalization and internalization.
- *Knowledge storage*: Knowledge storage is the process of storing knowledge in the form of a knowledge repository consisting of documents, reports and databases.
- *Knowledge application*: The knowledge created and captured is to be applied in different learning contexts to achieve competitive advantages such as creating KS.
- *Knowledge Evaluation*: Learners will be assessed on a regular basis to verify that knowledge must be relevant and accurate.

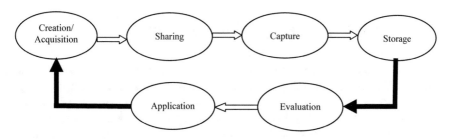

Fig. 2 Knowledge management process life cycle

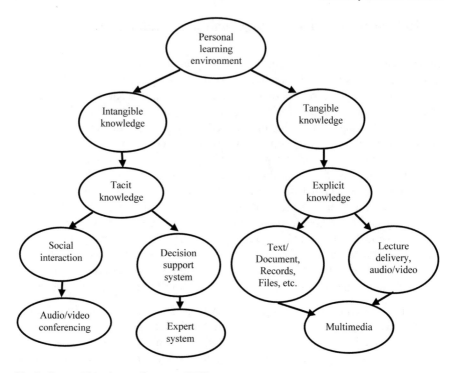

Fig. 3 Personal learning environment (PLE)

These KM processes take place in individual's personal learning environment (PLE) which consists of tacit and explicit knowledge. Tacit knowledge is the uncoded knowledge that remains in the human brain also called intangible knowledge, and explicit knowledge is the codified knowledge available for use of others called tangible knowledge. The knowledge acquired by each individual is known as tacit knowledge, and when that knowledge is delivered to others it is known as explicit knowledge. In this way, collaboration takes place to form knowledge society (KS).

4.1   The Personal Learning Environment (PLE)

PLE is not a technology, but also an approach/process that is formulated with the individual in mind. So, it varies from learner to learner. The PLE with the knowledge transfer modes is shown in Fig. 2. Tacit knowledge is the intangible or uncoded knowledge which has no physical existence. The knowledge which we acquire is the tacit knowledge. Explicit or tangible knowledge is the codified knowledge which has physical existence. When the acquired knowledge is ready for delivery, for example, texts, video and audio, it is the explicit knowledge (Fig. 3).

It has been observed that both types of knowledge are the essential components of the KM process in order to empower collaboration in the digital learning environment. The reason behind it is that the knowledge acquired by the individual is the tacit knowledge and is delivered to the society through explicit form.

4.2 Framework of Knowledge Management

Basing upon the above personal learning environment (PLE), the following framework of knowledge management has already been proposed by Pattnayak et al. [8] for the adoption of digital learning (Fig. 4).

It has been observed that the above framework consists of three logical divisions: people, process and technology. People are the various types of users of the system such as learner, instructor, instructional designer, subject matter expert and administrator. It may vary depending upon the requirement of the system. Process framework has already been suggested. In the technology part, the e-learning framework has been stated in Fig. 1. The other parts are left for the later part.

With the integration of people, process and technology within the digitized environment, knowledge takes the form of societal knowledge which becomes the warehouse of knowledge. This knowledge warehouse forms the knowledge society (KS). From the analysis of Digital India, the success of KS can be clarified.

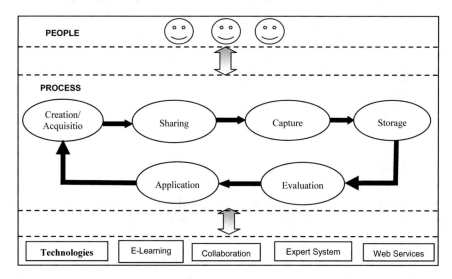

Fig. 4 Proposed knowledge management framework [8]

5   A Brief Analysis of Digital India

5.1   Benefits

Following are the important benefits of Digital India:

1. People all over the country can avail all the government services by the Digital India mission through common service delivery outlets. People can get chance for online education if they cannot afford school/colleges.
2. Availability and accessibility of online data to citizens of the country make the data more transparent.
3. E-governance can help to reduce corruption, and things can be done in a rapid manner.
4. Digital locker facility will help citizens to store their important documents such as PAN card, passport and mark sheets digitally.
5. For example, a person can give official details of his/her digital locker, to open an account where they can verify the documents. In this way, time can be saved and the pain of standing in long queues for document verification would be reduced.
6. Documentation and paperwork can be reduced to a significant level which will create Green India.
7. Digital India mission can restrict tax cheating by allowing cashless transaction.
8. Small-scale businesses people can be benefitted by using online tools to expand their business.
9. It will play a significant role in GDP growth. As per analyst, by 2015, the Digital India could raise GDP up to $1 trillion. World Bank report says that per capita GDP can be increased by 0.81 and 1.31%, respectively, in developing countries with a 10% increase in mobile and broadband penetration [9].
10. A large number of jobs in IT, electronics and telecommunication sector will be generated directly or indirectly through this programme.

5.2   Challenges

Digital India mission has been announced since 2015. But it is facing multiple challenges in successful implementation. Few of the challenges are:

1. *High level of digital illiteracy*: The biggest challenge in the success of Digital India programme is the requirement of high level of digital literacy. Low digital literacy is a key hindrance in adaptation of technologies. According to ASSOCHAM-Deloitte report on Digital India, November 2016, around 950 million Indians are still not on the Internet [10]. Here, digital learning plays an important role as it becomes the biggest challenge for the government.

2. *Changing the mindset*: People are accustomed to years of traditional pen and paper practice and not ready to change. Making people aware of paperless practice is also a big challenge.
3. *Connectivity to remote areas*: It is a huge task to connect each and every village, town and city. It is not so easy to connect 250,000 gram panchayats through National Optical Fibre Network.
4. *Internet Connectivity*: Providing high speed of the Internet is a core utility for online delivery of different services. India has low Internet speed. Thus, making a database to store such huge information is a challenge.
5. *Mobile Connectivity*: At present, 55,000 villages are lack of mobile connectivity because providing mobile connectivity in those locations is not commercially possible for the private service providers.
6. *Finance*: There is a large digital divide between urban and rural India. Sufficient funds have not been sanctioned till now to meet the infrastructural cost in rural areas.
7. *Infrastructure: Due to lack of the private participation in the government sector*, infrastructure development is very slow to meet the challenges of Digital India.
8. *Compatibility with centre state databases*: India comprises 1600 languages and dialects. Digital services are not available in local languages, which is a great obstacle in digital literacy.
9. *To make the common people technology savvy, to visualize and avail benefits of Smart India*: Here, the digital learning plays an important role as it becomes the biggest challenge for the government.

5.3 Suggestions

Digital India campaign cannot be successful on its own. Changes in policy are required to achieve Digital India a reality. Below are a few suggestions:

1. To enable citizens, digital literacy is the first step. People should learn how to secure their online data.
2. A large awareness programme must be conducted to make this programme successful. To increase the growth of the Internet usage, rural citizens have to be motivated about the benefits of the Internet services.
3. Addressing of digital divide is required.
4. Content manufacturing is not the government's strength. It requires partnerships between content and service with telecom sectors and other firms.
5. Exploration of PPP models must be done for the sustainable development of digital infrastructure.
6. Involvement and encouragement of the private sector are much essential for the development of infrastructure in rural and remote areas. There must be flexibility in taxation policies and quick clearance of projects to encourage the private sector.

7. To achieve Digital India, maximum connectivity with minimum cyber security risks is necessary for which anti-cybercrime team should be very strong to protect the data.
8. Cyber security course is to be introduced through various international certification bodies at graduate level to improve skill in cyber security.
9. Effective participation of different departments and demanding commitment and efforts are needed.
10. There should be amendments in various legislations for successful implementation which have prevented the growth of technology in India for a long time.

5.4 Threats

Cyber security is the biggest threat in adoption of digital technologies. As the country moves towards a digital economy, consumer and citizen data will be increased and will be stored digitally; as a result, a large number of online transactions will be carried out by companies, individuals as well as government departments. That will create a bigger target for cyber-criminals and hackers.

Most of the cyber security technologies including tools are imported. India does not have desired skills to analyse these for hidden malwares. Also at present, the country does not have top-level experts for these high-end jobs. As per NASSCOM report, 1 million trained cyber security professionals are required in India by 2025 where the current estimated number is 62,000 [11].

6 Conclusion

Digital India is a big step to build an actual empowered nation. If successful, it enables citizens to avail to multimedia information, content and services. It has been observed that digital learning takes a very important role to achieve the nine initiatives of Digital India. Digital learning allows students to learn collaboratively in groups where knowledge is globalized through knowledge management (KM) processes. An e-learning framework has been suggested to implement digital learning. Digital learning is embedded in KM environment to achieve collaboration. Users, knowledge processes and technologies are integrated to create knowledge society (KS). The various benefits of Digital India will surely move India towards a knowledge-based society and knowledge economy. Because there are a number of big challenges for the implementation of Digital India, they must be addressed in a proper manner to avoid the failure of the project. In fact, every citizen must be mentally prepared to accept the changes in order to build Smart India. Remedial measures have been suggested to overcome the threat of Digital India. So, the digital learning framework if successfully implemented will certainly recharge the concept of Digital India and will act as a booster for Smart India to be achieved in near future.

Digital Learning: The Booster for Smart India

References

1. https://www.eduba.com/digital-learning
2. https://whatis.ciowhitepapersreview.com/definition/digital-learning/
3. http://www.panworldeducation.com/2017/03/23/benefits-of-digital-learning-over-traditional-education-methods
4. Pattnayak, J., Pattnaik, S., Dash, P.R.: User centric model of E-learning to build up virtual learning environment. Int. J. Emerg. Trends Technol. Comput. Sci. (IJETTCS) 2(4), 16–22, ISSN: 2278-6856 (2013)
5. Sammour, G.: The role of knowledge management and learning. Int. J. Knowl. Learn. 4(5), 465–477 (2008)
6. Jacobfeuerborn, B., Muraszkiewicz, M.: E-Learning Methodology for High-Tech Organizations. e-learning, hitech organizations, knowledge management, corporate learning, staff competence development, e-learning system development (n.d.)
7. Yilmaz, Y.: Knowledge management in E-learning practices. TOJET: Turk. Online J. Educ. Technol. 11(2), 150–155 (2012)
8. Pattnayak, J., Pattnaik, S., Dash, P.R.: Knowledge management in E-learning: a critical review. Int. J. Eng. Comput. Sci. (IJECS) 6(5), 21528–21533 (2017)
9. Digital India Programme: Importance and Impact. Retrieved from http://iasscore.in/national-issues/digitalindia-programme-importance-and-impact
10. Digital India. Unlocking the Trillion Dollar Opportunity: ASSOCHAM—Deloitte report, November 2016. Retrieved from www.assocham.org
11. Avinash, K. (2015). Why cyber security is important for digital India. Retrieved from http://www.firstpost.com/business/why-cyber-security-is-important-for-digital-india-2424380.html

Hairpin Structure Band-Pass Filter for IoT Band Application

Kaibalya Kumar Sethi, Ashirwad Dutta, Gopinath Palai and Patha Sarkar

Abstract The present communication investigates on hairpin-line band-pass filter pertaining to GHz region with the help of finite element method. Further, the numerical techniques manipulate with a seven-pole hairpin-line microstrip band-pass filter on a 0.035-mm-thick substrate, which divulges a narrow bandwidth of 1.6 GHz with a notch at 14 GHz, VSWR of 1.2. The present simulation result affirms that the proposed hairpin structure can be a suitable candidate to realize different application related to IoT band, Ku band transmission, VSAT up linking, satellite broadcast services, etc.

Keywords VSWR · IoT · VSAT · Ku band

1 Introduction

In the communication system, a filter is performing as a two-port system, which is used as a frequency discriminating component. We required a specified frequency band for a particular application from the spectrum of electromagnetic wave [1]. At the receiver end, an antenna can receive a wide band of frequencies, but little of them may not be required for that particular application. So it should be filtered out properly before processing of the signal. Filter network has been used in juxtaposition with antennas to accomplish the preferred task. It plays a significant role in various RF/microwave applications. These are used to separate or combined dissimilar frequencies. Up-and-coming applications in wireless communications prolong to defy RF/microwave filters by means of ever more rigorous requirements such as higher concert, lesser size, lighter weight, and lower cost. Ideal strip lines and microstrip lines are for low selectivity wide-band applications. For higher

K. K. Sethi (✉) · A. Dutta · G. Palai
Department of Electronics and Communication Engineering, GITA, Bhubaneswar, India
e-mail: kaibalyaa@rediffmail.com

P. Sarkar
Department of CAPGS, BPUT, Rourkela, India

© Springer Nature Singapore Pte Ltd. 2020
S. Patnaik et al. (eds.), *New Paradigm in Decision Science and Management*,
Advances in Intelligent Systems and Computing 1030,
https://doi.org/10.1007/978-981-13-9330-3_40

bandwidth parallel coupled with very stiff coupling structure is preferred. For additional compactness, resonator sections are to be positioned alongside structures like interdigital, combo line and hairpin, etc., for MIC or MMIC filter design method hairpin structure is most preferred [2–4]. When ceramic substrate is selected at that time ground is not required. To dwindle the dimension higher dielectric constant material can be worn. For wide-band application high dielectric constant of 80 or 90 can be used in hairpin filter [5]. This type filter mainly is a folder adaption of a half wave parallel coupled line filter. Due to the availability of various full EM simulator software, the parallel coupled band-pass filter can be preferred to design. The full EM simulator software provides quicker optimization of disparate propagation velocities, bend and short-end coupling, etc.

In compared with combo line and digital filter, the hairpin-line filters are more in size, because inter-arm spacing is large. As the length-to-width ratio becomes small, the frequency for a particular substrate will increase. The relationship of numerous types of hairpin filter is specified by Wei et al. [6]. To reduce the dimension of hairpin filter than the parallel-coupled filter, the finite element method (FEM) is used in obsessed fatal solution type to investigate and deliberate the desired result for the frequency of 14 GHz.

2   Filter Parameter Design Equations

Kindly assure that the A prototype low-pass filter is chosen to intend a coupled band-pass filter. A seven-pole filter is required for good rejection. By using typical transformation equation, the premeditated topology is transferred to proposed band-pass filter. The coupling coefficient can be found out by using the Eq. (1) [7]

$$K = \frac{f_l - f_h}{f_O} \tag{1}$$

The hairpin-line structure is preferred to uphold the compact dimension of the proposed work. By flopping the resonators of parallel coupled half wavelength resonators into 'U' form practically [8]. The above 'U'-shaped resonator shown in Fig. 1 is called as hairpin-line resonator. The coupling effect between the resonator sections is created due to inductance [8]. The external quality factor can be calculated by using Eqs. (2) and (3) [8].

$$Q_{el} = \frac{g_0 g_1}{FBW} \tag{2}$$

$$Q_{en} = \frac{g_n g_n + 1}{FBW} \tag{3}$$

where Q_{el} and Q_{en} are the external quality factor and n is the no. of pole. For this design $n = 7$.

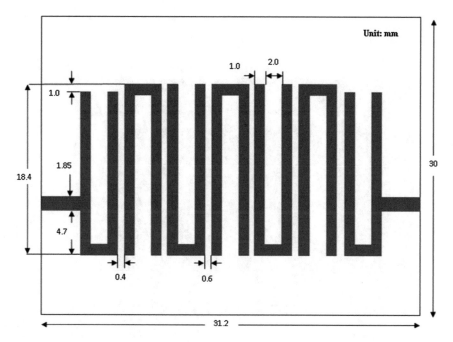

Fig. 1 Layout of seven-pole, hairpin-line microstrip band-pass filter on a 0.035-mm-thick substrate with a relative dielectric constant of 6.15

The design equation for estimating the taping point 't' for proposed filter is in (4) [8]

$$t = \frac{2L}{\pi} \frac{\pi Z_0/Z_r}{\sin^{-1}\left(\sqrt{2Q_e}\right)} \tag{4}$$

3   Filter Design Methodology

The proposed filter design has been developed by choosing a Taconic RF-60 substrate of dimension 31.2 × 30.0 mm². The thickness or the height of the substrate is of 0.0.035 mm with very low, loss tangent value. The substrate is having relative permittivity and loss tangent values as 6.15 and 0.0028, respectively. The ground plane is having same length as substrate. The front view of the structure is depicted in Fig. 2. In Fig. 3, the design filter is exposed to input and output port. The dimensions are shown in Fig. 1. All the units of dimensions are in millimeter.

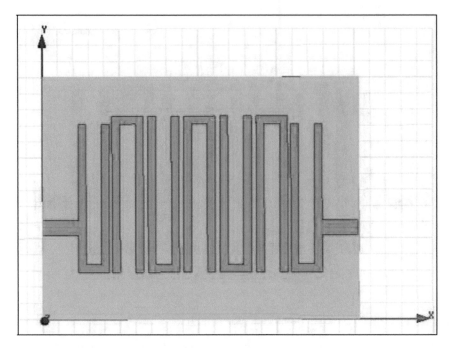

Fig. 2 Design of seven-pole, hairpin-line microstrip band-pass filter on a 0.035-mm-thick substrate with a relative dielectric constant of 6.15

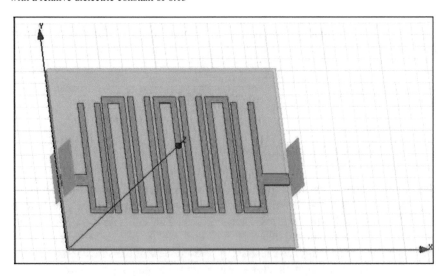

Fig. 3 Design of seven-pole, hairpin-line microstrip band-pass filter with i/p and o/p port

4 Result Analysis

We have obtained both S11 and S21 through simulation using full-wave electromagnetic simulation HFSS, and the corresponding result is plotted in Fig. 4. In this waveform, X-axis represents frequency in GHz and S-Parameters (S_{11} and S_{21}) in dB presented in Y-axis. The results, what we have got from the simulation, indicate a notch or dip at 14 GH$_z$ with a reflection loss close to -20.1 dB. This is normally considered to be a poor transmission which results in a band-stop behavior. In addition, we also have performed VSWR. The VSWR plot is quite sufficient to justify a poor transmission about 14 GH$_z$ which is shown in Fig. 5.

The frequency response result through HFSS is for the expected Ku band. The characteristic of band-pass filter is having center frequency 14 GHz and bandwidth 1.6 GHz.

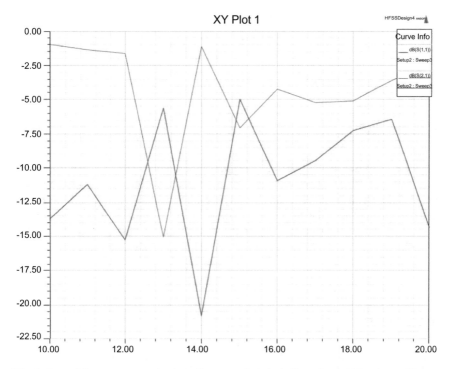

Fig. 4 S_{11} and S_{21} parameters of proposed seven-pole, hairpin-line microstrip band-pass filter

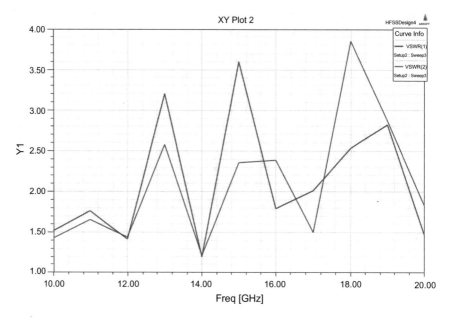

Fig. 5 VSWR of proposed hairpin-line microstrip band-pass filter

5 Conclusion

We have proposed a hairpin-line band-pass filter with suitable dimensions. The design will provide a superior result for IoT band applications. It could be generated different pass band frequency on varying the dimension of 'U' shape resonator and the gap between the resonators also to get other wireless license bands. So at the end, we landed to a conclusion that one particular structure can be used in multiple applications just by inserting more number of poles, by changing shape, gap, etc., to it.

References

1. Cristal, E.G., Franknel, S.: Hairpin-line and hybrid hairpin-line/half-wave parallel coupled-line filters. IEEE Trans. Microw. Theory Tech. **20**(11), 719–728 (1972)
2. Cristal, E.G., Franknel, S.: Design of hairpin-line and hybrid hairpin-parallel-coupled-line filters. In: IEEE GMTT International Microwave Symposium Digest, pp. 12–13 (1971)
3. Shivhare, J., Jain, S.B.: Design and development of a compact and low cost folded-hairpin line bandpass filter for L-band communication. In: IEEE (ELECTRO-2009), pp. 360–363 (2009)
4. Shaman, H., Almorqi, S., Haraz, O., Alshebeili, S.: Hairpin microstrip bandpass filter for millimeter-wave applications. In: IEEE (MMS-2014). ISBN No. 978-1-4799-7391-0. pp. 1–4 (2014)

5. Weller, T.M.: Edge-coupled coplanar waveguide band pass filter design. IEEE Trans. Microw. Theory Techn. **48**(12), 2453–2458 (2000)
6. Wei, J., Wu, J., Wang, L.: Analysis and design of a hairpin-line bandpass filter. COMPEL **29**(2), 355–361 (2010)
7. Hong, J., Lancaster, M.J.: Microstrip Filters for RF/Microwave Applications. Wiley, New York (2001)
8. Singh, T., Chacko, J., Sebastian, N., Thoppilan, R., Kotrashetti, A., Mande, S.: Design and optimization of microstrip hairpin-line bandpass filter using DOE methodology. In: IEEE ICICIT, pp. 1–6 (2012)

Author Index

Printed in the United States
By Bookmasters